# 1st International Conference on Rheology and Modeling of Materials 2013



# (IC-RMM1)

Journal of Physics: Conference Series Volume 602

## Miskolc-Lillafured, Hungary
## 7 – 11 October 2013

ISBN: 978-1-5108-0262-9
ISSN: 1742-6588

# 1st International Conference on Rheology and Modeling of Materials 2013

# (IC-RMM1)

Journal of Physics: Conference Series Volume 602

## Miskolc-Lillafured, Hungary
## 7 – 11 October 2013

ISBN: 978-1-5108-0262-9
ISSN: 1742-6588

IOP Institute of Physics

**Printed from e-media with permission by:**

Curran Associates, Inc.
57 Morehouse Lane
Red Hook, NY 12571

Some format issues inherent in the e-media version may also appear in this print version.



Printed by Curran Associates, Inc. (2015)



Institute of Physics
Dirac House, Temple Back
Bristol BS1 6BE UK

Phone: 44 1 17 929 7481
Fax: 44 1 17 920 0979

techtracking@iop.org

**Additional copies of this publication are available from:**

Curran Associates, Inc.
57 Morehouse Lane
Red Hook, NY 12571 USA
Phone: 845-758-0400
Fax: 845-758-2634
Email: curran@proceedings.com
Web: www.proceedings.com

# TABLE OF CONTENTS

Journal of Physics: Conference Series **602** (2015) 011001 doi:10.1088/1742-6596/602/1/011001

# ic-rmm1

## 1$^{ST}$ INTERNATIONAL CONFERENCE ON RHEOLOGY AND MODELING OF MATERIALS

October 7-11, 2013/ Miskolc-Lillafüred, Hungary

**Organized by** the International Scientific Advisory Board (**ISAB**),

the International Organization Board (**IOB**) and by

the Session chairs

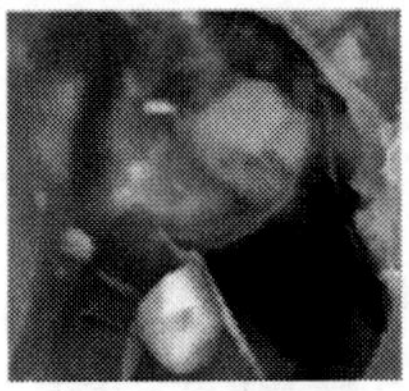

***ic-rmm1***

Journal of Physics: Conference Series **602** (2015) 011001 doi:10.1088/1742-6596/602/1/011001

# PREFACE

Understanding the rheological properties of materials and their rheological behaviors during their manufacturing processes and in their applications in many cases can help to increase the efficiency and competitiveness not only of the finished goods and products but the organizations and societies also. The more scientific supported and prepared organizations develop more competitive products with better thermal, mechanical, physical, chemical and biological properties and the leading companies apply more competitive knowledge, materials, equipment and technology processes. The idea to organize in Hungary the 1st International Conference on Rheology and Modeling of Materials we have received from prospective scientists, physicists, chemists, mathematicians and engineers from Asia, Europe, North and South America including India, Korea, Russia, Turkey, Estonia, France, Italy, United Kingdom, Chile, Mexico and USA.
The goals of **ic-rmm1** the 1st International Conference on Rheology and Modeling of Materials are the following:

- Promote new methods and results of scientific research in the fields of modeling and measurements of rheological properties and behavior of materials under processing and applications.
- Change information between the theoretical and applied sciences as well as technical and technological implantations.
- Promote the communication between the scientist of different disciplines, nations, countries and continents.

The international conference **ic-rmm1** provides a platform among the leading international scientists, researchers, PhD students and engineers for discussing recent achievements in measurement, modeling and application of rheology in materials technology and materials science of liquids, melts, solids, crystals and amorphous structures. Among the major fields of interest are the influences of material structures, mechanical stresses temperature and deformation speeds on rheological and physical properties, phase transformation of foams, foods, polymers, plastics and other competitive materials like ceramics, nanomaterials, medical- and biomaterials, cosmetics, coatings, light metals, alloys, glasses, films, composites, hetero-modulus, hetero-viscous, hetero-plastic complex materials, petrochemicals and hybrid materials, … etc.

Multidisciplinary applications of rheology and rheological modeling in material science and technology encountered in sectors like alloys, ceramics, glasses, thin films, polymers, clays, construction materials, energy, aerospace, automotive and marine industry. Rheology in food, chemistry, medicine, biosciences and environmental sciences are of particular interests.

In accordance to the program of the conference **ic-rmm1** more than 160 inquiries and registrations were received from 51 countries. Finally the scientists and researchers have arrived to our conference from 42 countries. Including co-authors, the research work of more than 300 scientists are presented in this book.

Prof. Dr. László A. Gömze
*chair, ic-rmm1*

## THE INTERNATIONAL SCIENTIFIC ADVISORY BOARD (ISAB)

*Prof. Dr. Igor EMRI* - President of the International Society of Rheology, (Slovenia)

*Prof. Dr. Gerald G. FULLER* - Stanford University, (USA)

*Prof. Dr. Grispulo GALLEGOS* - President of the European Society of Rheology, (Spain)

*Prof. Dr. László A. GÖMZE* - University of Miskolc, (Hungary)

*Prof. Dr. Jae Chun HYUN* - Korea University, (Republic of Korea)

*Prof. Dr. Ekhard SALJE* - University of Cambridge, (UK)

## THE INTERNATIONAL ORGANIZATION BOARD (IOB)

*Dr. Fernando BAUTISTA* - University of Guadalajara, (Mexico)

*Dr. Fares N, BEAINY* - University of Oklahama, (USA)

*PhD cand. Ivica DURETEK* - Montanuniversity Leoben, (Austria)

*Dr. Özlem Caglar DUVARCI* - Izmir Institute of Technology, (Turkey)

*PhD stud. Ádám EGÉSZ* - GE Appl. and Light. Budapest, (Hungary)

*Prof. Dr. Gerald G. FULLER* - Stanford University, (USA)

*Prof. Dr. Basavaraja GURAPPA* - Indian Institute of Technology, (India)

*A Prof. Dr. Markus HÜTTER* - Endhoven University of Technology, (The Netherlands)

*Prof. Dr. Jae Chun HYUN* - Korea University, (Republic of Korea)

*PhD kand. Albertas KLOVAS* - Kaunas University of Technology, (Lithuania)

*PhD kand. Malika KHODJA* - Boumerdes University Sonatrah, (Algeria)

*Dr. Giséle L. LECOMTE* - ENSCI, Limoges University, (France)

*Dr. Modesto T. LOPEZ* - University of Granada, (Spain)

*Dr. Edvin NUGIS* - Estonian University of Life Sciences, (Estonia)

*Prof. Dr. Marcello ROMAGNOLI* - University of Modena, (Italy)

*Prof. Dr. Ekhard SALJE* - University of Cambridge, (United Kingdom)

*Dr. Jérome SARRAZIN* - ExxonMobil Chemical EU Inc., (Belgium)

*Prof. Dr. Dennis A. SIGINER* - University of Santiago de Chile, (Chile)

## CHAIR OF THE INTERNATIONAL ORGANIZATION BOARD

*Prof. Dr. László A. GÖMZE* - University of Miskolc, (Hungary)

## THE ORGANIZER SESSION CHAIRS

**Session 1. Applied Rheology and Modeling of Clays, Minerals and Rocks**

*Dr. Ludmila N. GÖMZE* - Igrex Engineering Service Ltd, (Hungary)

**Session 2. Applied Rheology of Polymer Melts**

*PhD cand. Ivica DURETEK* - Montanuniversity Leoben, (Austria)

**Session 3. Applied Rheology and Material Structure of Foods**

*Dr. Thomas A, WAIGH* - University of Manchester, (United Kingdom)

**Session 4. Avalanche Dynamics in Rheology and Phase Transitions**

*Prof. Dr. Ekhard SALJE* - University of Cambridge, (United Kingdom)

**Session 5. Experimental Study and Modeling of Complex Flow**

*Dr. Fernando BAUTISTA* - University of Guadalajara, (Mexico)

**Session 7. Multiscale Mechanics of Materials**

*A Prof. Dr. Markus HÜTTER* - Eindhoven University of Technology, (The Netherlands)

**Session 8. Rheological Fluid Mechanics**

*Prof. Dr. Dennis A, SIGINER* - University of Santiago de Chile, (Chile)

**Session 9. Rheology and Modeling of Hetero-modulus Hetero-viscous Complex Materials**

*Prof. Dr. László A. GÖMZE* - University of Miskolc, (Hungary)

**Session 12. Rheology of Biological Systems**

*Prof. Dr. Blais NSOM* - University of Bretagne Occidentale, (France)

**Session 13. Rheology of Construction Materials and Concretes**

*PhD can. Albertas Klovas* - Kaunas University of Technology (Lithuania)

**Session 14. Rheology of Emulsions, Suspensions and Foams**

*Dr. Giséle LECOMTE* - ENSCI, University of Limoges, (France)

**Session 15. Rheology of Ferrofluids and Magneto-rheological Materials**

*Dr. Modesto T. LOPEZ* - University of Granada, (Spain)

*Prof. Dr. Sergei N. KULKOV* - Institute of Strength Physics SB RAS (Russia)

**Session 16. Rheology of Fluid Interfaces and Modeling**

*Dr. Basavaraja M. GURAPPA* - Indian Institute of Technology, (India)

**Session 17. Rheology of Melted Matals and Glasses**

*PhD cand, Csilla PARÓCZAI* - Guardian Orosháza Glass Ltd, (Hungary)

**Session 18. Rheology of Nanofluids**

*Prof. Dr. Marcello ROMAGNOLI* - University of Modena, (Italy)

**Session 19. Rheometric Measurement – Methods, Equipment and Errors**

*Dr. Oktay AKTAS* - University of Cambridge, (United Kingdom)

**Session 21. Rheology of Drug and Cosmetic Materials**

*Dr. Thomas A. WAIGH* - University of Manchester, (United Kingdom)

## ACKNOWLEDGEMENT

In the name of **ic-rmm Conference Boards** I would like acknowledge and say many thanks to our following sponsors for their support in press-campaign and contributions the news and information about **The 1st International Conference on Rheology and Modeling of Materials** between their members and in their media:

**AIPEA** – Internatonal Association for the Study of Clays
**AIR** – Associazione Italiana di REOLOGIA, (Italy)
**biomat.net** – The Biomaterials Network, (Portugal)
**ESR** – The European Society of Rheology
**GFC** – Groupe Francais de la Céramique, (France)
**ICG** – International Commission of Glass
**Igrex –** Engineering Service Ltd, (Hungary)
**Journals:** Academic Journals
APPLIED RHEOLOGY
cfi - Ceramic Forum International
ÉPÍTŐANYAG, Journal of Silicate Based and Composite Materials
refractiries WORLDFORUM
Rheology Bulletin
TECHNOLOGIJA HRANE
**MGU** – Moscow State University (Russia)
**NRC** – Nordic Rheology Society (Europe)
**RAS** – Institute of Strength Physics and Materials Science (Russia)
**SOR** – The Society of Rheology (USA)
**ME** – Miskolci Egyetem – University of Miskolc (Hungary)

Many thanks to colleagues of *ISAB, IOB* and to *chairs* of the SESSIONS for their support and help in organization work and in successful transaction of conference **ic-rmm1** the 1st International Conference on Rheology and Modeling of Materials

I would like to say many thank personally to *Dr. Özlem Caglar Duvarci* (Izmir Institute of Technology), *Prof. Dr. Igor Emri* (the president of the International Society of Rheology), *Prof. Dr. Gerald G. Fuller* (Stanford University), *Prof. Dr. Jae Chun Hyun* (Korea University), *Prof. Dr. Sergey N. Kulkov* (Tomsk State University), *Prof. Dr. Giséle L. Lecomte* (ENSCI Limoges University), *Prof. Dr. Ekhard SALJE* (University of Cambridge) and *PhD candidate Malika Khodja* (Boumerdes University Sonatrah) for their excellent organization works in Korea, Russia, Turkey, USA and other countries of Africa, America, Asia and Europe. Especially many thanks to editors of journals **Applied Rheology** (Switzerland), **Építőanyag** (Hungary) and **Rheology Bulletin** (USA) for their strong support in scientific press and media.

Prof. Dr. László A. Gömze
*chair of conference board*

# Peer review statement

All papers published in this volume of *Journal of Physics: Conference Series* have been peer reviewed through processes administered by the proceedings Editors. Reviews were conducted by expert referees to the professional and scientific standards expected of a proceedings journal published by IOP Publishing.

Journal of Physics: Conference Series **602** (2015) 012001 doi:10.1088/1742-6596/602/1/012001

# Selection of appropriate polyoxymethylene based binder for feedstock material used in powder injection moulding

**J Gonzalez-Gutierrez[1], G B Stringari[1], Z M Megen[1], P Oblak[1], B S von Bernstorff[2] and I Emri[1,3]**

[1] Center for Experimental Mechanics, Faculty of Mechanical Engineering, Univerity of Ljubljana – 1000 - Ljubljana - Slovenia
[2] BASF AG – 67063 – Ludwigshafen – Germany
[3] Institute for Sustainable Innovative Technologies – 1125- Ljubljana - Slovenia

joamin.gonzalez@fs.uni-lj.si, engenheirogus@gmail.com, zeruka@yahoo.com, pavel.oblak@fs.uni-lj.si, bernd-steffen.bernstorff-von@basf.com, igor.emri@fs.uni-lj.si

**Abstract**. Polyoxymethylene (POM) has found applications as a binder material in Powder Injection Moulding (PIM) due to its ability to depolymerize rapidly under acidic conditions. Such ability represents an advantage during the binder removal step of PIM. However, currently available POM has high viscosity that can complicate the injection moulding process of parts with complex geometry. For this reason it is necessary to investigate methods of lowering the viscosity of POM-based binders, but without affecting their solid mechanical properties (i.e. creep compliance). In this investigation, the addition of a low molecular weight polymer, and the reduction of the average molecular weight of POM were investigated as possible ways of decreasing the viscosity of PIM binders. The addition of the low molecular weight additive (WAX) caused a small decrease in the viscosity of the POM-based binder and a small increase in its solid creep compliance. On the other hand, lowering the average molecular weight of POM caused a large decrease in viscosity, but also an acceptable increase in creep compliance. Therefore, by selecting an appropriate molecular weight of POM, it is possible to improve the performance of POM-based binders for PIM.

## 1. Introduction

Polyoxymethylene (POM) is an engineering polymer that has also found applications as a binder material in powder injection moulding (PIM), which is a versatile mass production method for small, complex shaped components of metal or ceramic [1-4]. In PIM, POM acts as a carrier medium for metal or ceramic powders during the injection moulding process and it is later removed to obtain a metallic or ceramic piece after sintering. The main advantage of POM in powder injection moulding comes from the instability of acetal linkages, which results in rapid hydrolysis. Since the main chain of POM is composed of $-CH_2$-O- bonds, the methyl-oxygen bonds are easy to break under heat and oxygen [5]. The process is accelerated in the presence of acid vapours. This breakage results in a continuous depolymerization reaction yielding formaldehyde gas. The released formaldehyde and formic acid formed through oxidation of formaldehyde can accelerate the depolymerization reaction even further; this process is usually called the zipper mechanism [6]. The unzipping process represents a major advantage during the debinding process and it is generally called catalytic debinding [7].

Thus, catalytic debinding occurs at a significantly higher rate than other dedinding techniques, such as solvent and thermal debinding, and greatly speeds up the PIM process [8]. Additionally, POM in the solid state has good mechanical properties and thus makes easier handling of moulded parts, which can be fragile or easily deformable in other binder systems. However, POM based-feedstock materials have higher viscosity than other feedstock materials based on polyolefins. Such high viscosity can complicate the injection moulding of parts with complex geometry. Therefore, it is important to reduce the viscosity of POM-based feedstock materials.

In this study, the effect of a low molecular weight additive and reduction of the average molecular weight of POM were investigated as possible ways of decreasing the viscosity of POM-based binders used in PIM. Since it is important to maintain high solid mechanical properties, creep compliance was measured on the modified binder systems.

## 2. Materials and methods

### *2.1. Material preparation*

Polyoxymethylene copolymers of different average molecular weight ($M_w$) were synthesized at BASF (Ludwigshafen, Germany). The nomenclature and average molecular weight of all the POM materials used in this study is shown in table 1. Molecular weights were measured by the supplier using gel permeation chromatography. For the first part of this investigation POM MW092, which is the POM used in commercially available feedstock materials by BASF, was blended with a low molecular weight polymer. The blend was prepared by extruding together the two components in a twin-screw extrusion machine (PolyLab, Thermo Scientific, Germany) seven times. The details of the process are described elsewhere [8, 9]. The low molecular weight additive is simply referred as WAX due to confidentiality terms. Blends of POM and wax were prepared at different weight percentage of WAX as shown in table 1.

Table 1. Average molecular weight of POM copolymers and description of POM-WAX blends

| Material ID | Average Molecular Weight, $M_w$, [g/mol] | Description |
|---|---|---|
| MW010 | 10240 | Virgin POM copolymer, laboratory scale synthesis |
| MW024 | 24410 | |
| MW052 | 52750 | |
| MW081 | 81100 | |
| MW092 | 92360 | Virgin POM copolymer, industrial scale synthesis |
| MW109 | 109000 | |
| MW129 | 129300 | |
| MW204 | 204400 | |
| WAX | 36400 | Low molecular weight additive |
| B-01 | N/A | Extruded MW092 + 1wt% WAX |
| B-02 | N/A | Extruded MW092 + 2wt% WAX |
| B-04 | N/A | Extruded MW092 + 4wt% WAX |
| B-08 | N/A | Extruded MW092 + 8wt% WAX |
| B-16 | N/A | Extruded MW092 + 16wt% WAX |

### *2.2. Viscosity measurements*

Viscosity measurements in oscillatory mode were performed in a MARS-II rotational rheometer (Thermo Scientific, Germany). Rheological tests were performed at 210 °C, which is within the rage of temperatures at which POM is generally processed. A truncated cone-plate measuring-geometry with a 20 mm diameter and angle of 1° was used in all measurements. Two frequency sweeps were performed in each measurement; the first one increasing from 0.01 Hz (0.0628 rad/s) to 100 Hz

(628.32 rad/s), and the second one decreasing from 100 to 0.01 Hz. This was done to monitor any possible flow instabilities. All measurements were performed applying a shear stress of 200 Pa, previously determined to be within the linear viscoelastic region of all materials investigated. All viscosity measurements at a given temperature were performed six times per material. In this study, viscosity results are presented as the magnitude of the complex viscosity ($|\eta^*|$), which is related to the constant rotational viscosity ($\eta$) through the Cox-Merz rule [10], which has been previously determined to be applicable to POM copolymers and its WAX blends [11, 12].

*2.3. Creep measurements*

In order to perform creep compliance measurements cylindrical specimens with diameter $D = 5.8 \pm 0.1$ mm and length $l = 29.0 \pm 2.0$ mm were prepared by gravimetrical casting from materials listed in table 1. Details on the procedure and schematics of the device used are shown elsewhere [8, 9, 13]. After casting specimens were cut to the required length and glued with acrylate-based glue (F524 black and activator B, Kemis Plus, Slovenia) to custom made metal holders for gripping the cylindrical specimen to the measuring device. Three specimens were prepared for each type of material shown in table 1.

Creep measurements in solid state were performed following the methodology developed at the Center for Experimental Mechanics at the University of Ljubljana. Effect of wax addition on creep compliance was performed in the self-developed torsiometer [14], while the effect of molecular weight was analyzed using a MARS II rotational rheometer fitted with solid clamps. Creep compliance was measured in segmental form at various temperatures and shifted to a reference temperature of 110 °C using the principle of time-temperature superposition. Time-temperature superposition was performed using the self-developed Closed Form Shifting Algorithm [15]. The reference temperature of 110 °C was selected because it is one of the temperatures at which debinding is performed.

## 3. Results and Discussion

*3.1 Addition of Low molecular weight polymer (WAX)*

*3.1.1 Viscosity measurements*

The effect of WAX addition on the viscosity of POM is shown in figure 1. The general trend is that as the content of WAX increases, the viscosity linearly decreases. However, the decrease is not as large as needed. For example, adding 16wt% of WAX reduces the viscosity by approximately 20%, which is not low enough. The needed viscosity should be at least 10 times smaller (90% reduction) to reach the recommended viscosity for a PIM binder of approximately 10 Pa s [16]. Blends with a higher amount of WAX were tried, but they lead to phase separation just after extrusion. Thus, it can be concluded that addition of WAX is not an optimal solution to decrease the viscosity of POM-based binders for PIM.

*3.1.2 Creep compliance*

The effect of WAX addition on the creep compliance of POM is shown in figure 2. It appears that at the lowest content of wax there is a decrease in viscosity compared to unblended POM. After 1wt% content, creep compliance steadily increases until it reaches the maximum value at 16wt%. Adding 16wt% of WAX causes an increase of 16% in creep compliance, which is acceptable for our purposes. The observed increase in creep compliance could be the result of a decrease in crystallinity, which facilitates movement of polymeric chains when loaded. Such behaviour has also been observed with other polymers, for example LDPE [17]. In terms of binder selection, it appears that the best creep compliance properties could be achieved by adding 1wt% of WAX, but the drop in viscosity is not enough for purposes of PIM. Therefore, WAX addition is not the best solution to optimize the behaviour of binders used in PIM.

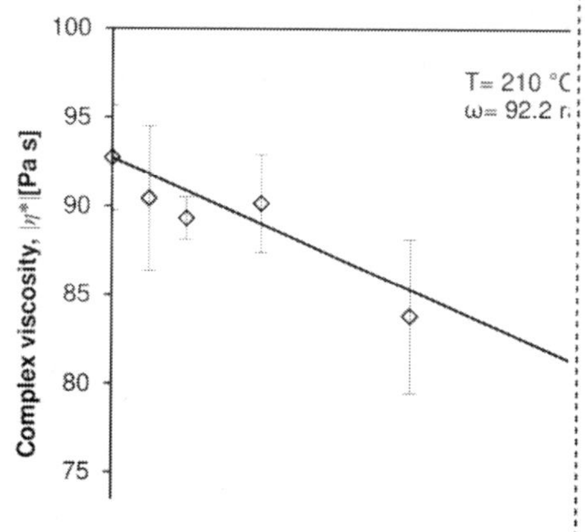


**Figure 1.** Viscosity of POM and WAX blends at 210 °C and angular frequency of 92 rad/s. The solid line represents a linear fit to the experimental data [18].

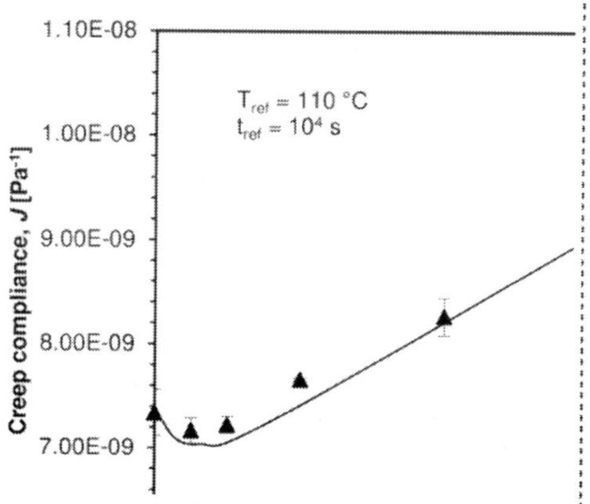


**Figure 2.** Shear creep compliance of POM and WAX blends at reference temperature 110 °C and time of 10000 s. The solid line represents a manual fit to the experimental data.

### *3.2. Reduction of average molecular weight*

#### *3.2.1. Viscosity measurements*

The magnitude of the complex viscosity as a function of angular frequency for all POM copolymers was measured at 210 °C. From the measured complex viscosity data ($|\eta^*|$) the Newtonian viscosity ($\eta_0$) was estimated from the plateau at frequencies below 10 rad/s. The results are presented as a function of the average molecular weight in figure 3. As with other polymers, POM copolymers show a rapid decrease in viscosity as the average molecular weight decreases following a power function as proposed by Fox and Flory [19].

With respect to the selection of an appropriate binder for PIM, one could choose between the first three molecular weights (MW010, MW024 and MW052), which are all below the recommended viscosity of 10 Pa s [16]. However, in order to make the correct decision, one should consider the results of the solid mechanical properties.

#### *3.2.2. Creep compliance measurements*

After the creep compliance segments were shifted to a reference temperature of 110 °C, isochronal curves at 10000 s (~3 h) were estimated and plotted as a function of the average molecular weight (figure 4). As it can be seen in figure 4, the creep compliance has a plateau at the highest molecular weights ($M_w$ = 81100 to 204400 g/mol). And as the molecular weight is decreased beyond $M_w$ = 81100 g/mol there is a rapid increase in the creep compliance. Since it is desirable not to have a drastic increase in creep compliance in the PIM binder, it is not recommended to choose the lowest molecular weight (MW010) available, but the second lowest (MW024). This will ensure an acceptable increase in creep compliance, but a very large decrease in viscosity (99% or 240 times) when compared to the current binder (MW092). Since there is a decrease in creep compliance, extra care should be taken when handling the moulded part with the suggested binder.

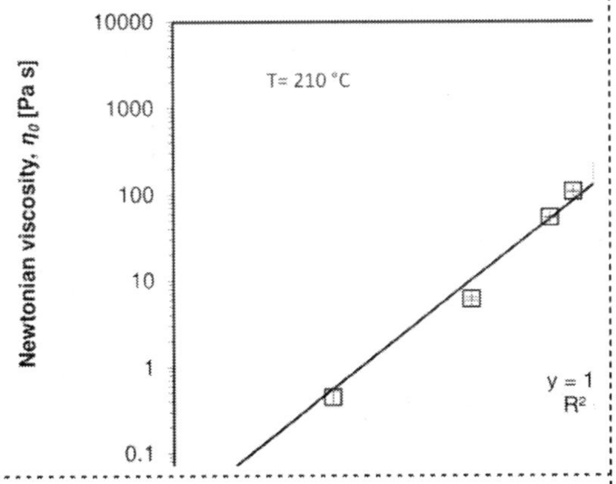


**Figure 3.** Newtonian viscosity of POM copolymers at 210 °C. The solid line represents a power law fit to the experimental data.

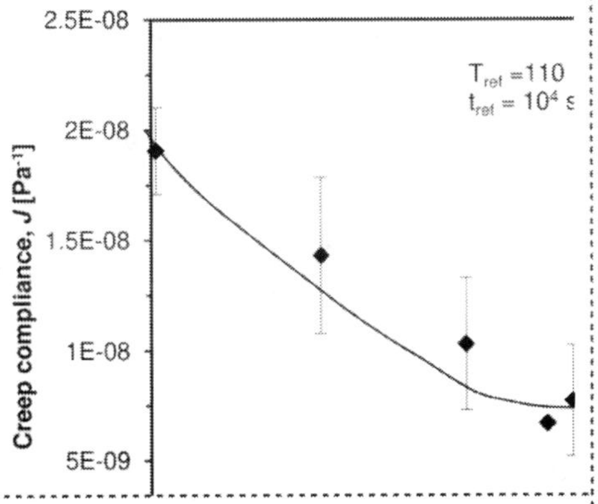


**Figure 4.** Shear creep compliance of POM copolymers at reference temperature 110 °C and time of 10000 s. Solid line is a manual fit to the experimental data.

## 4. Conclusion

Reduction of viscosity of PIM feedstock is very important, since it will facilitate the injection moulding of parts with complex geometry. However, it is important to maintain the solid mechanical properties in the moulded part. This will ensure shape retention and dimensional stability during demoulding and debinding steps of PIM.

In this study, it was observed that addition of 16wt% of WAX to POM copolymer reduces viscosity of the binder by approximately 20% and increases its shear creep compliance by 16%. The increase in creep compliance can be considered acceptable, but the reduction of viscosity can be considered insufficient, since the goal was to reduce viscosity by at least 10 times. Therefore, addition of WAX is not recommended to solve flowability problems of POM-based binder systems for PIM.

Furthermore, it was observed that viscosity of POM copolymers increases with average molecular weight following a power law relationship. On the other hand, shear creep compliance of POM copolymers decreases with average molecular weight reaching a plateau at molecular weights larger than 81100 g/mol. With these results on hand, one can recommend to use a POM-based binder with a molecular weight around 24410 g/mol. Using POM MW024 represents a reduction of 200 times in viscosity and an increase in creep compliance of 85% with respect to the current POM-based binder system. However, this can be considered a significant improvement, but care should be taken when handling the green part with the selected binder.

## 5. Acknowledgements

The authors would like to acknowledge the financial support of the Ad Futura Fund of the Republic of Slovenia, the Slovenian Research Agency and the Erasmus Mundus Program; as well as, the support of the staff at BASF Ludwigshafen, Germany for the synthesis and molecular weight characterization of POM copolymers.

## 6. References

[1] Krug S, Evans J R G, and ter Maat J R R 2001. *J. Am. Ceram. Soc.* **84** pp 2750-2766
[2] Krug S, Evans J R G, and ter Maat J R R 2002. *J. Eur. Ceram. Soc.* **22** pp 173-181
[3] Schneider J, Iwanek H, and Zum Gahr K H 2005. *Wear* **259** pp 1290-1298
[4] Attia U M, and Alcock J R 2012. *J. Mater. Process. Technol.* **212** pp 2148-2153

[5] Kongkhlang T, Tashiro K, Kotaki M, and Chirachanchai S 2008. *JACS* **130** pp 15460-15466
[6] Pielichowska K 2012. *J. Polym. Res.* **19** pp 9788-9798
[7] Krug S, Evans J R G, and ter Maat J R R 2001. *J. Eur. Ceram. Soc.* **21** pp 2275-2283
[8] Stringari G B, Zupančič B, Kubyshkina G, von Bernstorff B, and Emri I 2011. *Powder Technol.* **208** pp 590-595
[9] Gonzalez-Gutierrez J, Stringari G B, Zupančič B, Kubyshkina G, von Bernstorff B, and Emri I 2012. *J. Solid Mech. Mater. Eng.* **6** pp 419- 430
[10] Cox W P, and Merz E H 1958. *J. Polym. Sci.* **28** pp 619-622
[11] Stringari G B. 2010,*Time-dependent Properties of Bimodal POM- Applications in Powder Injection Molding, Master Thesis* (Ljubljana: Faculty of Mechanical Engineering, University of Ljubljana) chapter 3 pp 46-72
[12] Gonzalez-Gutierrez J, Oblak P, von Bernstorff B, and Emri I 2012. *Proc. 11th Global Conf. on Sustainable Manufacturing (Berlin)* (Berlin: Universitätsverlag der TU Berlin) pp 434-438
[13] Kubyshkina G, Zupančič B, Stukelj M, Grošelj D, Marion L, and Emri I 2011. *J. Biomater. Nanobiotech.* **2** pp 361-368
[14] Metlikovič P, and Emri I 1990, *Strojniški vestnik* **36** pp 101-104
[15] Gergesova M, Zupančič B, Saprunov I, and Emri I 2011 *J. Rheol.* **55**, pp 1-16
[16] Mutsuddy B C, and Ford R G 1995. *Ceramic Injection Molding* (London:Chapman and Hall) chapter 1 pp 1-26
[17] Jin H, Gonzalez-Gutierrez J, Oblak P, Zupančič B, Emri I, 2012 *Polym. Degrad. Stabil.* 97, pp. 2262-2272
[18] Gonzalez-Gutierrez J, Dorgham A, Stringari G, Zupančič B, von Bernstorff B, and Emri I 2012. *Intersectoral Almanac* (International and Russian Academy of Engineering) **5**, pp. 58-63
[19] Fox T G, and Flory P J 1951. *J. Phys. Colloid Chem.* **55**, pp 221-228

# Deformation, orientation and bursting of microcapsules in simple shear flow: Wrinkling processes, tumbling and swinging motions

**A Unverfehrt[1], I Koleva[2] and H Rehage,[1]**

[1] Technical University Dortmund, Department of Physical Chemistry, Germany
[2] Phoenix CBS GmbH, Bad Blankenburg, Germany

anja.unverfehrt@tu-dortmund.de, ivanka.koleva@tu-dortmund.de, heinz.rehage@tu-dortmund.de

**Abstract**. In a series of experiments we studied the deformation and orientation behaviour of microcapsules in simple shear flow. For a large number of capsules we observed folding processes which were induced by the bending resistance, by membrane pre-stresses or the mechanical asymmetry of the surrounding viscoelastic wall materials. Periodic oscillations of the inclination angle were detected for non-spherical particles. At low shear rates a tumbling motion occurred in which the capsule turned around its axis. A swinging mode at evaluated shear rates was accompanied by tank-treading motions, a rotation of the membrane around the capsule core. Between these two well-known motions we also observed an intermittent regime.

## 1. Introduction

Microcapsules with well-defined wall materials are widely used in technical, cosmetic and pharmaceutical applications. There is still a great demand for new designs of membranes with controlled release properties and bursting processes. Taylor-made capsules require, therefore, the knowledge of their mechanical properties. In scientific experiments capsules are not only used as transport carriers for active ingredients and substances but also as artificial model systems for biological cells, e.g. Red Blood Cells (RBC). In order to get more information on the flow induced deformation of biological cells, we investigated microcapsules in simple shear flow. For that matter the complex RBC membrane was replaced by a simple structured polymer material with adjustable viscoelastic properties. This procedure allowed to study basic principles and mechanisms with variable parameters.

A microcapsule generally consists of a liquid core enclosed by a deformable membrane. These particles are often synthesized by interfacial polymerization of emulsion droplets. The dynamic response of microcapsules in simple shear flow depends on many parameters. The initially shape [1], the viscosity ratio of the inner and outer phases [2], the membrane constitutive laws [3], the bending stiffness [4-7] and the membrane pre-stresses [8] have striking influence on the deformation and orientation behavior. In former experiments, we observed wrinkling instabilities for polysiloxane and polyamide microcapsules [9-11]. Osmotic pressures, membrane pre-stresses and the bending elasticity might be responsible for these phenomena [12-14].

[1] To whom any correspondence should be adressed

A model for initially spherical microcapsules in simple shear flow was first proposed by Barthès-Biesel [15]. A capsule with a membrane of negligible bending stiffness is deformed by hydrodynamic forces to an ellipsoid and this particle will be orientated at a constant angle of $\Theta = 45°$(figure 1).

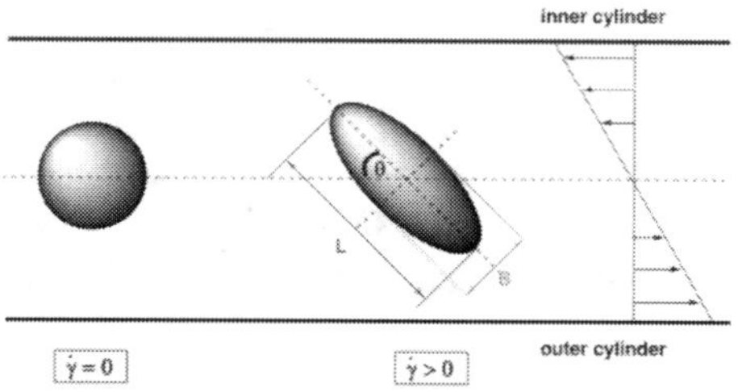


**Figure 1.** Deformation and orientation of an initially spherical microcapsule in shear flow.

The deformation is described in terms of the Taylor-Parameter $D$, in which $L$ and $B$ represent the major and minor axis of the elliptically deformed microcapsule.

$$D = \frac{L - B}{L + B} \tag{1}$$

The surface Young's modulus $E_s$ of the enclosing, viscoelastic shells can be determined using Hooke's law for small elastic deformations or Mooney-Rivlin's law for flexible membranes which exhibit non-linear elasticity. For purely elastic membranes, however, a linearity between the Taylor-parameter $D$ and the shear rate $\dot{\gamma}$ was detected in regime of small deformations. This enables the calculation of $E_s$ using the simple equation (2) [16,17]. Here $\eta_a$ denotes the viscosity of the outer phase, $r$ the capsule radius and $C_a$ the capillary number.

$$D = \frac{25\eta_a r\dot{\gamma}}{4E_s} = \frac{25}{4}C_a \tag{2}$$

Initially non-spherical capsules show an interesting orientation behavior. It depends mainly on the shear rate, the capsule shape and the viscosity ratio $\lambda$ of the inner and outer fluids. At low shear rates we observed a "tumbling motion" (TU), in which the inclination angle varied continuously between +90° and -90° [1,2,11]. The "swinging motion" (SW) was detected at evaluated shear rates and characterized by oscillations of $\Theta$ around positive values [1,2,11,18]. This mode was superimposed by a "tank-treading" (TT) motion. Tank treading describes the rotation of the membrane around its core. The tumbling motion and the swinging process were accompanied by periodic shape oscillations [1,2,9-11] The influence of the viscosity ratio was analyzed in model recently proposed by Keller and Skalak [19]. These authors observed that less viscous particles undertook a "tank treading" motion (TT) while more viscous ellipsoids tumbled. Skotheim and Secomb extended this model and included the "shape memory effect" [20]. They predicted that the transition from tumbling to swinging occurs via an intermittent regime (IR), in which a series of oscillation is interrupted by a tumble. Sui *et al.* presented a numerical simulation which also suggested the presence of an intermittent regime, but there was not sufficient evidence to confirm this special motion [1]. Recent studies of Yazdani *et al.* took the membrane viscosity of capsules into account [21]. They observed that a transition from SW to TU occurred when the membrane viscosity increased. Additionally they concluded that membrane viscosity had similar effects as the inner viscosity of the capsule.

In contrast to this theoretical work, less experimental results are published up to now [9-11,18, 22, 23]. To support these theoretical and numerical models we investigated microcapsules in simple shear flow by means of an optical flow cell (rheoscope).

## 2. Experimental

The capsule membranes were synthesized by interfacial polycondensation using self-assembled monolayers (SAM) of octadecyltrichlorosilane (OTS). In contact with water the oil soluble OTS molecules hydrolyzed and formed the corresponding alcohols (silanoles).

The formed alcohol was even more hydrophilic and adsorbed at the oil/water interface. In a second step silanole molecules condensated forming siloxanes. Silanoles could also react with the origin octadecyltrichlorosilane molecules.

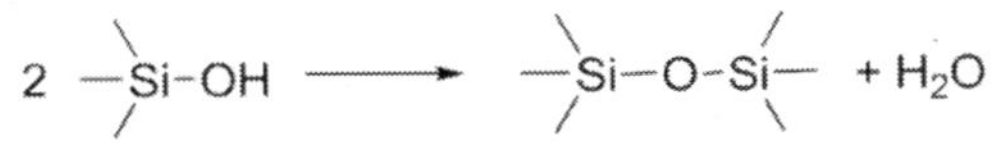

As OTS had three functional groups, cross-linking processes occurred, and we obtained a stable polysiloxane network with viscoelastic properties. The reaction took place at room temperature and slightly alkaline or acid conditions.

The polysiloxane network was characterized at the water/oil-interface by means of two-dimensional rheology. In these experiments we obtained basic information on the kinetics of surface gelation, the network type and the viscoelastic regime (LVR). The microcapsules were studied in an optical flow cell (rheoscope). For each experiment we used separate conditions. For interfacial shear rheometers the density difference should be high to give a planar interface. In an optical flow cell however, the densities of the organic and the water phase should be nearly the same in order to avoid sedimentation and creaming of capsules. These different conditions were realized by using mixtures of organic and polar solvents. To compare the results of the flat and curved polysiloxane membranes we adjusted the OTS bulk concentration in such a way that the interfacial monomer concentration in the different solvent mixtures was identical. We determined the mean surface area of the flat membranes and the microcapsules with diameters of 300 μm and adjusted the interfacial excess concentration $\Gamma$.

$$\Gamma = \frac{c \cdot V \cdot N_A}{A} \qquad (3)$$

$N_A$ denotes the Avogadro constant, $V$ represents the volume of the organic phase and $c$ denotes the bulk concentration of OTS. Equation (3) holds as long as all monomers adsorb at the interface and become part of the two-dimensional network. In previous experiments, we have shown that this condition is approximately fulfilled [23,24].

Relevant data of the volume and surface concentrations are summarized in Table 1.

**Table 1.** Interfacial excess concentration and volume concentration.

| $\Gamma$ / molecules nm$^{-2}$ | $c_{OTS}$ (flat networks) / mM | $c_{OTS}$ (capsules) / mM |
|---|---|---|
| 6.0 | 0.00137 | 0.2 |
| 12.0 | 0.00274 | 0.4 |

### 2.1. Flat membranes

The interfacial shear rheometer was based on a conventional rotational ARES-spectrometer (TA instruments). The ultra-thin membranes were investigated using a 2D-Couette-geometry. The outer cylinder consisted of a glass cup attached to the motor unit. A biconical titanium plate served as the measuring tool and was placed exactly at the interface between oil and water. 30 ml of the aqueous phase, consisting of 90 wt% glycerol and 10 wt% 1 M NaOH-solution ($\rho$ = 1.241 g/cm³), was poured into the glass cup. It was covered with 10 ml of an OTS solution dissolved in a mixture of p-xylene and 1,2,4-trichlorobenzene with a density of 1.1 g/cm³.

The two dimensional storage modulus $\mu'$ and the loss modulus $\mu''$ were calculated from experiments with sinusoidal oscillations. Three different measurements were carried out successively: A time-sweep-test to observe the process of surface gelation, a frequency-sweep-test to get information of the network-type and a strain-sweep-test to determine the viscoelastic regime.

### 2.2. *Synthesis of microcapsules*

In order to synthesize microcapsules with spherical shapes and equal sizes, we used a microfluidic device with T-shaped channel. In this instrument a disperse oil phase was injected vertically to the flow direction of a continuous aqueous phase. Due to the acting shear forces tiny oil droplets were generated which contained the reacting OTS monomers. The aqueous phase consisted of 90 wt% glycerol and 10 wt% of an aqueous 1 M NaOH-solution ($\rho$ = 1.241 g/cm³). To adjust the same density for the oil phase, OTS was solved in a mixture of p-xylene and 1,2,4-trichlorobenzene ($\rho$ = 1.243 g/cm³). We used automatic syringes in order to generate a constant solvent flow through the PVC-tube. We succeeded in producing capsules with diameters of 300 µm for aqueous solution velocities of $v_a$ = 85 ml/h ($d_{needle}$ = 2mm) and organic solution velocities of $v_i$ = 0.3 ml/h ($d_{needle}$ = 85µm). The polymerizing droplets were collected in a beaker where they remained stored until the membrane was formed. The density adjustment and the high viscosity (261 mPas) of the continuous phase prevented the capsules from sedimentation, coalescence and creaming processes. This procedure leaded to nearly monodisperse and spherical particles. A very small deviation from the true spherical shape was caused by the flow processes in the tube. After different polymerization times the microcapsules were inserted into the optical flow cell.

### 2.3. *Optical flow cell (rheoscope)*

The optical flow cell (figure 2) consisted of a coaxial cylinder system which was inserted into an inverse microscope. Due to the different radii with 42.4 and 41.0 mm a gap of 1.4 mm was formed. We induced a laminar shear field by rotating the cylinders in opposite directions. In the centre of the annular gap a stagnation zone was formed where the microcapsules did not translate. Each cylinder was connected to a DC motor drive instrument and could be controlled separately. For measurements an automatic control unit adjusted the shear rate and retained the capsule within the field of observation. The outer cylinder had glass windows to observe the particles within the gap. To avoid turbulences in simple shear flow a small Reynolds number was required. We therefore filled the gap with a highly viscous solvent, and all experiments were carried out in anhydrous glycerol.

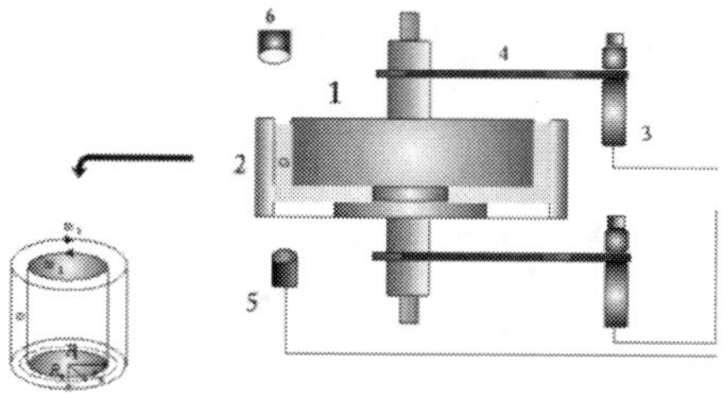


**Figure 2.** Schematic drawing of the rheoscope: 1- inner cylinder, 2 - outer cylinder, 3 - motor drive unit, 4 - belts, 5 - microscope objective, 6 - light source.

The microcapsules produced in the microfluidic device were placed via a syringe into the centre of the gap. In a series of experiments, we performed measurements with increasing, decreasing and constant shear rates. The capsule´s motion during flow was analyzed by means of a high speed camera (Kodak SR-Corder Motion Analyzer). A video file with 25 images/s and a log file with the value of the shear rate were recorded simultaneously. The capsule deformation and the inclination angle were calculated from the capsule contour and position using the NI VDM software. More details, concerning the optical measurements are summarized in [25].

### *2.4. Materials*

As octadecyltrichlorosilane is extremely reactive towards small amounts of water we used anhydrous solvents. P-xylene (99+%) and 1,2,4-trichlorobenzene (99+%) were purchased from Sigma-Aldrich and stored under argon. Their water content was determined by Karl-Fischer-Titration to 0.0025 wt% $H_2O$. OTS (ABCR) was purified via vacuum distillation (7.5 $*10^{-2}$ Torr,~155°C). Anhydrous glycerol was purchased from Merck.

## 3. Results and discussion

### *3.1. Interfacial shear rheology of the planar polysiloxane networks*

The kinetics of the surface gelation processes were measured by time-sweep-tests. In these experiments the evolution of the storage and loss moduli and, hence, cross-linking processes, could be observed as a function of time. The angular frequency $\omega$ and the strain amplitude $\gamma$ remained constant during these measurements. The striking increase of $\mu'$ and $\mu''$ points to the presence of viscoelastic networks. Typical results of these measurements are shown in figure 3.

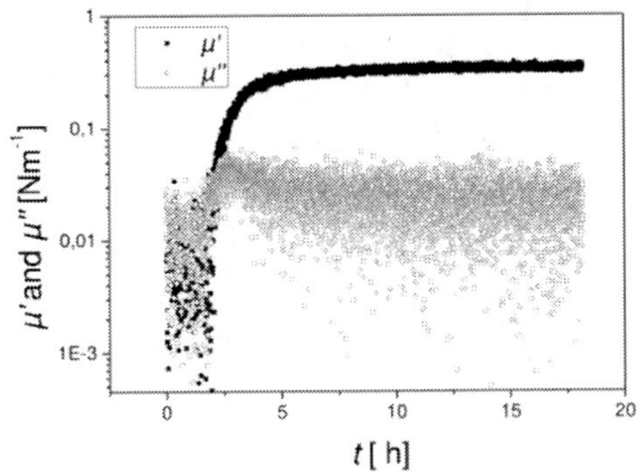


**Figure 3.** Time-sweep-experiment showing the kinetics of surface gelation (OTS solution with $\Gamma = 12$ molecules/nm², $c = 0.00274$ mM $\omega = 1$ rad/s, $\gamma = 0.2$ %).

After two hours polymerisation time the gelation point ($\mu'=\mu''$) was reached and the storage modulus increased strongly up to values of 0.33 $Nm^{-1}$. The loss modulus remained nearly constant with values about one decade smaller. The membrane showed striking viscoelastic properties. When both moduli $\mu'$ and $\mu''$ attained the plateau regime we supposed that the network formation was completed. In order to make sure that the capsules did not continue their polymerization after desired time scales, we stopped this chemical reaction by removing the water and transferring the capsules into anhydrous glycerol.

It turned out that many parameters had influence on the gelation process such as the OTS concentration and the mixture of the organic solvents. These processes were extensively investigated in our group [11, 23-25]. In this context it is interesting, to note, that the addition of 1,2,4-trichlorobenzene to p-xylene showed similar values of $\mu'$ in the range of $\Gamma = 6$ and 12 molecules per nm² as the pure p-xylene solution. On grounds of this special situation we performed the experiments at these conditions [23]. The elastic modulus slightly increased with rising OTS concentrations but it was not greatly influenced by the water content of the organic phase [11].

A typical frequency-sweep-test is presented in figure 4. In this experiment the angular frequency was varied between 0.01 and 100 rad/s and the strain amplitude was adjusted to $\gamma = 0.2$. The constant curve progression of $\mu'$ pointed to the existence of a chemical cross-linked, rubber-elastic membrane.

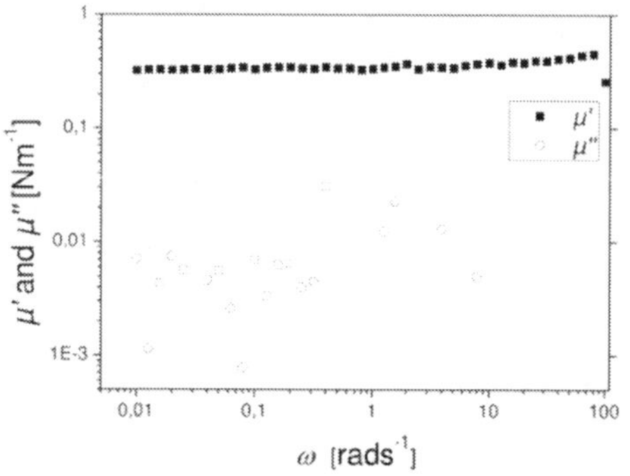


**Figure 4.** Frequency-sweep-test: $c$(OTS) = 0.00274 mM, $\gamma$ = 0.2%.

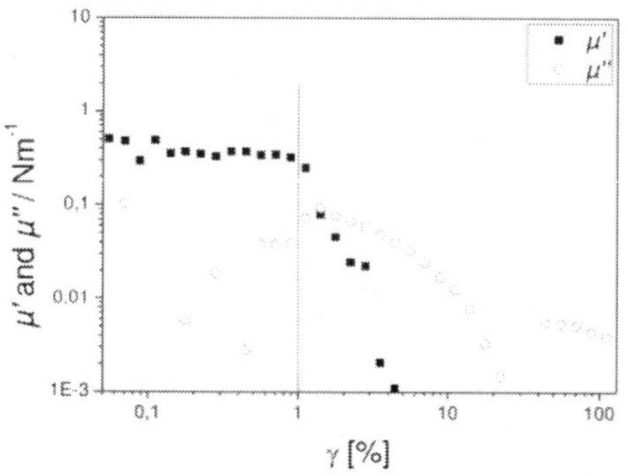


**Figure 5.** Strain-sweep-test: $c$(OTS) = 0.00274 mM, $\omega$ = 1 rad/s.

The relative large scattering of the loss moduli values was induced by the large differences between $\mu'$ and $\mu''$ and the bad signal to noise relation of the torque transducer for the evaluation of these experimental data.

The non-linear viscoelastic properties of the polysiloxane networks are represented in figure 5. The upper limit of the linear viscoelastic range (LVR) was determined from the descent of the elastic modulus. The synthesized membranes showed a rather small LVR of about 1 %. We therefore applied small deformations below this yield value so that the membrane response could still be described by linear differential equations. Above the LVR we also observed irreversible deformations and network rupture.

### *3.2. Microcapsules in simple shear flow*

*3.2.1. Folding and bursting processes.* In a series of experiments we investigated microcapsules with OTS concentrations of 0.2 and 0.4 mM at different polymerization times. In contrast to the time-sweep-time-tests (figure 3), network formation of the curved membranes occurred much faster. This could be determined by measurements in the optical flow cell regarding the deformation as a function of shear rate. In comparison to similar emulsion droplets the deformation curves of capsules showed lower slopes. This is easy to understand because instead of interfacial tension, the elasticity of the membrane creates resistance to deformation. Typical elastic moduli for polysiloxane capsules with $\Gamma$ = 6 and 12 molecules/nm² range between $E_s$ = 1.47 and 1.68 Nm$^{-1}$ (polymerization time of 60 min.) [11]. The surface Young's moduli $E_s$ were calculated from the linear *slope* between the measured deformation $D$ and the shear rate $\dot{\gamma}$ according to equation 2.

Figure 6 presents a typical result of rheoscopic experiments, which illustrates the difference between an emulsions droplet and a capsule (axis ratio $B/L$ = 0.99). For both suspensions the viscosity ratio was identical ($\lambda$=0.001). While the emulsion droplet elongated continuously with raising shear rate, the capsule deformed only slightly. At a critical shear rate of about 25-26 s$^{-1}$ however, the capsule deformation increased rapidly. At these conditions irreversible membrane deformation followed by network rupture occurred.

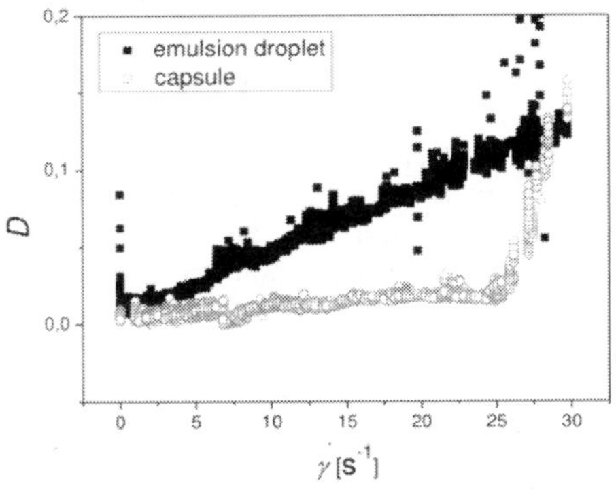


**Figure 6.** Comparison of the deformation behaviour of an emulsion droplet and a microcapsule in simple shear flow; polymerisation time = 4 h, $\Gamma$ = 12 molecules/nm$^2$, T = 26°C.

Close inspection revealed that the capsule showed wrinkling instabilities during flow. These folding processes always appeared parallel to the main axis of the capsule (figure 7). With increasing shear rate the folding was more pronounced, but reaching the membrane rupture point they became blurred and finally disappeared. After breaking the deformation of the capsule could easily be followed with the naked eye, because elongation proceeded rapidly. In this range we should rather talk about an emulsion droplet because the membrane was not intact anymore. It is evident that the slope of the broken capsule is steeper than the curve of the pure emulsion droplet. This phenomenon can be explained by membrane fragments remaining on the surface which lower the interfacial tension. It is worthwhile to mention that the disappearance of wrinkling instabilities was accompanied by rupture processes of the polysiloxane networks.

The appearance of folding processes is even more difficult to understand. They occurred when shear forces were acting on the surface of the capsules (figure 7).

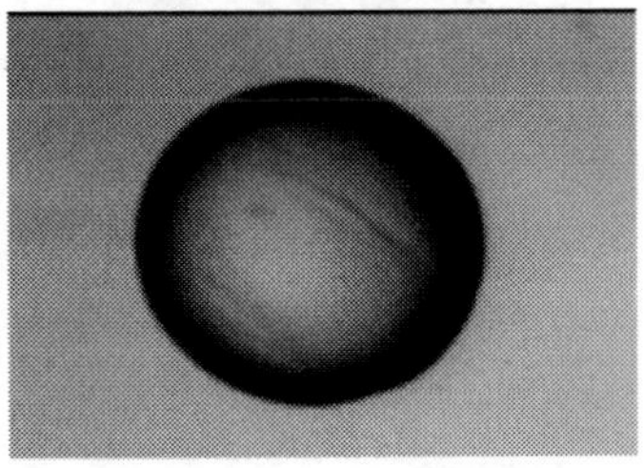

**Figure 7.** Wrinkling instabilities during shear flow for an microcapsule with $\Gamma$ = 12 molecules/nm$^2$; polymerisation time = 5 h, T = 26°C.

An explanation for this phenomenon might be the presence of negative membrane tensions. It was recently suggested that osmotic pressures could avoid folding processes by inducing positive membrane stresses [8]. Since the measurements were carried out under ambient air and at room temperature, we cannot exclude that tiny amounts of water were dissolved in the different phases. Especially glycerol is very hygroscopic. The water content of the organic solution should not be affected much because it was stored under argon and used immediately after preparing. In order to calculate the magnitude of osmotic pressures the solubility of water molecules and glycerol in the oil phase must be considered. As these solubilities were very low [11], the resulting pressure gradients can only be small. In a series of experiments we did not observe any volume changes of the capsules which point to the presence of transport processes across the interfaces. For these reasons the effects of osmotic pressures can be neglected.

It is also conceivable that wrinkling instabilities depend on the bending elasticity of the surrounding membranes. Minor changes during the surface gelation can lead to slightly deviations of membrane properties and might be responsible for the different behavior of capsules. It is also possible that the membranes are pre-stressed and not in an equilibrium state. This can be caused during the synthesis by small shear stresses, gravitational forces or diffusion processes.

According to recent studies of Yazdani *et al.* [21] the wrinkles should disappear in the presence of membrane viscosity when the viscosity ratio $\lambda$ increased. In our case both the viscosity ratio ($\lambda = 0.001$) and the membrane viscosity ($\mu''$) were very small. It is, hence, difficult to estimate in which degree the membrane viscosity is involved in folding processes. At the present state it seems also possible that the bending resistance or the mechanical anisotropy, which is caused by oriented membrane molecules, induce buckling instabilities. Polysiloxane membranes very often showed this phenomenon. Membrane folding processes might even exist in the quiescent state if their sizes are below the limit of the optical resolution of our microscope. Such small phenomena are, of course, not visible to the naked human eye.

A series of very small membrane wrinkling processes are shown in figure 7 (red rectangle). During capsule movement these processes are generally easier to observe in a video film. However, the interesting aspect here is another phenomenon: the bursting process. As soon as the membrane breaks, the wrinkling processes disappear and the capsule deformation increases.

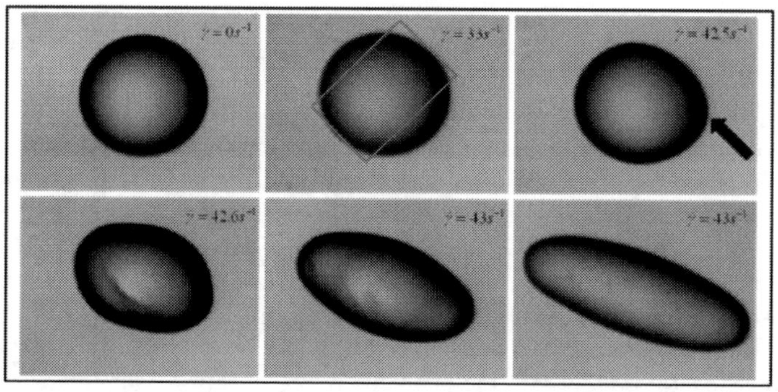

**Figure 8.** Bursting and wrinkling process of a microcapsule in simple shear flow. (Polymerisation time = 94 min., $\Gamma$ = 12 molecules/nm², T = 26°C , $B/L$ = 0.97).

During the whole measurement the capsule deformation remained nearly constant until a critical shear rate of 42-43 $s^{-1}$ was reached (see also figure 9). The snapshot images at shear rates of 0 and 33 $s^{1}$ illustrate that no remarkable capsule deformation occurred in this range. This small deformability agrees well with the results of the LVR.

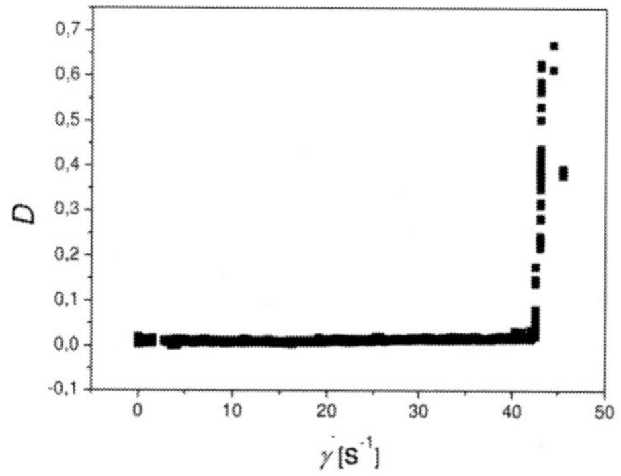


**Figure 9.** The deformation curve of an microcapsule in shear flow; polymerisation time = 94 min., $\Gamma$ = 12 molecules/nm², T = 26°C , $B/L$ = 0.97.

At $\dot{\gamma} = 42{,}5 s^{-1}$ the shear forces were sufficent to induce a bulge which initiated the bursting. The capsule membrane was destroyed within a few seconds. This means that the deformation proceeded so fast that even the now generated emulsion droplet (figure 8 last photo) soon teared apart. The almost vertically slope in figure 9 gives an impression of this very rapid bursting mechanism.

*3.2.2. Tumbling and swinging motions.* Non-spherical particles showed an interesting orientation behavior. This process was very different from the performance of spherical capsules. As the viscosity ratio was adjusted to be constant ($\lambda$ = 0.001), the shear rate was the only parameter which could be varied. In figure 9 different oscillation motions of a capsule with an axis ratio of B/L = 0.96 are presented. In regimes of small shear rates we observed tumbling motions. In this mode the inclination angle varied continuously between +90° to -90° (figure 10), which correlates to one half turn of the capsule rotation. Shape oscillations accompany this motion (figure 10). First inspections revealed that the capsule tumbled similar to a rigid body due to its small deformability. Obviously the hydrodynamic forces were not large enough to align the capsule in the flow direction. At evaluated shear rates, however, a swinging motion was detected (figure 11), superimposed by a tank-treading motion. Here, the inclination angle varied around positive values while the membrane rotated around its core. Tank-treading could be observed if tracer particles were used or inhomogeneities on the membrane surface were visible.

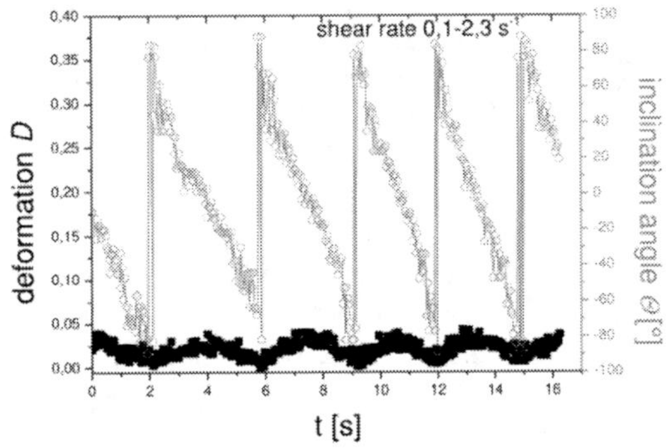


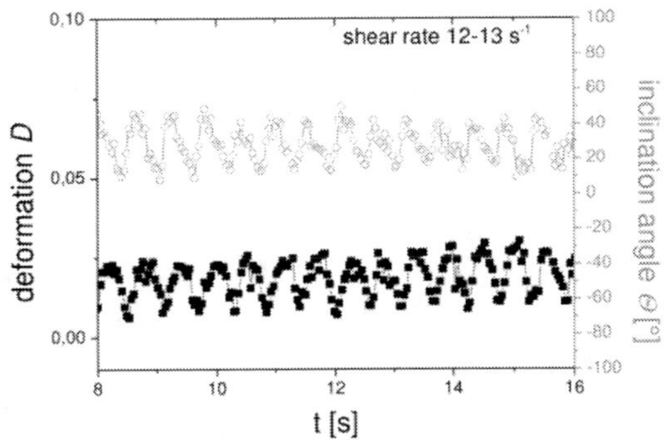


**Figure 10.** Tumbling motion of a microcapsule with an axis ratio of $B/L$= 0.96; polymerization time = 75 min., $\Gamma$ = 12 molecules/nm², T = 26°C.

**Figure 11.** Swinging motion of a microcapsule with an axis ratio of $B/L$= 0.96; polymerization time = 75 min., $\Gamma$ = 12 molecules/nm², T = 26°C.

Obviously the growth of shear forces induced TT and orientated the capsule to positive inclination angles. The oscillations around a constant value were caused by the membrane rotation and the "shape memory effect" [20] and therefore by periodic changes of the capsule shapes. The minor and major axis of the flexible particles varies continuously. It is worthwhile, to mention, that with rising shear rate the amplitude diminished and finally faded away. This oscillation decay was another indication for capsule bursting. It correlated very well with the steep rise of the deformation, and both phenomena occurred simultaneously (figure 12). The transition from TU to SW usually occurred via an intermittent motion. This regime was characterized by single oscillations ranging from +80° to -80°, +80° to -30°and +75° to -85° (figure 13).

At these conditions the capsule failed to make a complete tumble. Due to the growing shear forces the capsule underwent shape oscillations in which a nearly spherical contour was reached. This globular shape only existed for an instant, but made it difficult to determine an axis and an inclination angle. It was evident that the different capsule shapes during the oscillation period caused these varying orientation fluctuations.

For the capsule presented in figure 13 the IR was slightly different from the motion described by Skotheim, Secomb and Sui et al. [1,20]. In this investigation the SW mode was only interrupted by one single tumble motion. The inclination angles changed between positive and negative values, but did not reach -90° and +90°. For a quantitative comparison between theory and experiments it must be considered that the numerical simulations [1] were carried out with a viscosity ratio of $\lambda = 1$, and we have, instead, a very low value of $\lambda = 0{,}001$.

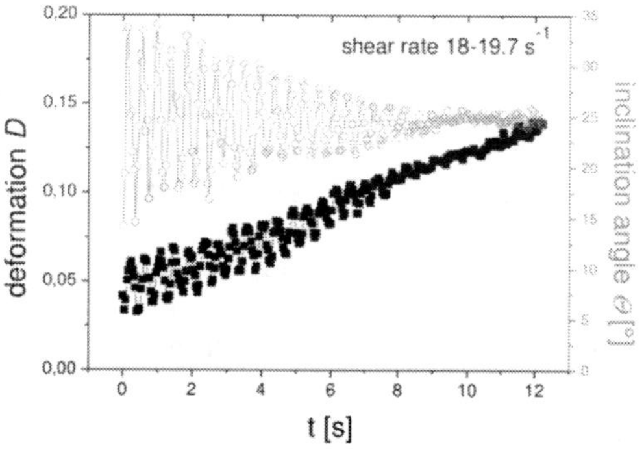


**Figure 12.** Swinging regime of a microcapsule with an axis ratio of $B/L = 0.98$; polymerization time = 70 min., $\Gamma = 6$ molecules/nm², T=25°C.

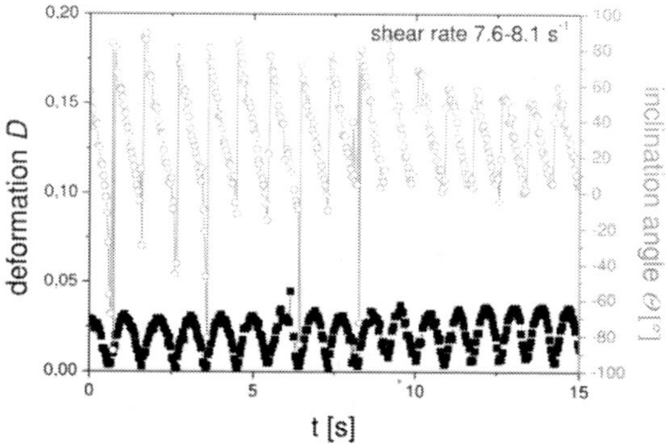


**Figure 13.** Intermittent and swinging regime of a microcapsule with an axis ratio of $B/L = 0.98$; polymerization time = 70 min., $\Gamma = 6$ molecules /nm², T=25°C.

## 4. Conclusion

In a series of experiments we systematically measured the shear induced deformation and orientation behaviour of polysiloxane microcapsules. These particles showed a rather small linear-viscoelastic regime up to deformations of about 1%. Above this yield value non-linear properties and membrane busting occurred.

For a large number of capsules we observed shear induced membrane wrinkling instabilities. These folding processes were probably induced by pre-stressed membranes, the bending resistance or the mechanical anisotropy of the surrounding polymer layers. Network rupture immediately quenched the presence of wrinkling processes.

For non-spherical capsules we observed at least three different motions and periodic oscillations. In the regime of low shear rates tumbling motion occurred similar to the behaviour of a rigid body. At increasing shear forces we detected a swinging mode which was accompanied by a tank-treading membrane rotation. The transition from tumbling to swinging occurred via an intermittent regime.

## References

[1] Sui Y, Low H T, Chew Y T and Roy P 2008 *Phys. Rev* E **77** 016310
[2] Bagchi K and Kalluri R M 2009 *Phys. Rev.* E **80** 016307
[3] Barthès-Biesel D, Diaz A and Dhenin E 2002 *J. Fluid Mech.* **460** 211
[4] Pozrikidis C 2001 *J. Fluid Mech.* **440** 269
[5] Le D V 2010 *Phy. Rev* E **82** 016318
[6] Sui Y, Chew Y T, Roy P, Chen X B and Low H T 2007 *Phys. Rev.* E **75** 066301
[7] Noguchi H 2010 *Phys. Rev.* E **81** 056319
[8] Lac E and Barthès Biesel D 2005 *Phys. Fluids* **17** 072105
[9] Rehage H, Husmann M and Walter A 2002 *Rheol Acta* **41** 292
[10] Walter A, Rehage H and Leonhard H 2001 *Colloid Surf.* A **183-185** 123
[11] Koleva I and Rehage H 2012 *Soft Matter* **8** 3681
[12] Barthès-Biesel D 2011 *Curr. Opin. Colloid Interface* **16** 3

[13] Finken R and Seifert U 2006 *J. Phys.: Condens Matter* **18** L185
[14] Lac E, Bartès-Biesel D, Pelekasis N A and Tsamopolous J 2004 *J. Fluid Mech.* **516** 303
[15] Barthès Biesel D 1980 *J. Fluid Mech.* **100** 831
[16] Barthès-Biesel D 1991 *Physika* A **172** 103
[17] Barthès-Biesel D 1998 *Progr. Colloid Polym. Sci.* **111** 58
[18] Abkarian M, Faivre M and Viallat A 2007 *Phys. Rev. Lett.* **98** 188302
[19] Keller R S and Skalak R 1982 *J. Fluid Mech.* **120** 27
[20] Skotheim J M and Secomb T W 2007 *Phys. Rev. Lett.* **98** 078301
[21] Yazdani A and Bagchi P 2013 *J. Fluid Mech.* **718** 569
[22] Chang K S and Olbricht W L 1993 *J. Fluid Mech.* **250** 609
[23] Koleva I and Rehage H 2012 *Soft Matter* **8** 7672
[24] Husmann M 2001 PhD Thesis, Polyorganosiloxanfilme zwischen fluiden Phasen: Rheologisches Verhalten von ultradünnen Membranen, Mikrokapseln und Kapseldispersionen University of Essen
[25] Koleva I 2011 PhD-Thesis, Untersuchungen der mechanischen Eigenschaften von Polysiloxanmembranen und der Deformations- und Orientierungsdynamik von Polysiloxanmikrokapseln im Scherfeld, University of Dortmund

# Rheological properties of purified illite clays in glycerol/water suspensions

**I Dusenkova**[1,*], **J Malers**[1], **L Berzina-Cimdina**[1]

[1]Institute of General Chemical Engineering, Riga Technical University, Azenes street 14/24, Riga, LV-1048, Latvia

*E-mail: inga.dusenkova@rtu.lv

**Abstract**. There are many studies about rheological properties of clay-water suspensions, but no published investigations about clay-glycerol suspensions. In this work apparent viscosity of previously purified illite containing clay fraction < 2 μm and glycerol/water suspensions were investigated. Carbonates were removed by dissolution in hydrochloric and citric acids and other non-clay minerals were almost totally removed by centrifugation. All obtained suspensions behaved as shear-thinning fluids with multiple times higher viscosity than pure glycerol/water solutions. Reduction of clay fraction concentration by 5% decreased the apparent viscosity of 50% glycerol/water suspensions approximately 5 times. There was basically no difference in apparent viscosity between all four 50% glycerol/water suspensions, but in 90% glycerol/water suspensions samples from Iecava deposit showed slightly higher apparent viscosity, which could be affected by the particle size distribution.

## 1. Introduction

Clays and clay minerals are widely used in cosmetics, health care and therapeutic products as excipients and active ingredients. The most important properties for these applications are high specific area and adsorptive capacity, chemical inertness and rheological properties. Most cosmetic and health care products are designed for topical application, which require appropriate rheological properties. These products must flow easily when deformed and retain their shape in repose. The addition of clay minerals improves the rheological properties of suspensions and emulsions, therefore they are used as excipients and thickening agents in many semi-solid preparations [1-4].

An important factor for application of clay minerals in cosmetics and health care is the purity. Most naturally occurring clays typically are mixtures of clay and non-clay minerals [5]. The value and application of certain clays depends on the required purification, which accordingly depends on the type and amount of non-clay minerals. In Latvia the most abundant clay mineral is illite, but it can also contain kaolinite and mixture of non-clay minerals, such as quartz, feldspar, calcite, dolomite, metal oxides and hydroxides and muscovite. Latvia has large clay reserves, but they are traditionally used in building materials and pottery. To use Latvian clays as excipients in cosmetic and therapeutic products, it is necessary to perform simple, but effective purification process and to investigate the rheological properties of purified illite. In this study the rheological properties of illite and glycerol/water suspensions were investigated, because glycerol is often used in semi-solid cosmetic and health care products due to its moisturizing, soothing and healing properties [6].

## 2. Materials and methods

### *2.1. Materials*

Clay samples were obtained from two deposits in Latvia – Laza and Iecava. They were matured in distilled water for 1 month and fractionated by wet sieving method through 230 mesh (63 μm) sieve. Both samples have different colour (Figure 1) and mineralogical composition (Table 1).

**Figure 1.** Powders of clay samples a) Laza, b) Iecava.

**Table 1.** Mineralogical composition of clays (wt%) with standard deviation 2-4 %

| *Deposit* | *Illite* | *Kaolinite* | *Chlorite* | *Quartz* | *Feldspar* | *Calcite* | *Dolomite* | *Muscovite* |
|---|---|---|---|---|---|---|---|---|
| **Iecava** | 51 | - | - | 11 | 26 | - | 4 | 8 |
| **Laza** | 34 | 10 | 8 | 13 | 16 | 8 | 5 | 6 |

### *2.2. Dissolution of carbonates*

Calcite and dolomite were removed by dissolution in 1M hydrochloric and citric acid solutions. 50g of clay samples was mixed with 200mL distilled water and stirred. The acid was added stepwise, keeping the pH above 4.5, in order to avoid damaging the structure of clay minerals [7]. All carbonates were dissolved in 2-4 days (depending on the sample and acid), when the pH was constant at values 5-6. Afterwards the clay suspensions were washed with distilled water until the electrical conductivity was under 10 μS/cm.

### *2.3. Preparation of clay and glycerol/water suspensions*

After removal of carbonates clay fraction < 2 μm was obtained by centrifugation of the suspensions at 1100 rpm for 3 minutes, then dried in 105°C and crushed into powder. For the preparation of suspensions, a chosen amount of clay powder was mixed with certain amount of distilled water until the obtained paste was homogenous. Then a pure glycerol was added to obtain 50% or 90% (w/w) glycerol/water concentration. The suspensions of 90% glycerol were homogenized and matured for 1 week, but 50% - for 2 weeks. The chosen concentrations of clay samples were obtained by diluting suspension with the highest concentration.

### *2.4. Characterization of the samples*

X-ray powder diffraction (PANalytical X'Pert PRO diffractometer with spinning sample holder) was used to identify the mineralogical composition of clay samples before and after removal of carbonates (sample fraction < 63 μm) and the purity of obtained clay fraction < 2 μm.

The measurements were taken with rotational viscometer (Fungilab, Expert series) at room temperature 20±1°C and shear rates 1, 5, 10, 20, 50 and 100 rpm. At each shear rate the viscosity data were recorded every 3 seconds for totally 60 seconds (except for 1 rpm where it was 180 seconds), before changing to higher shear rate. At each shear rate from all recorded viscosity values the average value was used in further research.

The particle size distribution in the suspensions was determined by laser particle size analyser Analysette 22, with measuring range 0.01-53 μm and laser wavelength 650 nm.

## 3. Results and Discussion

The mineralogical composition of clay samples after the purification process is shown in Table 2. All carbonates are completely dissolved by both acids. Small amounts of quartz and feldspar are still present in clay samples from Laza deposit, but samples from Iecava deposit contain basically pure illite.

**Table 2.** Mineralogical composition of obtained clay fraction (wt%) with standard deviation 2-4 %

| *Deposit* | *Used acid* | *Sample* | *Illite* | *Kaolinite* | *Quartz* | *Feldspar* |
|---|---|---|---|---|---|---|
| **Iecava** | HCl | Ie_HCl | 97 | - | 2 | 1 |
| | Citric acid | Ie_C | 96 | - | 2 | 2 |
| **Laza** | HCl | L_HCl | 64 | 20 | 8 | 8 |
| | Citric acid | L_C | 66 | 18 | 9 | 7 |

Pure 50% and 90% glycerol/water solutions are Newtonian fluids with viscosity 0,006 and 0,219 Pa*s at 20°C, respectively. The results in Figure 3 show that the addition of clay fraction increase the apparent viscosity multiple times and the obtained suspensions behave as Non-Newtonian fluids, where the apparent viscosity decreases by increasing the shear rate. The magnitude of increase depends on the clay sample concentration and shear rate. In 50% glycerol/water suspensions samples

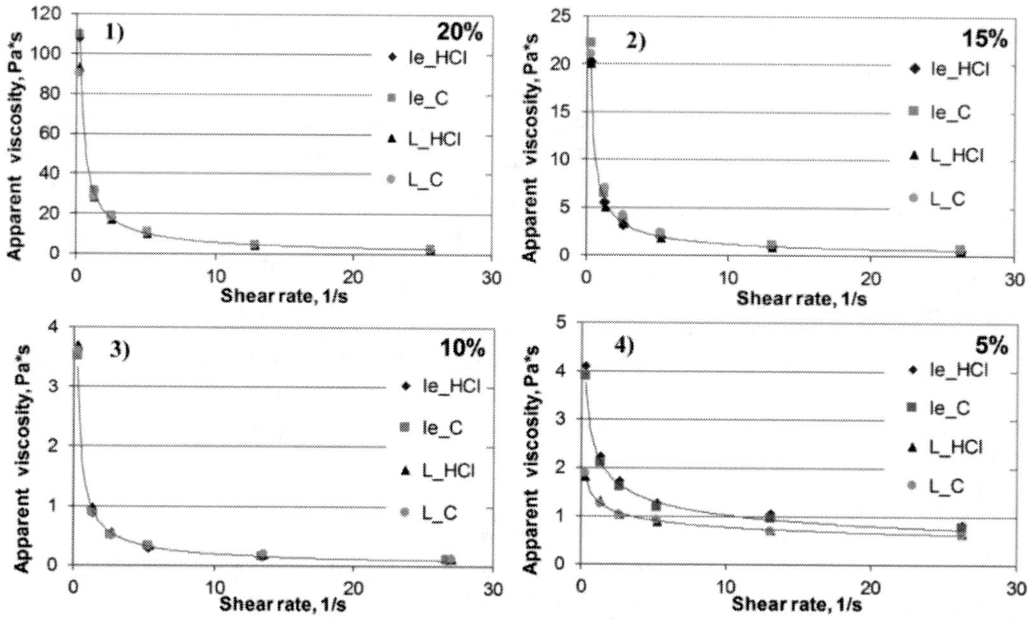


**Figure 2.** Apparent viscosity of 50% (1-3) and 90% (4) glycerol/water suspensions with 20, 15, 10 and 5 wt% of clay fraction

from both deposits show almost identical visocosity values, only with small differences at the lowest shear rates in suspensions containing 20 and 15wt% of clay fraction. Reduction of clay fraction concentration by 5% the apparent viscosity decreases approximately 5 times. From all 90% glycerol/water suspensions, clay samples from Iecava deposit shows the highest apparent viscosity (Figure 2 (4)). This could be connected with the particle size distribution (Figure 3), where Ie_HCl and L_C containing suspensions show finer particles than L_HCl and L_C containing suspensions. At the same time the difference between the particle size distribution in 50% glycerol/water suspensions is approximately 2 times smaller, which obviously does not affect the apparent viscosity (Figure 2 (1-3)). The differences in particle size distribution between suspensions containing clay samples from Iecava and Laza deposits could be connected to the fact, that samples from Laza contain quartz and feldspar particles, which act as cementing agents [7].

To describe the rheological behavior of suspensions, Herschel-Bulkley (H-B) model was used:

$$\tau = \tau_0 + m\gamma^n \qquad (1)$$

where $\tau$ is the shear stress, $\tau_0$ is the yield stress, $\gamma$ is the shear rate, m is the consistency coefficient and n is the flow behavior index. The H-B model can be used for shear-thinning and shear-thickening fluids [8]. The H-B model fits well the $\tau$ versus $\gamma$ data, giving correlation coefficients between 0,960 and 0,999. The obtained H-B model parameters are reported in Table 3. As all flow behaviour indices n are less then 1 ($n < 1$), all suspensions behave as shear-thinning fluids. All suspensions also exhibit yield stress, where a significant increase is observed at clay concentration 20 wt% in 50% glycerol solution.

**Table 3.** Herschel-Bulkley model parameters of clay suspensions

| **Clay fraction concentration, wt%** | | **Ie_HCl** | **L_HCl** | **Ie_C** | **L_C** |
|---|---|---|---|---|---|
| **50% glycerol** | **90% glycerol** | | | | |
| Yield stress ($\tau_0$), Pa | | | | | |
| | 5% | 0,25 | 0,23 | 0,37 | 0,32 |
| 10% | | 0,78 | 0,80 | 0,56 | 0,70 |
| 15% | | 1,88 | 0,87 | 0,64 | 0,68 |
| 20% | | 21,10 | 18,80 | 21,90 | 19,10 |
| Flow behavior index (n) | | | | | |
| | 5% | 0,71 | 0,82 | 0,72 | 0,84 |
| 10% | | 0,48 | 0,5 | 0,34 | 0,52 |
| 15% | | 0,35 | 0,23 | 0,24 | 0,21 |
| 20% | | 0,18 | 0,20 | 0,19 | 0,20 |
| Flow consistency coefficient (m), Pa*s$^n$ | | | | | |
| | 5% | 2,11 | 1,11 | 1,86 | 1,01 |
| 10% | | 0,35 | 0,43 | 0,59 | 0,44 |
| 15% | | 3,90 | 5,80 | 7,26 | 8,52 |
| 20% | | 38,08 | 34,87 | 39,5 | 35,64 |

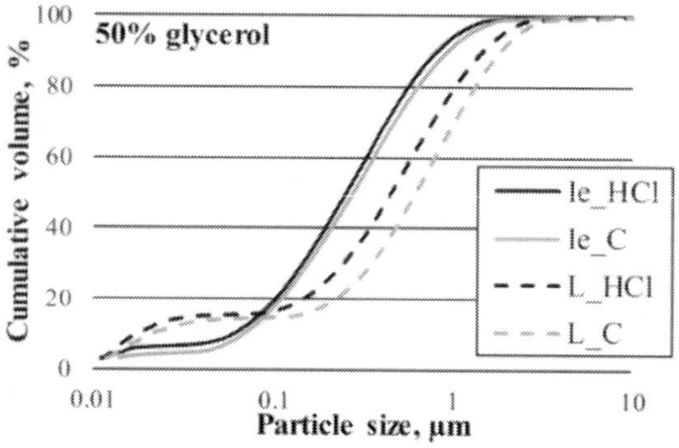


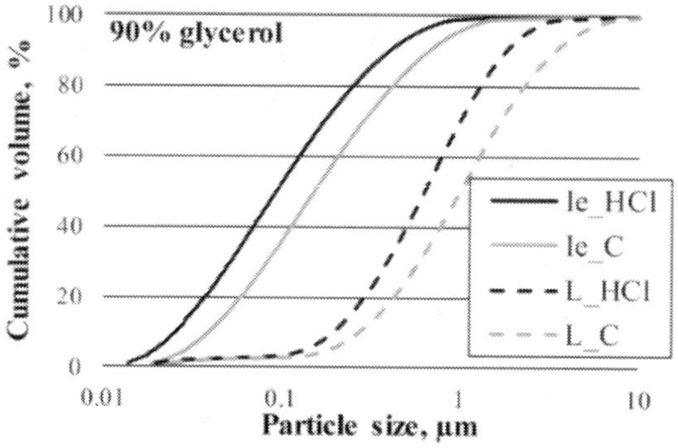


**Figure 3**. Particle size distribution in suspensions

## 4. Conclusions

The addition of clay fraction < 2 µm to glycerol/water solutions increase the viscosity multiple times and the obtained suspensions behave as shear-thinning fluids. The magnitude of increase in 50% glycerol/water suspensions is influenced by the added clay fraction concentration. Based on the obtained results, purified illite containing clay fraction can be used as excipients and thickeners in glycerol containing semi-solid production.

**Acknowledgements**
This work has been supported by the National Research Program of Latvia 2010-2013 "Sustainable use of local resources (underground resources, wood, food and transport) – new products and technologies (NatRes)" within the project V7904.

**References**

[1] Viseras C, Aguzzi C, Cerezo P and Lopez-Galindo A 2007 *Appl. Clay Sci.* **36** 37-50
[2] Carretero M I and Pozo M 2009 *Appl. Clay Sci.* **46** 73–80
[3] Carretero M I 2002 *Appl. Clay Sci.* **21** 155-163
[4] Carretero M I, Gomes C S F and Tateo F 2006 Clays and human health. *Handbook of Clay Science.* Elsevier Ltd., Amsterdam, **1** 717–741
[5] Lopez-Galindo A, Viseras C and Cerezo P 2007 *Appl. Clay Sci.* **36** 51-63
[6] Tan W H, Abdul Aziz A R and Aroua M K 2013 *Renew. Sust. Energ. Rev.* **27** 118-127
[7] Carrado K A, Decarreau A, Petit S, Bergaya F and Lagaly G 2006, Synthetic clay minerals and purification of natural clays. *Handbook of Clay Science*. Elsevier Ltd., Amsterdam, **1** 115-139
[8] De Blasio F V 2011 Non-Newtonian Fluids, Mudflows, and Debris Flows: A Rheological Approach. *Introduction to the Physics of Landslides*, Springer 89–130

Journal of Physics: Conference Series **602** (2015) 012004 doi:10.1088/1742-6596/602/1/012004

# Rheological investigations of tailings of kimberlite ore dressing and numerical simulation of its behaviour in PLAXIS

**A Korshunov**[1], **A Nevzorov**[2]

[1] Associate Professor, Northern (Arctic) Federal University, Department of Geotechnics – Arkhangelsk – Russia

[2] Professor, D.E., Northern (Arctic) Federal University, Department of Geotechnics – Arkhangelsk – Russia

E-mail: a.a.korshunov@yandex.ru, a.l.nevzorov@yandex.ru

**Abstract**. The article presents the results of analysis of rheology properties of sandy-clay tailings (wastes) of kimberlite ore dressing of the diamond deposit (Russia, Arkhangelsk region). The coefficient of secondary compression as main parameter of soil's creep is defined by implementing standard one-dimensional consolidation test. The linear correlation between initial void ratio and coefficient of secondary compression of sandy-clay tailings were obtained. For numerical simulation of tailing's behaviour subject to its rheology properties in Soft Soil Creep (SSC) model (time independent behaviour) of PLAXIS software was used. According to laboratory tests calibration of SSC model was implemented. It allows predicting dam's safety and reliability for long-term outlook.

## 1. Introduction

Development of diamond deposits in the Arkhangelsk region requires to build special storage for industrial wastes, so-called tailing dumps. One of the main engineering structures of tailing dump is multi-tiered dam (Fig. 1). In the operation of the tailing dump a staged construction of dam and washing of tailings on upstream slope of the dam are implemented. Tailings, which are presented by sandy- clay soils with specific physical and mechanical properties, formed the foundation for the 2nd and following tiers of dam. Taking into consideration of the long-term operation and rate of responsibility of dam, specific properties of tailings and its time-to-time variability, the estimation of dam's reliability subject to rheological properties of the foundation is an actual objective.

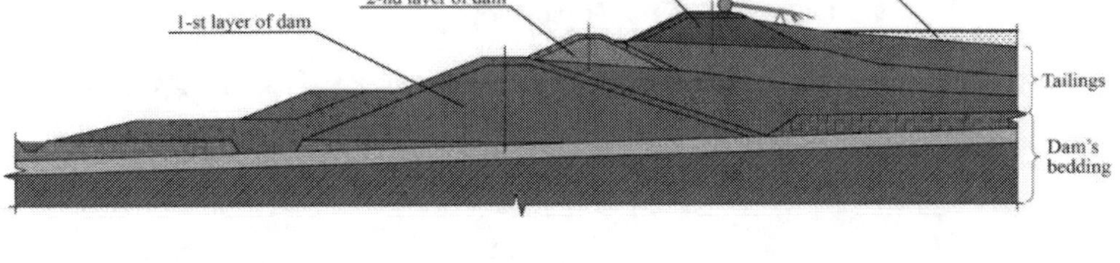


**Figure 1.** Cross-section of 3-tiered dam.

## 2. Laboratory testing

One of the rheological properties is a creep - the process of deformations in time under constant pressure. The main parameters of creep are compression index ($C_c$) and coefficient of secondary compression ($C_\alpha$). The first parameter characterizes the compressibility of the soil on phase of primary consolidation, the second - the deformability of the soil skeleton in time under constant pressure. Coefficient of secondary compression was determined by one-dimensional compression tests of fully saturated sandy-clay tailings and subject to two-side drainage in automated consolidation apparatus LoadTrac-II (GeoComp Corp, USA) ( Fig. 2).

**Figure 2.** Automated consolidation apparatus LoadTrac-II

The sample had the following dimensions: diameter - 73 mm, height - 20.85 mm. We applied 300 kPa in tests, which corresponds to the representative stresses on sandy- clay tailings during the operation of the dam. To calculate the start time of the secondary compression we plotted the curve with axes: strain (ε) - time (t, min) in a logarithmic scale. On the intersection of two tangents to the part of the curve of the primary consolidation and secondary compression we determined the start point of the secondary consolidation ( Fig. 3).Coefficient of secondary compression is defined as slope of straight line of strain (ε)-logarithm of time (lg t) curve (line 2, Fig. 3). Coefficient of secondary compression calculated on the test results of 12 samples of sandy-clay tailings with different initial void ratio. Initial void ratio varied between 0.72....0.91. This range includes all void ratios obtained from the results of laboratory tests of sandy-clay tailings during geotechnical monitoring [3].

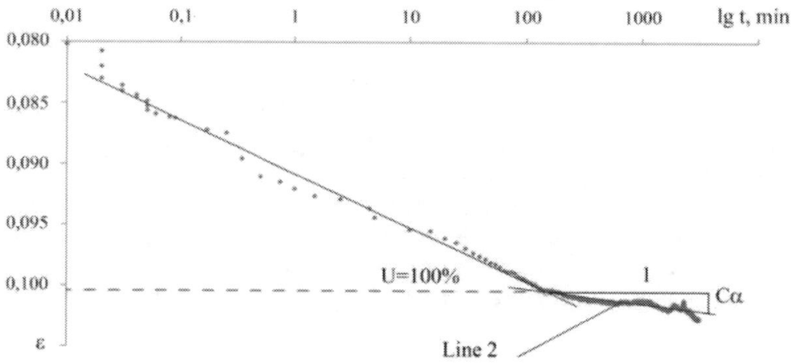


**Figure 3.** Strain-time curve of sandy-clay tailings

Linear relationship between the initial void ratio $e_0$ and $C_\alpha$ under constant pressure (300 kPa) was determined:

$$C_\alpha = 4,3937(1-0,9789e_0)\cdot 10^{-3} \quad (1)$$

Analysis of the consolidation of sandy-clay tailings showed that the primary consolidation is completed within 100-150 minutes after application of pressure. Due to the fact that the primary consolidation period significantly less secondary consolidation period (5000-6000 min) , compression index can be defined as characteristic of primary consolidation under applied load.

To determine the compression index we applied pressure of 12.5, 25, 50, 100, 200, 300 kPa to the sample. Compression index $C_c$ was determined for each value of the initial void ratio by plotting a straight line between initial void ration and applied pressure in the logarithm scale (Fig. 4). Relationship between compression index and initial void ration for industrial tailings is presented in Figure 5.

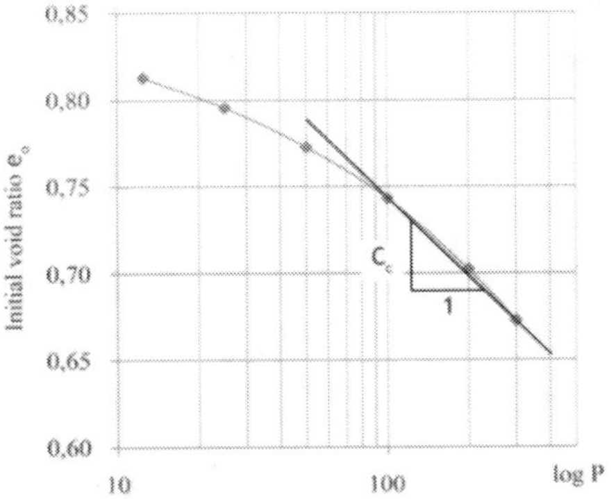

**Figure 4.** Compression Index $C_c$

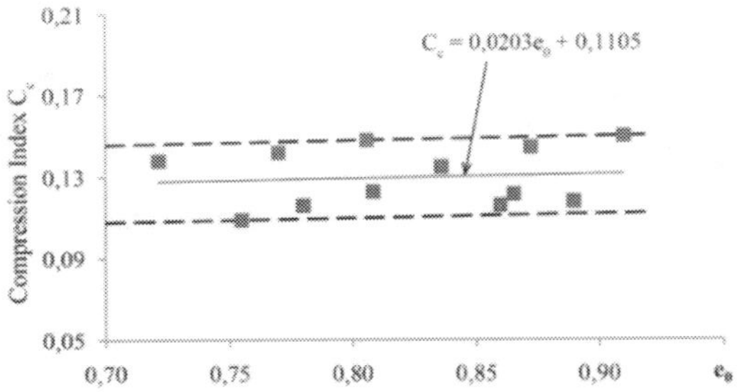

**Figure5.** Linear relationship $C_c$ and $e_0$

A linear relationship between the initial void ratio $e_0$ and $C_c$ was defined:

$$C_c = 0{,}0203 \cdot e_0 + 0{,}1105 \tag{2}$$

## 3. Numerical Simulation

To estimate the reliability of the dam subject to the rheological properties tailings we used PLAXIS 2D software and numerical Soft Soil Creep model. The main parameters of the model are modified compression index ($\lambda$*), modified swelling index ($\kappa$*), modified creep index ($\mu$*). The first two parameters are determined by plotting the logarithm of stress as a function of strain. The plot can be approximated by two straight lines. The slope of normal consolidation line gives the modified compression index ($\lambda$*) and the slope of unloading (recompression) line gives us modified swelling index ($\kappa$*). Modified creep index $\mu$* are determined as slope of straight line in strain ($\varepsilon$) - logarithm of time (*ln t*) curve.

As an alternative parameters PLAXIS uses the compression index ($C_c$), recompression index ($C_r$), the index of secondary compression ($C_\alpha'$). Index $C_\alpha'$ is determined as slope of straight line of initial void ratio ($e_0$)and the logarithm of time (*lg t*) curve and can be calculated by multiplying the coefficient of secondary compression ($C_\alpha$) by ($1+e_0$). Alternative parameters associated with the basic characteristics of creep model by the following expressions (3) - (5) [2]:

$$\lambda^* = \frac{C_c}{2{,}3(1+e)} \tag{3}$$

$$\kappa^* = \frac{2}{2{,}3} \cdot \frac{C_r}{(1+e)} \tag{4}$$

$$\mu^* = \frac{C_a'}{2{,}3(1+e)} \tag{5}$$

The mean of initial void ratio of sandy- clay tailings was equal 0.86 according to results of geotechnical monitoring. Indexes $C_c$ and $C_\alpha$ related to defined initial void ration are shown in Table 1. In the virtual laboratory PLAXIS were modeled odometer tests taken into consideration indexes from Table 1. The results are shown in Figure 6a. The simulated and observed compression curves are differed. To model the behavior of sandy-clay tailings properly it is necessary to calibrate the parameters of the numerical model of Soft Soil Creep according to laboratory tests.
Calibration of the parameters is made in a virtual laboratory of PLAXIS software for the following parameters: $\lambda^*$, $\mu^*$, $c$ - cohesion and $\varphi$ - friction angle. Observed and simulated results have a good agreement (Fig. 6b).

**Table 1.** $C_c$ and $C_\alpha$ indexes ($e_0$=0.86)

| | $C_c$ | $C_\alpha$ x$10^3$ |
|---|---|---|
| Mean | 0,128 | 0,695 |
| Min | 0,111 | 0,586 |
| Max | 0,149 | 0,830 |

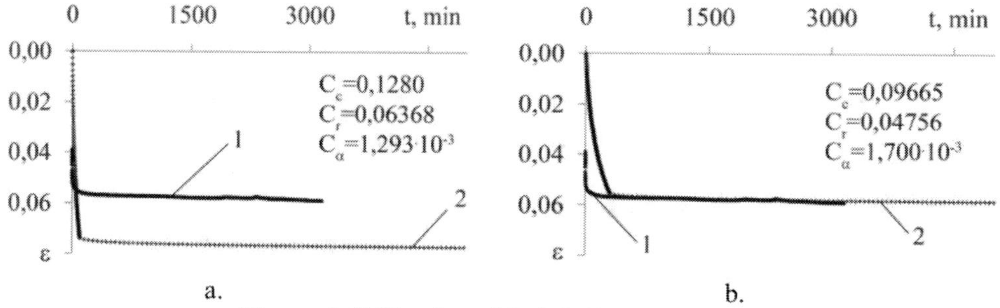


**Figure 6.** Calibration of Soft Soil Creep Model:
1- observed; 2 – simulated
a – before calibration; b – after calibration

To estimate the stress-strain state of the dam and determine its stability we modeled dam's behavior in the PLAXIS 2D (Fig. 7). Hardening soil model was applied for simulating foundation's and dam body's soils. Soft Soil Creep model was used for simulating rheological behavior of tailings.

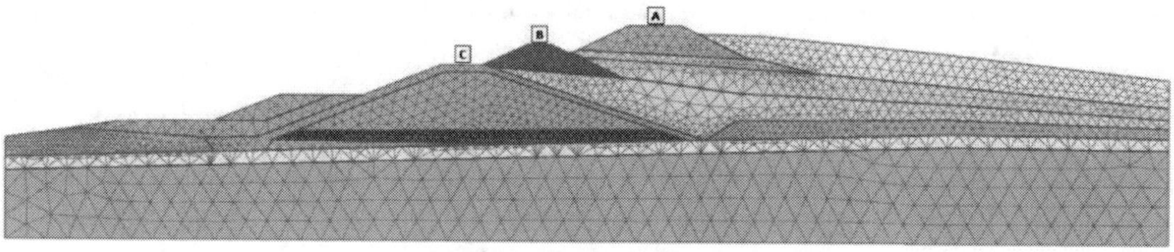


**Figure 7.** FE model of dam

Staged construction method was applied and we simulated the following stages:
1. Building of 1-st tier of dam and washing of tailing on upstream slope;
2. Building of 2-st tier of dam and washing of tailing on upstream slope and etc;
3. Evaluation of reliability of the dam at 1 year and 15 years of operation after the completion of third tier of dam

Safety factors were 1,369 and 1,373 in 1 year and 15 years respectively. Various parameters of the numerical model have a different influence on the safety factor. Assessment of the influence of parameters is key objective in terms of planning and carrying out additional researches of soils and sandy-clay tailings.

We performed parameter mapping of Soft Soil Creep model on rate of influence in the short-term and long-term outlook in PLAXIS software. Distribution of the rate of influence $\lambda^*$, $\mu^*$ on the safety factor is shown in Figure 8. For further analysis 4 parameters were selected: $\lambda^*$ - for tailings of 2 and 3 tiers; $\mu^*$ - for tailings of 1 and 2 tiers which had the maximum rate of influence.

So we calculated the maximum and minimum safety factor with specified variation of the rheological properties of sandy-clay tailings. Maximum safety factor was 1,375 after 15 years, the minimum - 1,348.

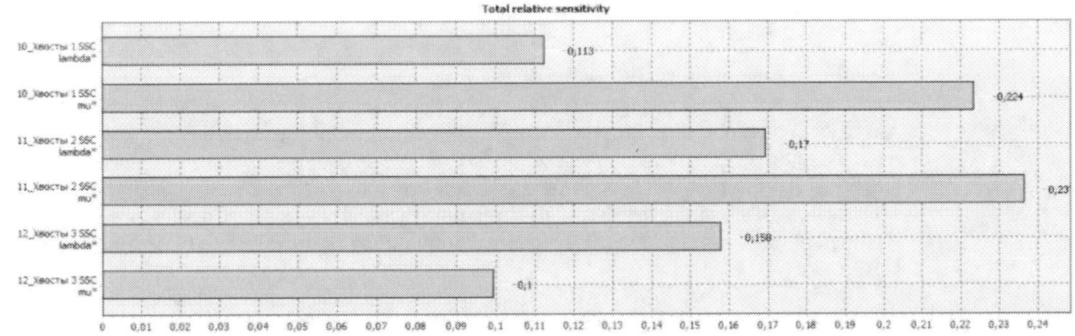


**Figure 8.** Rate of influence $\lambda^*$, $\mu^*$ on safety factor

## 4. Conclusion

Calibration of characteristics Soft Soil Creep model in PLAXIS must be performed taking in consideration the results of laboratory tests. The initial parameters of the numerical model have been changed: $C_c$ decreased by 24.5%, $C_\alpha$ increased by 31.5 %.

Modeling of building and operation of dams subject to variation of the rheological characteristics of the soils and tailings, obtained by laboratory testing is appropriate. It provides us to get qualitative and quantitative assessment of the safety factors over time, as well as to establish the limits of their variation depending on the initial rheological properties of soils.

Such approach will allow to evaluate the influence of various factors independently (e.g., strength parameters and filtration characteristics) on the reliability of dams.

## References

[1] Murthy V 2003 *Geotechnical Engineering: principles and practices of Soil Mechanics and Foundation Engineering* ,(NY, Murthy-Marcel Dekker Inc.)

[2] *User Guide Plaxis 2D 2012*, 2012 (Netherlands, Plaxis bv).

[3] Nevzorov A, Koptyev V, Korshunov A. (2011). "Results of geotechnical moni-toring for the erection and operation of the tailing dam at the diamond deposit named after M.V.Lomonosov" Proceedings of the 15th European conference on Soil Mechanics and Geotechnical engineering, ISSMGE, Athens, Greece, pp. 1177-1184.

# A new approach for development of rheological relations for saturated porous media

**E Khramchenkov**[1,2], **M Khramchenkov**[1]
[1] Kazan Federal University – Institute of Mathematics and Mechanics– Kazan – Russia
[2] Joint Supercomputer Center – Kazan department – Kazan – Russia

ekhramch@gmail.com and mkhramch@gmail.com

**Abstract**. Equations of hydro-geo-mechanics in filtrating porous media with porous variable-mass skeleton are examined. Variation of skeleton mass occurs due to heterogeneous chemical reactions. Regularities of mass transfer and deformations in such kinds of media are analyzed. Peculiarities of obtaining of rheological relations are investigated. A new approach to obtain rheological relations is proposed. Numerical simulations in order to verify obtained model are performed.

## 1. Introduction

Hydrogeomechanical models are important for solution of various problems in hydrogeology, hydrogeoecology, oil production and geophysics. Main modern concepts of the hydrogeomechanics are presented, for example, in [1, 2]. Oil depletion on some of Russian oil fields necessitates formulating models that take into account the variation of the stress-strain state of the rock mass caused by chemical interactions between components of underground fluid and the material of the porous skeleton to perform effective enhanced oil recovery. Those models are also essential in problems of hydrogeology, such as filtration of solutions in clay layers, suffosion processes and karst processes.

Above-mentioned chemical interactions usually cause the variation of the mass of the porous matrix [3]. That is why it is important to perform an additional research on the influence of this variation on rheological relations, which are required to obtain a closed model of deformations of filtrating porous media.

It is also necessary to perform systematic development of main equations of underground mass-transfer in this case. Whereas those questions did not receive exhaustive explanation in specialized literature, it makes sense to obtain required equations and examine most important applications.

## 2. Mathematical model

### *2.1. Mass balance equations of porous skeleton and percolating liquid*

First, it is essential to develop set of equations of filtration in a deformable porous medium with porous variable-mass skeleton. From the definition of the volume strain of the porous medium $\theta$:

$$\theta = (V - V_0)/V_0 \tag{1}$$

Assuming values of $\theta$ are small, we can obtain the following ratio:

$$V = V_0 \exp\theta \tag{2}$$

V is the representative volume of porous medium, lower index "0" stands for initial values in the zero time. Therefore, for the mass of the porous medium we have:

$$M_s = (1-m)\rho_s V_0 \exp\theta \tag{3}$$

Where: $\rho_s$ is the density of the solid phase; $m$ is the porosity of the rock. Last equation was differentiated with respect to time.

$$\frac{\partial m}{\partial t} = \frac{(1-m)}{\rho_s}\frac{\partial \rho_s}{\partial t} + (1-m)\frac{\partial \theta}{\partial t} - \frac{(1-m)}{M_s}\frac{\partial M_s}{\partial t} \tag{4}$$

Mass balance of the solid material of the porous skeleton is described by the equation

$$\partial[(1-m)\rho_s]/\partial t + \mathrm{div}[(1-m)\rho_s \mathbf{W}] = j. \tag{5}$$

Here $\mathbf{W}$ is the velocity of the solid phase; $j$ denotes the source/drain of the mass of the porous skeleton caused by interface interaction. The porous skeleton is assumed to loose its mass during interface interaction, so hereinafter $j$ will represent the drain of the mass. The mass of the material of the porous skeleton in the representative volume there can be defined as

$$\rho_s V_s = \rho_s (1-m) V = M_s \tag{6}$$

Here $V_s$ is the volume of the solid phase in the representative volume. Drain in the equation (5) is the loss of the mass of the porous skeleton caused by processes like dissolution, leaching or suffosion. It can be written as

$$j = \frac{1}{V}\frac{\partial M_s}{\partial t} \tag{7}$$

From the equation (5), after the expanding of derivatives we have

$$-\frac{\partial m}{\partial t}\rho_s + (1-m)\frac{\partial \rho_s}{\partial t} + (1-m)\rho_s \mathrm{div}\mathbf{W} + \mathbf{W}\mathrm{grad}[(1-m)\rho_s] = j \tag{8}$$

Taking into account the equations (4) and (7) and assuming the last member of the (8) to be the second order infinitesimal:

$$\partial\theta/\partial t = \mathrm{div}\mathbf{W} \tag{9}$$

The last assumption is explained in the traditional way of the poromechanics [4]. From the Terzaghi's Principle $P = \sigma^f + p$, where $P$ is the external load, $\sigma^f$ is the effective stress, and $p$ is the water pressure in pores, it follows that $\mathrm{grad}\sigma^f = -\mathrm{grad}p$ if $\mathrm{grad}P = 0$. Since the porosity of the rock $m$ and the density of the solid phase $\rho_s$ are functions of $\sigma^f$ and $p$ and taking into account the last equation it follows that $\mathbf{W}\mathrm{grad}[(1-m)\rho_s]$ from the (7) is in proportion to $\mathbf{W}\mathrm{grad}p$. According to the Darcy's law, this member is also in proportion to the product of the velocity $\mathbf{W}$ and the filtration velocity $\mathbf{q}$. Values of velocities in examined processes are relatively low, so the second degree of the velocity can be ignored as infinitesimal value [4].

Mass balance of the dissolving fluid in saturated rock is described by the equation:

$$\partial(m\rho)/\partial t + \mathrm{div}(m\rho\mathbf{V}) = 0. \tag{10}$$

Here $\rho$ is the density of the fluid, $\mathbf{V}$ is the velocity of the fluid. Relative velocity of the fluid in rock (filtration velocity) is $\mathbf{q} = ms(\mathbf{V} - \mathbf{W})$. Then, from the equations (9) and (10), we have:

$$m\partial\rho / \partial t + \rho\partial m / \partial t + \mathrm{div}(\rho\mathbf{q}) + \mathrm{div}(\rho m\mathbf{W}) = 0. \tag{11}$$

After transformation of the last equation, while $\mathbf{q}\mathrm{grad}\rho$, $\mathbf{W}\mathrm{grad}(m\rho)$ are considered infinitesimal values, it leads to the following equation:

$$m\rho^{-1}\partial\rho / \partial t + \partial m / \partial t + \mathrm{div}\mathbf{q} + m\partial\theta / \partial t = 0. \tag{12}$$

Combined with (4) the equation (12) gives the following expression:

$$[m\rho^{-1}\frac{\partial\rho}{\partial t} + (1-m)\rho_s^{-1}\frac{\partial\rho_s}{\partial t}] + \frac{\partial\theta}{\partial t} + \mathrm{div}\mathbf{q} = \frac{(1-m)}{M_s}\frac{\partial M_s}{\partial t}. \tag{13}$$

It is obvious from $\rho_s = M_s / V_s$ that:

$$(1-m)\rho_s^{-1}\frac{\partial\rho_s}{\partial t} = \frac{(1-m)V_s}{M_s}\frac{\partial(M_s / V_s)}{\partial t}. \tag{14}$$

Expanding the derivative in the right side and taking (13) into the account:

$$(1-m)\rho^{-1}\frac{\partial\rho}{\partial t} + \frac{\partial\theta}{\partial t} + \mathrm{div}\mathbf{q} = \frac{(1-m)}{V_s}\frac{\partial V_s}{\partial t}. \tag{15}$$

The right side of (15) can be expressed as:

$$\frac{(1-m)}{V_s}\frac{\partial V_s}{\partial t} = (1-m)\frac{\partial\varepsilon}{\partial t} \tag{16}$$

Here $\varepsilon$ is the volume strain of the skeleton material (sum of the diagonal elements of the strain tensor). Then (15) can be written as:

$$(1-m)\rho^{-1}\frac{\partial\rho}{\partial t} + \frac{\partial\theta}{\partial t} + \mathrm{div}\mathbf{q} = (1-m)\frac{\partial\varepsilon}{\partial t}. \tag{17}$$

Water is a slightly compressible fluid, so first member in the left side of (17) can be neglected:

$$\frac{\partial\theta}{\partial t} + \mathrm{div}\mathbf{q} = (1-m)\frac{\partial\varepsilon}{\partial t}. \tag{18}$$

Time integration of the last equation gives following result:

$$\theta + \int_0^t \mathrm{div}\mathbf{q} \cdot d\tau = \int_0^t (1-m)\frac{\partial\varepsilon}{\partial t} \cdot d\tau. \tag{19}$$

Partial integration of the right side leads to the equation:

$$\theta + \int_0^t \mathrm{div}\mathbf{q} \cdot d\tau = (1-m)\varepsilon + \int_0^t \varepsilon\frac{\partial m}{\partial t} \cdot d\tau. \tag{20}$$

Using the equation (4) for the time derivative of the porosity and (14) for the member $(1-m)\rho_s^{-1}\partial\rho_s / \partial t$ we can transform (20) to the following equation:

$$\theta + \int_0^t \mathrm{div}\mathbf{q} \cdot d\tau = (1-m)\varepsilon + \int_0^t \varepsilon \left[ (1-m)\frac{\partial \theta}{\partial t} - (1-m)\frac{\partial \varepsilon}{\partial t} \right] \cdot d\tau. \tag{21}$$

It is obvious that $\varepsilon \le \theta$ so for the last member of the right side of (21) it is true that:

$$\int_0^t \varepsilon \left[ (1-m)\frac{\partial \theta}{\partial t} - (1-m)\frac{\partial \varepsilon}{\partial t} \right] \cdot d\tau \le \int_0^t \left[ (1-m)\theta\frac{\partial \theta}{\partial t} - (1-m)\varepsilon\frac{\partial \varepsilon}{\partial t} \right] \cdot d\tau \tag{22}$$

Last integral contains time derivatives of the squares of $\varepsilon$ and $\theta$ so it can be neglected. Therefore, finally we receive:

$$\theta + \int_0^t \mathrm{div}\mathbf{q} \cdot d\tau = (1-m)\varepsilon. \tag{23}$$

The obtained equation has clear physical sense: total strain of the porous rock is made up of two terms. The first term is the water that was forced out during the filtration. The second term is deformation of the porous skeleton.

### 2.2. *Rheological relations for the filtrating porous media with mass-variable porous skeleton*

Usually rheological relations are obtained from the expression for the free energy of the porous medium [1]. Dissipation caused by chemical reactions is not included in this approach, and that is why it can hardly be applied in our case. In order to obtain rheological relations we consider that solid skeleton is elastic, then from the [5, 6] we have:

$$\sigma_{ij}^{(s)} = -(K - \frac{2}{3}G)\varepsilon\delta_{ij} - 2G\varepsilon_{ij},\ \varepsilon = \sum_i \varepsilon_{ii} \tag{24}$$

$\varepsilon_{ij}$ is the strain tensor of the skeleton, $G$ is the shear modulus. In one-dimensional case, vertical compressive stress $\sigma_{zz}$ can be expressed as:

$$\sigma_{zz}^{(s)} = -(K + \frac{4}{3}G)\varepsilon \tag{25}$$

Denoting $\alpha = (K + \frac{4}{3}G)$, we can write (23) as:

$$\alpha \left[ \theta + \int_0^t \mathrm{div}\mathbf{q} \cdot d\tau \right] = -(1-m)\sigma_{zz}^{(s)}. \tag{26}$$

From the definition of actual stresses [5], the right side of (26) can be written as:

$$-(1-m)\sigma_{zz}^{(s)} = P - mp \tag{27}$$

From the Terzaghi's Principle and (27), it follows that:

$$\sigma^f = \alpha\theta,\quad \alpha\int_0^t \mathrm{div}\mathbf{q} \cdot d\tau = (1-m)p. \tag{28}$$

The first equation is the rheological relation for the volume strains of the filtrating medium. Shear deformations obviously coincide with deformations of the porous medium, so rheological relations can be written as:

$$\sigma_{ij}^{f} = -(K - \frac{2}{3}G)\theta\delta_{ij} - 2G\varepsilon_{ij} \tag{29}$$

Introducing the porosity coefficient $e = m/(1-m)$ and differentiating last equation of the system (28) with respect to time, we receive the equation for the pressure in the filtrating porous medium with mass-variable porous skeleton:

$$(1-m)\partial p/\partial t - p\partial m/\partial t = -\alpha \cdot \mathrm{div}\mathbf{q} \tag{30}$$

If the porosity is constant, this equation turns into the equation of the filtrational consolidation of the porous medium with the porous skeleton of constant mass. This equation is widely used in civil engineering [7].

## 3. Results and Discussion

### *3.1. Numerical simulation of filtrational dissolution*

When the density of the solid phase is constant during the chemical interaction, the following set of equation can be written:

$$\partial m/\partial t = (1-m)\partial\theta/\partial t - [V_0(1+\theta)]^{-1}\partial V_s/\partial t$$

$$\partial m/\partial t + m\partial\theta/\partial t + \mathrm{div}\vec{v} = 0, \quad \vec{v} = -k\mu^{-1}\nabla(p+\rho g z) \tag{31}$$

$$\partial(mc)/\partial t + [V_0(1+\theta)]^{-1}\partial V_s/\partial t + mc\partial\theta/\partial t = \mathrm{div}(D\nabla c - \vec{v}c),$$

$$[V_0(1+\theta)]^{-1}\partial V_s/\partial t = \beta(c-c_0)^k;\ \beta, k, c_0 = \mathrm{const}$$

This set also includes the equation for the concentration of active component in the solution.

Here $c$ is the concentration of the skeleton material; $k$ is the permeability coefficient; $\mu$ is the viscosity of the filtrating fluid. The viscosity of the solution is constant; the permeability is dependent on the porosity in compliance with the Archie's law [1].

First we performed a comparison with one-dimensional analytical solution [10] to verify results of the numerical simulation. The analytical solution was calculated with the same parameters as the numerical solution above.

The results of the comparison are presented below. Graphics represent the dependence of the correspondent value on distance to the borehole (in meters). The dash line stands for numerical solution.

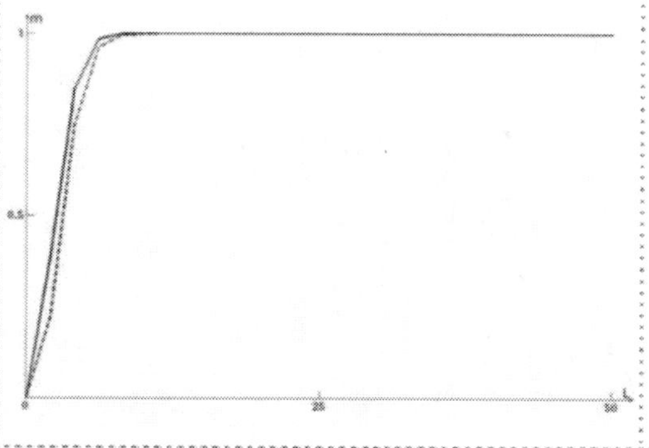

**Figure 1.** Distribution of the concentration of the dissolved material in the fluid phase.

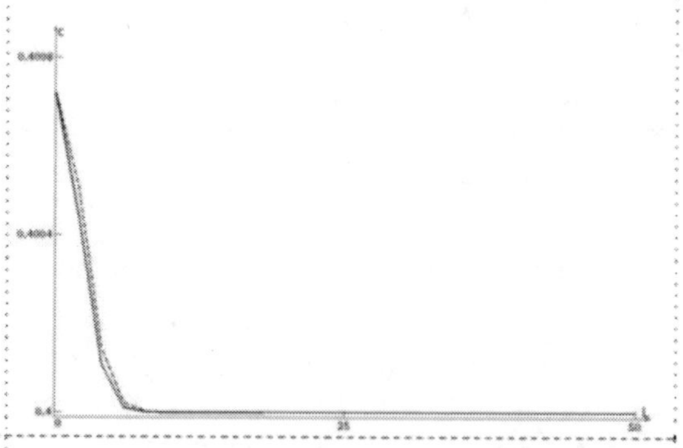

**Figure 2.** Distribution of the porosity

### *3.1. Simulation of injection of dissolving fluid*

We performed three-dimensional numerical simulation of an injection of dissolving agent into homogeneous porous layer with vertical borehole as an example of the real physical problem. The carbonate rock was considered as the layer, water was considered as the dissolving agent. Data for speed of dissolution, solubility and other parameters were taken from [8].

Implicit finite difference scheme with regular computational grid was used to perform discretization of the (31). Simulated area had dimensions 100x100x40 meters, dimensional step was 2 meters, time step 10 seconds. Due to the large size of the area, PETSc solver was used to perform efficient parallel computations on those big systems [9]. GMRES method in combination with LU preconditioner showed the best performance for our case. This method showed good convergence in studied case. The residual was set to $10^{-8}$.

Visualization of obtained results was performed in ParaView open-source software. (Contributors: National Center for Supercomputing Applications (NCSA) at the University of Illinois at Urbana-Champaign (UIUC), Lawrence Livermore National Laboratory (LLNL), Sandia National Laboratories (SNL), Los Alamos National Laboratory (LANL), Jean-Loup Gailly and Mark Adler (gzip library)).

We performed calculations with different configurations and properties of the clot. Permeability in the layer for one of the tests is presented on Fig. 1. Dark area stands for low permeability region, bright area designates extremely high permeable channel, which has a low permeable gap.

**Figure 3.** Permeability of the layer. Test configuration.

Journal of Physics: Conference Series **602** (2015) 012005 doi:10.1088/1742-6596/602/1/012005

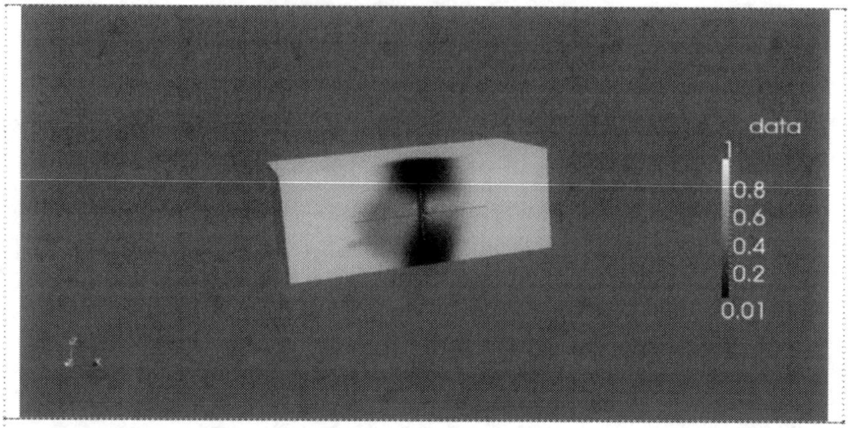


**Figure 4.** Distribution of concentration. Dark area designates the expansion of the injected fluid

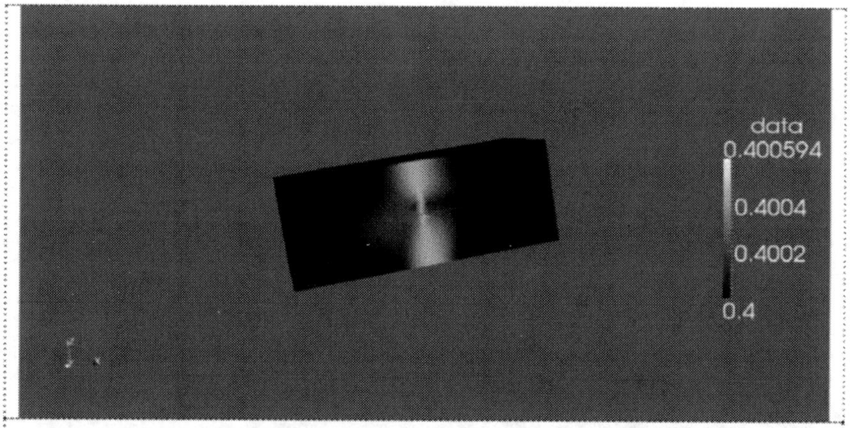


**Figure 5.** Distribution of porosity. Bright area is the region with larger porosity

## 4. Conclusion

New approach to the derivation of equations of the underground mass transfer for filtrating porous medium with the variable-mass skeleton is proposed. Obtained equations can be used to solve important practical problems, which require taking into account variations of the stress-strain state of the rock mass. Those variations are caused by chemical interactions between components of the underground fluid and the material of the porous skeleton. The applicability domain of the model includes problems of enhanced oil recovery, simulation of filtration of brines through clay layers, simulation of suffosion and karst processes.

Rheological relations for saturated porous media were obtained in the same way as equations of mechanics of continuum for solid body with dislocations [11,12]. Tensor of dislocational polarization is introduced in that method. Its analog in our case is the integral $\int_0^t \mathrm{div}\mathbf{q} \cdot d\tau$ in (19) and other equations. Fluid in pores is isotropic, so the tensor is reduced to the contraction. Our approach gives an opportunity to obtain rheological relations for saturated porous media in the new way and to avoid some difficulties of the traditional method.

**Acknowledgements**
Work is conducted under the RFBR project №12-07-00007.

**References**

[1] Fjaer E, Holt R M, Hoursrud P, Raaen A M, Risnes R 2008 Petroleum Related Rock Mechanics **2$^{nd}$** ed. *Elsevier, Amsterdam and Oxford*
[2] Jaeger J.C., Zimmerman R.W. 2007 Fundamentals of Rock Mechanics **4$^{th}$** ed. *Oxford, Wiley*
[3] Taron J., Elsworth D., Ki-Bok Min 2009 *Int. J. Rock Mechanics and Mining Sci.* **46** 842
*[4]* Nikolaevskiy NV. 1996 Geomechanics and fluid dynamics, *Dordrecht: Kluwer Academic Publishers*
[5] Lopatnikov S L, Gillespie J W (Jr.) 2010 *Transp. Porous Med.* **84** 471
[6] Lopatnikov S L, Gillespie J W (Jr.) 2012 *Transp. Porous Med.* **92** 597
[7] Verruijt A 1984 The theory of consolidation // In: Fundamentals of transport phenomena in porous media. Part **2**: Deformation of porous media. *Martinas Nijhoff Publishers* 351 – 368.
[8] Siemers J, Dreybrodt W 1998 *Water Resour. Res.* **34** (3) 409
[9] S Balay W D, Gropp L C, McInnes B F 1997 Smith Efficient Management of Parallelism in Object-Oriented Numerical Software Libraries. Modern Software Tools in Scientific Computing, *Birkhäuser Press, Boston,* 163-202
[10] Kosterin A V, Khramchenkov M G, Chugunov V . 1991 *Inz.-Fiz. Zhurnal* **60** (6) 942
[11] Landau L D, Lifshitz E M 1986 Theory of Elasticity **7** (**3$^{rd}$** ed.) *Butterworth–Heinemann*
[12] Kosevich A M 1999 Theory of Crystal Lattice. *WILEY-VCH, Berlin, New York*

Journal of Physics: Conference Series **602** (2015) 012006 doi:10.1088/1742-6596/602/1/012006

# Colloidal behavior of aqueous montmorillonite suspensions in the presence of non-ionic polymer

**M Gareche[a,1], N Azri[1], A Allal[2] and N Zeraibi[1]**
[1] UMBB, LGPH Laboratoire – Faculté des Hydrocarbures et de la Chimie – 35000 Boumerdes – Algeria
[2] UPPA, IPREM, EPCP – Université de Pau et des Pays de l'Adour – 64053 Pau Hélioparc - France

m_gareche@yahoo.fr, nadjet.n@gmail.com, ahmed.allal@univ-pau.fr, no_zeraibi@yahoo.fr

**Abstract.** In this paper we characterized at first, the rheological behavior of the bentonite suspensions and the aqueous solutions of polyethylene oxide (PEO), then we were investigated the influence of this polymer in a water-based drilling fluid model (6% of bentonite suspension). The objective is to exhibit how the non ionic polymer with molecular weight $6x10^3$ g/mol. of varying concentration mass (0.7%, 1%, 2% et 3%) significantly alter the rheological properties (yield stress, viscosity, loss and elastic modulus) of the bentonite suspensions. The rheological measurements made in simple shear and in dynamic on the mixture (water-bentonite-PEO), showed rheological properties of bentonite suspensions both in the presence and absence of non-ionic polymer. The PEO presents an affinity for the bentonite particles slowing down their kinetic aggregation. The analysis by X-rays diffraction also allowed understanding the structure of this mixture. It had revealed the intercalation between of the clay platelets on one hand, and the links bridges assured by the chains of polymer between bentonite particles beyond a critical concentration in PEO on the other hand. The Herschel-Bulkley rheological model is used for the correlation of our experimental results.

## 1. Introduction

Clay minerals are extensively used in a wide range of applications. They are key component in the formulation of ceramic products, cement, drilling fluids, moulding sands, paints and paper, among others [1]. An important characteristic that clay minerals are able to provide in such applications is adequate particle dispersion, which is necessary to obtain a uniform and stable system. Under certain conditions the clay particles may become aggregated, which leads to the variation of important properties required for a particular function. In drilling fluids, for example, the flow behavior of the system is of utmost importance due to its circulation a round the wellbore [2]. The aggregation of the clay particles under varying conditions of temperature and electrolytes leads to strong variations of the flow properties. It becomes necessary therefore to add certain additives, or polymers, to stabilise the clay particles and prevent this behavior. In soil science, non-ionic polymers that are water soluble have been found to impart stability to natural soil aggregates. On contact with the particle these molecules tend to collapse and spread out over the clay surface, improving the physical conditions of soil [3].

---

[a] Mourad GARECHE: UMBB, LGPH Laboratory – Faculté des Hydrocarbures et de la Chimie – 35000 Boumerdes – Algeria

Since non-ionic polymers appear to be well suited for stabilising a clay structure, there lies fundamental interest to comprehend the structure and flow behavior of the clay / non-ionic polymer systems. In the literature, we find numerous works treating the ternary system water-clay-additive, where the additive is often a non-ionic polymer. By being interested in the interactions particles additives and to understand the microstructure, the authors often couple rheological measures with physico-chemical techniques such as X-rays diffraction (XRD), dynamic and static light scattering (DLS/SLS), transmission electronic microscope (TEM) etc. [4-11].

The purpose of this article is to highlight the rheological properties of bentonite suspensions both in the presence and absence of non ionic polymers, the polyethylene oxide (PEO). The stability of the suspension was enhanced with increasing polymer adsorption. The influence of this additive was investigated in terms of viscosity, yield stress, loss and elastic modulus which are principal characteristics of the drilling fluids.

## 2. Experimental

### *2.1. Materials*

In our work, we used the calcic bentonite (B3378) supplied by Sigma-Adrich. Its average composition is (% mass): $SiO_2$-48.35, $Al_2O_3$-12.15, $Fe_2O_3$-8.26, CaO-6.68, MgO-5.47, $Na_2O$-3.65, $K_2O$-2.39. It possesses a specific surface of 39.3 $m^2/g$, a capacity cationic exchange of 0.88 meq/g, pH=9 and a specific gravity of 2.4 $g/cm^3$.

The polyethylene oxide (PEO) supplied by Acros organics, is non ionic polymer of the polyether's family. Its chemical formula: $HO-CH_2-(CH_2-O-CH_2-)n-CH_2-OH$. It possesses a molecular weight $6.10^3$ g/mol, a number monomer/chain 136, a volumic mass 1173 $kg/m^3$ and radius gyration 30 Å. It was added to the clay suspensions.

### *2.2. Samples preparation*

The bentonite suspensions having mass concentrations of 2, 4, 5, 6 and 8% were prepared by pouring a volume of clay into demineralised water under a magnetic glass-rod during 24 hours. We noticed, after formulation of suspensions, at a more or less long rest, that is particles sedimentation for 2% and consistency gel formation with increases concentration another suspension. The same protocol is used to preparation for aqueous solutions PEO (1-5%).

For the preparation of mixtures (water-bentonite-PEO), the experimental protocol is the following: in the quantity of demineralised water, we add the bentonite. After 6 hours of agitation, the PEO of molecular mass $6x10^3$ g/mol is introduced into the basic bentonite suspension (6%). The homogenization is obtained by magnetic a glass-rod during 48 hours. Before making the rheological measures, every sample is mixed during 45 minutes. After sample's placed in geometry, it's subjected to a preshear, then, we proceed to the measure.

The type of experiment consists in determining the flow curves in simple shear and to study the viscoelatic properties of the different suspensions. The measurements were made with a geometry of coaxial cylinders (Φ32 mm/Φ29,5 mm and L : 44,3 mm) for rheometer SR-5000, cone plan (Φ60 mm/2°) for Bohlin C-VOR-150. By being interested in interaction particles-additives, we coupled with the rheological measurements, the analysis by X-rays diffraction realized on Inel XRG-3000.

## 3. Results and Discussion

### *3.1. Rheological measurements of bentonite suspensions*

The results available on such suspensions are often very divergent. This can be explained by the diverse qualities of bentonite used (origin, manufacturing, etc.), as well as the practised experimental methods. Figure 1 presents the flow curves for the bentonite suspensions of varying concentrations. An increase is noticed in the apparent $\mu$ viscosity and the yield stress which becomes more significant with the increase of concentration. The rheological behavior of these suspensions is thus Newtonian for the concentration value of 2 % and shear-thinning with yield stress beyond this latter.

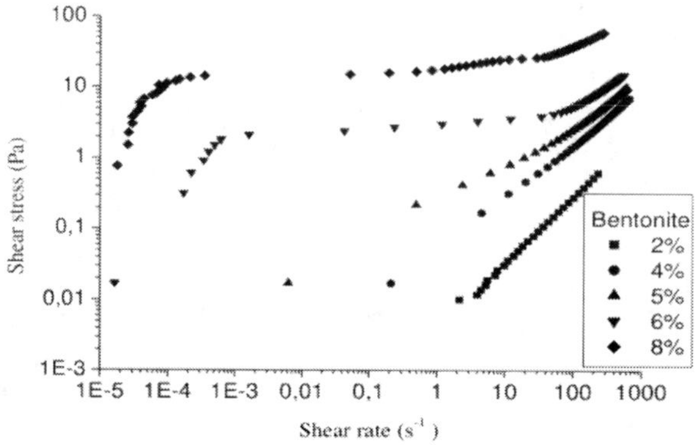


**Figure 1.** Flow curves for bentonite suspensions at different concentrations

The increase of the viscosity can explain by the crystalline structure of the mineral group to which belongs the montmorillonite. In the water, the particles of bentonite which have a strong affinity for the environment in which they are scattered, form suspensions possessing a structure in house of cards [12]. They are constituted by minerals the shape of which is generally flattened. These materials are silicates of aluminum the crystalline composition of which is a pile of leaves. The phenomenon of inflation, which is translated by a modification of the rheological characteristics of the suspension, occurs by the penetration of the water between the leaves which provokes their space [13].

The existence of yield stress of flow is associated with the Van der Waals forces, which facilitate the formation of flocs or particles aggregates which provoke a resistance to the flow [14]. It is only significant at relatively concentration beyond 2% of bentonite. Similar results were obtained by [2] on sodic montmorillonite above 3% and by [15] on mass fractions from it bentonite beyond value of 2%. When the shear stress is lower than the "yield stress", the bentonite suspension has the behaviour of solid type. There is no flow. But, when the shear stress is sufficient to maintain a flow, the shear rate abruptly increases. The material quickly fluidifies. During this phase, the aggregates of particles were practically destroyed and particles were directed according to current lines.

### *3.2. Rheological measurements of aqueous solutions PEO*

The development and the environmental protection, make that the water soluble polymers take an important place to restrict the use of the organic solvents. The PEO associated with particles of clay in an aqueous solution allows to obtain solutions with very interesting rheological properties such as the stabilization of the colloidal suspensions in the drilling fluids. The figure 2 presents the flow curves of the PEO solutions for varying mass concentrations (1%, 2%, 3%, 4% and 5%). We notice an increase of the viscosity according to the concentration. The behaviour of the PEO solutions is quasi-Newtonian.

### *3.3. Rheological measurements of mixtures suspensions bentonite-PEO*

To exhibit, the influence of the PEO on the rheological properties of bentonite suspensions, we have selected a basic suspension of concentration 6%, into which we introduced the PEO of molecular weight $6x10^3$ g/mol. by varying its concentration (0.7%, 1%, 2% and 3%).

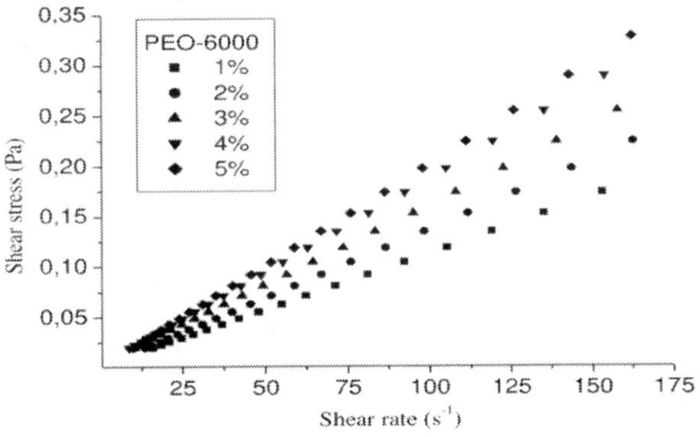


**Figure 2.** Flow curves for aqueous solutions PEO at different concentrations

*3.3.1. Shear stress-shear rate measurements*

A shear stress ($\tau$ ) - shear rate ($\gamma$) sweep was applied to each suspension by varying the shear stress over a wide range between 0,017 and 16 Pa. The flow curves were analyzed and fitted to the Herschel-Bulkley model, which has previously been used to describe the flow behavior of clay suspensions applied to drilling fluids [16]. The Herschel-Bulkley equation is given by:

$$\tau = \tau_c + K \gamma^n$$

Where $\tau_c$ is yield shear stress, K is the consistency, n is the flow behavior index and $\mu_{HB}$ is the high shear viscosity Herschel-Bulkley.

The rheological behavior of the bentonite suspension, in the presence of the polymer (PEO) to different concentrations is practically shear-thinning with yield stress (fig. 3).

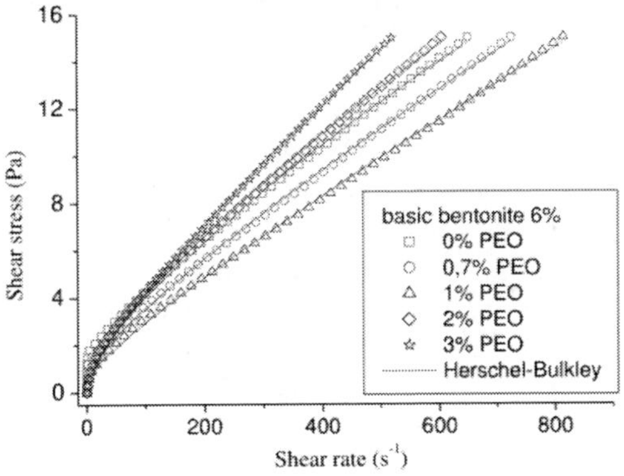


**Figure 3.** Modelling of flow curves for mixture bentonite-PEO at different concentrations

However, the PEO provokes, the decrease of the yield stress of the mixture, even the disappearance of this last one for high concentrations of PEO ($\tau_c$ is determined by the intersection point for the first slope with the second slope of the deformation curve on the X-axis in fig. 4).

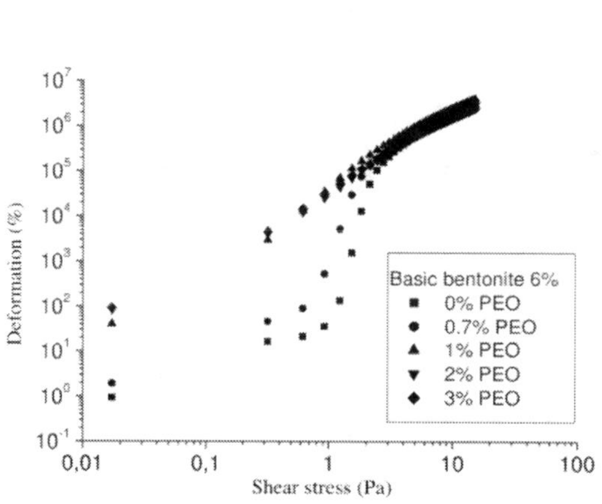


**Figure 4.** Deformation shear stress curves for mixture bentonite-PEO at different concentrations

The addition of the PEO in the bentonite suspension allows the progressive covering of the bentonite particles, so reducing the interactions particle-particle and favouring the interactions particle-additive [17]. Fig. 5a, 5b illustrate the parameters evolution of the Herschel-Bulkley model at the different mass concentration in PEO. A light reduction of yield stress $\tau_c$ is observed, accompanied with a significant decline of the high shear viscosity ($\mu_{HB}$), when the quantity of polymer on the surface increases. Beyond a polymeric concentration of 1% in PEO, τc is essentially constant while $\mu_{HB}$ increase significantly. This last effect is due to the increase of the continuous phase of the viscosity by the presence of the polymer. Similar observations were reported by Rossi and al. [6,18] on the suspensions of Na+montmorillonite (5%) in the presence of non ionic polymer (PEO-4000).

The same observations are made on the consistency (K) which evolves practically in the same direction as the high shear viscosity ($\mu_{HB}$) and the index of flow (n) conversely to this last one. It means, that the presence of the PEO decreases the interaction between the particles of clay (rigid) for the benefit of the interactions particles-additive (less rigid) which indeed explains the drop of $\tau_c$.

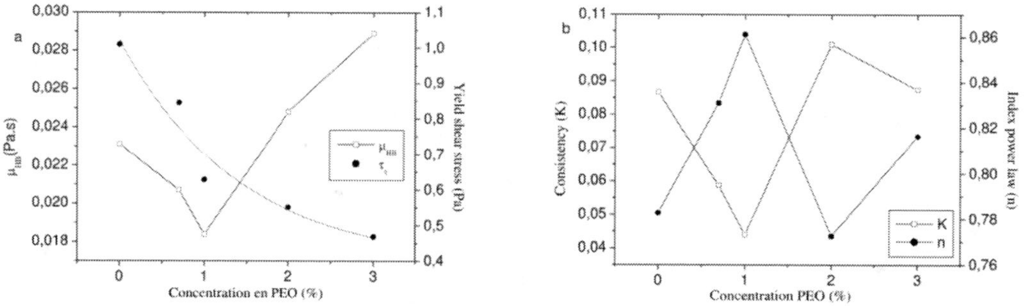


a b

**Figure 5.** Parameters of Herschel-Bulkley model as function of PEO concentrations:(a) $\mu_{HB}$,$\tau_c$ (b) K,n

*3.3.2. Dynamic oscillatory measurements*

In dynamic state, we approached the linear viscoelasticity of our mixtures suspensions by making essays of sweeping in frequency and sweeping as function time:

After four hours of rest under the geometry, so that the suspensions can reach their equilibrium state, essays of sweeping in frequency revealed that the elastic modulus (G') are superior to the loss modulus (G") for all the concentrations. It is concerned about the viscoelastic solid behaviour (fig. 6).

However, modulus G' and G" for mixtures always remain lower than the modulus of basic suspension (6% bentonite). That explains the reducing effect of the PEO on these modulus. Besides, we notice that the elastic module is independent from the frequency while the loss module increases in high and low frequency.

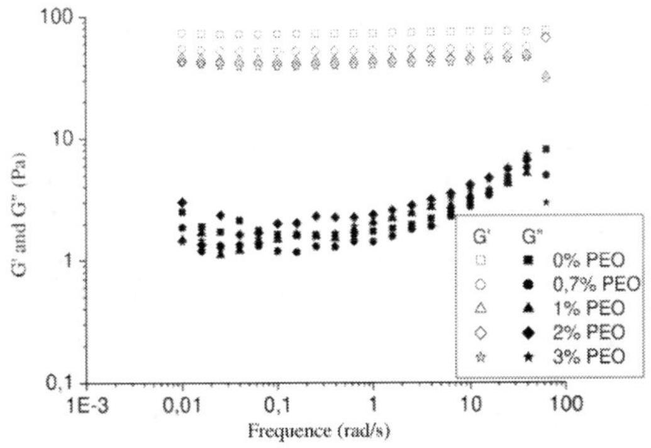


**Figure 6.** Elastic and loss modulus frequency sweep as function of PEO concentration for a 6% bentonite suspension

Besides, elastic (G') and loss (G") modulus – time sweep was applied, to highlight the influence of the concentration in PEO on the kinetic of gel to each mixture suspension. Measures of the temporal evolution of elastic and loss modulus (at ω=1 rad/s) for very low deformations were recorded for eight hours in average (fig. 7). We notice that the most diluted system gels the least fast. These data were re-normalised by the time of gel which was determined for every corresponding system to the crossing of modulus (G' = G") [19]. It is interesting to notice that once re normalised, modules are little dependent on the concentration, in particular beyond the point of gel. In this phase of the gel formation, it is thus more the kinetic of aggregation of the particles which determines the modules that the concentration.

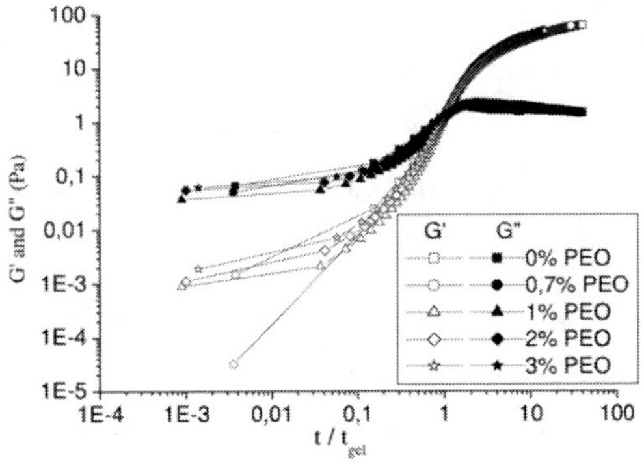


**Figure 7.** Elastic and loss modulus time sweep as function of PEO concentration for a 6% bentonite suspension

The analysis of the table 1 shows the evolution of the gel time according to the concentration of the mixture bentonite-PEO. This time of gel always remains superior to that of the basic suspension. Consequently, we can say that the PEO slows down the kinetic of aggregation clay particles. Recently, similar studies on the laponite suspensions in the presence of PEO of low molecular weight were obtained by Mongondry [20]. Besides, it exists a critical concentration (1%), where the gel time presents an optimum (Max): before this critical concentration, the time gel increases and after that it decreases.

**Table 1.** Gel Time for mixture bentonite-PEO

| System | Bentonite | Bentonite + PEO-6000 | | | |
|---|---|---|---|---|---|
| Concentration (%) | 6 | 0,7 | 1 | 2 | 3 |
| $T_{gel}$ (s) | 800 | 843 | 3363 | 3004 | 2163 |

There are then two observed characteristic rheological behaviours: the first one is that corresponding to the concentrations lower than 1%, where the particles of clay do not interact between them directly or by intervention of the polymer. The electrostatic repulsion being strong, the PEO is chosen then between one or several of the particles by adsorbing without being able to bridge. In that case, the PEO allows stabilizing the particles of clay. The second behavior, beyond the critical concentration where is the time of gel decreases without exceeding that of the basic suspension (800 s). We can think that there is possible appearance of the phenomenon of bridge between the particles of clay which favours system flocculation while increasing the concentration of the polymer. These chains of polymer adsorbed on the surface of the particle can interact with those which are in solution. Also, if the chain is long enough and the saturation of particles is not reached, they can adsorb on several particles at the same time (particles are now suspended by the chains of polymer as long as the saturation of particles is not reached).

The change of behavior observed on the previous figures, could be due to the microstructural relations between the polymer chains and the particles of clay. Indeed, the non ionic polymers can coat the particles of clays or can intercalate the layers of the clay particles [8,21].

### *3.4. Analysis by X-rays diffraction*

X-ray diffraction (XRD) studies could specify the nature of this rheological behaviour. Indeed, this analysis by X-rays diffraction of our samples bentonite-PEO mixture, dried in the ambient temperature under hood using a radiation CuK filtered by iron (source 40 kV and 20 mA), had revealed the d-spacing of the (001) peak following (table 2).

**Table 2.** d-spacing of the (001) peak by X-ray diffraction (XRD)

| System | 6% Bentonite + PEO-6000 | | | | |
|---|---|---|---|---|---|
| Concentration of PEO (%) | 0 | 0,7 | 1 | 2 | 3 |
| d-spacing (001) peak (Å) | 11,80 | 13,83 | 14,03 | 17,57 | 17,57 |

The mechanism envisaged possible is the following: for low concentrations in polymer (≤ 1%), the adsorbed chains will have a flattened configuration so that the polymer does not exceed the Debye length of the particle. In that case the electrostatic repulsion will continue to impose the stability of the system. If we increase a little concentration in polymer (> 1%), the conformation flattened by chains is not possible because the number of adsorbed polymer increases. The thickness of polymers on the surface of the particles of bentonite becomes then superior to the Debye layer where the electrostatic repulsion quickly loses of their strength. So the particles bridge probability by a chain of polymer becomes much stronger as well as the surfaces are not saturated in polymer; what involves the flocculation of the suspension (the d-spacing is practically stable, while the time of gel decreases).

## 4. Conclusion

In the present paper, we investigated on one hand, the rheological behavior of the solutions of PEO which was almost Newtonian and on the other hand, the rheological behavior of the bentonite suspensions in the absence of non-ionic polymer which was practically, shear-thinning with a yield shear stress of flow at relatively high concentration (> 2%). In the presence of this additive (PEO) in the water based drilling fluids (6% bentonite), the influence was essentially discussed in terms of viscosity, yield shear stress of flow and viscoelastic modulus. It had revealed that the PEO which has an affinity for the bentonite particles by adsorbing, slows down their kinetics of aggregation (gel structure formation) by adopting two possible mechanisms: intercalation between the platelets of clay or then favouring the flocculation by forming bridges assured by the chains of polymer between the clay particles beyond critical concentration in PEO (>1%), moreover proved by the essays of viscoelasticity and the analysis by X-rays diffraction. The stability of the suspensions was improved by the polymer which gradually coated the clay particles and so involves the reduction of particle-particle interaction.

**References**

[1] Grim R E 1968 *Clay mineralogy* **2nd** ed. *MC Graw-Hill, NY* 596.
[2] Darley H C H, and Gray G R 1988 *Composition and properties of drilling and completion fluids* **5th** ed. *Gulf company, Texas* 643
[3] Then, B K G 1979 *Developments in Soil Science* **9** *Elsevier Sci. Pub. Co. NY* 362
[4] Luckham P F and Rossi S 1999 *Adv. Colloid Interface Sci.* **82** 43-92
[5] Shen Z, Simon G P and Cheng Y B 2002 *Polymer* **43**, 4251-4260
[6] Rossi S, luckham PF, Green N, and Cosgrove T 2003 *Col. and Surf.* **A 215**, 11-24
[7] Tombacz E, Szekeres M 2004 *Appl. Clay Sci.* **27** 75-94
[8] Almedar A and Gungor N 2005 *J. of Mater. Sci.* **40** 171-177
[9] Lagaly G. and Ziesmer S 2006 *Colloid Polymer Sci.* **284** 947-956
[10] Ebagninin K W Benchabane A et Bekkour K 2007 *Proc.**18** ème Congrès Français de Mécanique, France* 271-274
[11] Allal A., Gareche M., Dupin J.C., Roby F. et Zeraibi, N 2009 *Rhéologie* **16** 35-44
[12] Van Olphen H 1964 *J. Col. Interf. Sci.* **19** 313-322
[13] Bekkour K, Ern H et Lounis M 2001 *Proc. XV ème Congrès français de mécanique, Nancy*
[14] Guzik J M, Pantet A, et Monnet P 1994 *Les cahiers de Rhéologie* **13** (1-2) 150-159
[15] Berthier P et Flaud P 1992 *Les cahiers de la Rhéologie* **X** (3-4) 263-270
[16] Alderman N J, Gavignet A, Guillot D and Maitland G C 1988 *Proc. **63** rd ANN. Tech. Conf. Houston, 10/88, SPE* 18035, 187-195
[17] Rossi S, luckham P F, and Tadros T 2002 *Colloids and Surfaces* **A 201** 85-100
[18] Rossi S, luckham P F and Tadros T 2003 *Colloids and Surfaces* **A 215** 1-10
[19] Tung C.Y.M, and Dynes P J 1982 *J. Appl. Polym. Sci.* **27** 569-574
[20] Dau J, and Lagaly G 1988 *Croatica Chem. acta, CCACAA* **71** (4),983-1004
[21] Mongondry P, Nicolai T, and Tassin F 2004 *J. of Coll. and Interface Sci.* **275**, 191-196

Journal of Physics: Conference Series **602** (2015) 012007 doi:10.1088/1742-6596/602/1/012007

# Mechanical properties of materials obtained via alkaline activation of illite-based clays of Latvia

**I Sperberga[1], M Rundans[1], A Cimmers[1], L Krage[1] and I Sidraba[2]**

[1]Riga Technical University, Institute of Silicate Materials, Azenes Str. 14/24, Riga, LV 1007, Latvia

[2] University of Latvia, Faculty of Geography and Earth Sciences, Raina Blvd. 19, Riga, LV 1586, Latvia

**Abstract.** Materials has been synthesized in the temperature range from 60-100 $^0$C from two illite based clays of Latvia under activation of KOH and NaOH solutions (4-6 M). Compressive strength and apparent porosity were measured. The effect of concentration of KOH and NaOH solutions on the material mechanical properties was investigated by means of infrared spectroscopy (IR). Compressive strength data of the materials showed that via such activation could obtain building materials with good quality.

## 1. Introduction

Alkaline activation is a chemical process in which a powder material of an alumosilicate nature is mixed with an alkaline activator to produce a paste that is able to set and harden in a short time [1,2]. The reaction process of this system is characterized by dissolution of alumino-silicate oxides and followed by the polymerization reaction of those dissolved species, in the presence of alkali ions for charge balancing in framework cavities nearby $AlO_4$ constituents. The reaction produces $SiO_4$ and $AlO_4$ tetrahedral frameworks linked by shared oxygens as polysialates or polysialate-siloxo or polysialate-disiloxo depending on the $SiO_2/Al_2O_3$ ratio in the system [2]. It is three-dimensional alumino-silicate binder obtained using inorganic raw material having pozzolanic properties such as some aluminosilicate natural minerals or clays [3]. Geopolymeric materials possess excellent mechanical properties, including fire and acid resistance [4,5]. Mentioned properties make geopolymers as alternative construction material. Kaolinite is the most often used natural raw material because of its high content of alumina [3,6]. The search for alternative easily available materials in Latvia have lead among another materials to illite-based clays. From geological point of view – Latvia is rich in clay resources to be used for material production currently and at least for centuries [7].

## 2. Materials and methods

Two Quaternary illite clays of Latvia (Q) and (D) were used for geopolymer synthesis were characterized by means of classic chemical analysis (table 1).

**Table 1.** Chemical composition of the studied clays, weight %

| Deposit / Oxides | $SiO_2$ | $Al_2O_3$ | $Fe_2O_3$ | CaO | MgO | $TiO_2$ | $K_2O$ + $Na_2O$ | LOI |
|---|---|---|---|---|---|---|---|---|
| Q clay | 47.32 | 17.44 | 5.34 | 9.21 | 1.78 | 0.56 | 3.44 | 14.16 |
| D clay | 43.05 | 14.10 | 4.24 | 11.46 | 5.05 | 0.64 | 3.65 | 15.36 |

Particle size distribution was determined by sedimentation analysis and X-ray diffraction by model Rigaku, Japan, with $CuK_{\alpha}$ radiation at a scanning interval from $2\theta = 10\text{-}60^{\circ}$ and speed 4 °/min. Results are shown in Table 2.

**Table 2.** Particle size distribution and mineralogical composition of the studied clays

| Deposit | Particle size distribution, % | | | Mineralogical composition |
|---|---|---|---|---|
| | Sand fraction > 50 µm | Silt fraction 5 – 50 µm | Clay fraction < 5 µm | |
| Q clay | 3.1 | 18.4 | 78.5 | Illite, kaolinite, quartz, calcite, dolomite, feldspar |
| D clay | 6.3 | 23.5 | 70.2 | |

In order to investigate the effect of the activators (KOH and NaOH) on the properties of the material, two series (Q and D) were tested. The potassium and sodium hydroxide solutions were obtained by dissolving dried pellets of 99 % purity in distilled water. Thus the activators were added to the clay powders and mechanically mixed. Afterwards cylindric samples (h=25 mm, d=16 mm) were made in plastic way with 25-30 % of average moisture content. Samples were cured at 60, 80 and 100 °C temperature for 4 hours and were allowed to mature at room temperature for 28 days to increase their mechanical strength. Final samples were used for the mechanical strength and apparent porosity measurements.

Compressive strength was determined by means of the "Compression Test Plant ToniNorm" Toni-Technic by Zwick (300 kN).

Fourier transform infrared spectroscopy was carried out on powders of hardened materials using "21 Prestige, Shimadzu Corp" FTIR spectrophotometer in transmittance mode. FTIR spectra were gathered between 400 and 1600 $cm^{-1}$.

## 3. Results and discussion

Table 1 shows differences in chemical composition between studied clays. Content of CaO is higher for clay Q than that of D clay. The CaO content of source material appears to strengthen the geopolymer by forming amorphously structured Ca-Al-Si gel during geopolymerisation. The $Al_2O_3$ and CaO are regarded as more significant elements than the others, because $AlO_4^-$ and silicate oligomers combine to form geopolymers.

Moreover alkali metal cations ($K^+$ and Na play a catalytic role and could contribute to strong structure formation.

Fig. 1 presents the role of KOH as chemical activator on the compressive strength of final materials obtained from both studied clays and cured at different temperatures. KOH solution with higher used concentration (6M KOH) did not promote the strength enhancement for both clays.

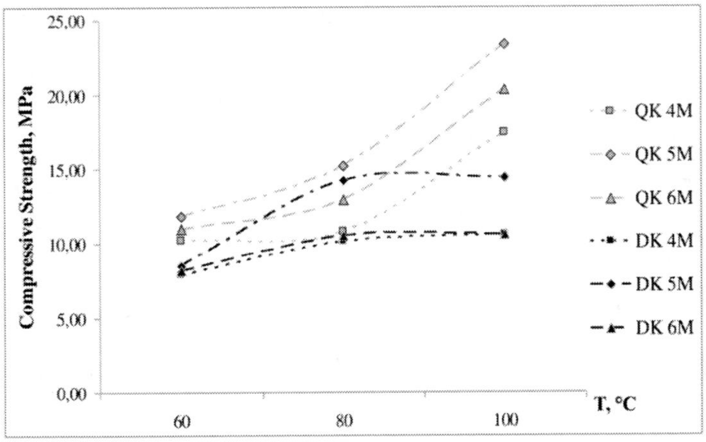


**Figure 1.** Compressive strength evolution vs. KOH concentration at different curing temperatures.

The higher compressive strength (23 MPa at 100 $^0$C) showed Q clay activated with 5 M KOH solution. By activation with 4 M and 6 M compressive strength of materials obtained from Q clay was lower (17 and 20 MPa, accordingly). Obviously higher concentration of KOH solution facilitates the dissociation of different silicate and aluminate species, thus promoting further polymerization resulting in higher compressive strength. However, if a very high alkaline environment is used, the connectivity of silicate anions may be reduced resulting thus in poor polymerization and lower compressive strength. Compressive strength increases with an increasing of curing temperature in all cases. Contrary to Q clay compressive strength is lower when D clay is treated with KOH solution reaching the best value (14 MPa) when curing at 80 $^0$C temperature took place.

Results presented in Fig. 2 shows effect of NaOH solution on the compressive strength of materials obtained from both used clays. Strength increased along with decrease of NaOH concentration for Q clay.

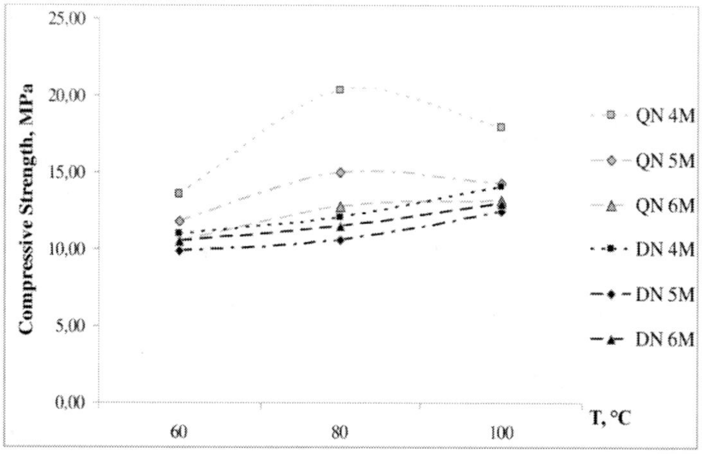


**Figure 2.** Compressive strength evolution vs. NaOH concentration at different curing temperatures.

In this case the lower $Na_2O/Al_2O_3$ and $Na_2O/SiO_2$ ratios promoted the strength improvement than higher ones in the case of Q clay. Obviously using higher than 4M NaOH concentration for chemical activation, excess hydroxide ion concentration caused alumosilicate gel precipitation at the very early stages, resulting in lower compressive strength. The highest compressive strength (20 MPa) showed material activated with 4M NaOH and cured at 80 $^0$C. Chemical treatment of D clay with different NaOH concentrations gave lower compressive strength results comparing with Q clay at similar curing and chemical conditions.

FTIR analysis shows increased sensitivity for structures of short-range structural order and is considered as an appropriate technique for studying the structural evolution of amorphous alumosilicates exhibiting high heterogeneity. Infrared absorption bands enable identification of specific molecular components and structures.

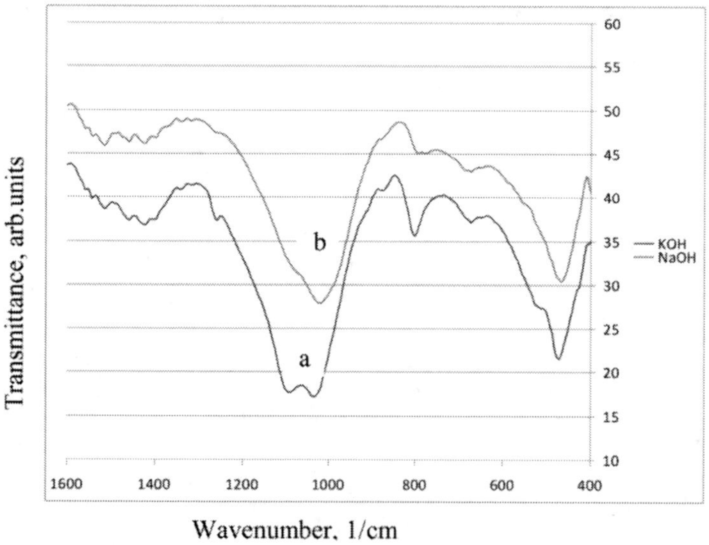


**Figure 3.** FTIR spectra of materials from Q clay cured at 100 $^0$C: a – activated with 5 M KOH solution; b – activated with 5 M NaOH solution.

The difference in absorption frequencies between differently activated Q and D clays predicts transformations taking place during material synthesis. The most characteristic difference observed between the FTIR spectra of differently activated Q clay is concerning the band at around 800 $cm^{-1}$ which is related to the $AlO_4$ vibrations [8].

The bands at 470 $cm^{-1}$ seen for both specimens are due to bending vibrations of Si-O-Si and O-Si-O bonds indicating that quartz is almost chemically inert in the strong alkaline activating solutions [9,10]. The small band seen at ~ 530 $cm^{-1}$ only for material obtained by activating with KOH solution is mainly due to the symmetric stretching vibrations of Si-O-Si and Al-O-Si [8], which are referred to the formation of amorphous to semi-crystalline alumosilicate materials. Small bands at 680 $cm^{-1}$ represent the Si-O symmetrically stretching showing that 6-coordinated Al changed into 4-coordinated one, and participates in the formation of framework structure [11]. The very small band seen at 870 $cm^{-1}$ corresponds to dissolved silicate and/or alumosilicate species and indicates that dissolution of raw materials takes place [12]. Characteristic band at 1033 $cm^{-1}$ have been assigned to asymmetric stretching of Al-O and Si-O bonds originating from within individual tetrahedra. All bands at around 1088 $cm^{-1}$ are a major fingerprint of the inorganic polymer matrix and define the extent of polysialation or aluminium incorporation; they are also due to asymmetric stretching of Al-O and Si-O bonds originating from within individual tetrahedra [9]. The very small band seen at 1270 $cm^{-1}$ may be due to Si-O-Si or Al-O-Si asymmetric stretching vibration as a result of $SiO_4$ and $AlO_4$ reorganisation that takes place during synthesis [12].

## 4. Conclusions

Geopolymers are materials rapidly developed during last decades and they are quite new ones in Latvia. Illite-based geopolymer materials through activating of two Quaternary clays of Latvia by KOH and NaOH solution were synthesized. Compressive strength of the geopolymers were greatly dependent both on curing conditions and the concentration of activation solution. It was shown that higher compressive strength demonstrated activated Q clay containing higher clay and lower sand fraction. Better compressive results showed Q clay cured at 100 °C and activated with KOH. Obviously KOH provides more inorganic polymer precursors compared to NaOH since the larger $K^+$ favours the formation of larger silicate oligomers with which $Al(OH)_4^-$ prefers to bind, thus higher compressive strength is acquired in this temperature. Curing in lower temperatures gave decreased compressive strength of materials in the case of activation with KOH solution. Using NaOH as activator better results reached Q clay as well.

## References

[1] Duxon P, Fernández-Jiménez A, Provis JL, Lukey GC, Palomo A and van Deveter JSJ 2007 *J. Mater. Sci.* **42** 2917
[2] Steins P, Poulesquen A, Diat O and Frizon F 2012 *Langmuir* **28** 8502
[3] Kuenzel C, Vandeperre LJ, Donatello S, Boccaccini A and Cheeseman C 2012 *J. Amer. Ceram. Soc.* **95** 3270
[4] Davidovits J 1991 *J. Therm. Anal.* **37** 1633
[5] Davidovits J 2008 *Geopolymer chemistry & applications.* 2nd edition, Saint-Quentin: Institute Géopolymère 333
[6] Elimbi A, Tchakoute Kouamo H and Njopwouo D *Constr. Build. Mat.* 2011 **25** 2805
[7] Seglins V 2008 *Scientific Proceedings of Riga Technical University, Material Science and Applied Chemistry* Riga **17** 62
[8] Bakharev T 2005 *Cem. Concr. Res.* **35** 1224
[9] Van Jaarsveld JGS, Van Deventer JSJ, Lukey G C 2002 *Chem. Engin. J.* **89** 1-3 63
[10] Lee W K W, Van Deventer JSJ 2002 *Colloids Surf. A: Physicochemical and Engineering Aspects* **211** Issues 2–3 49
[11] Zhang Y, Wei S, Zongjin L 2012 *Appl.Clay Sci.* **47** 271
[12] Rees CA, Provis JL, Lukey GC, Van Deventer JSJ 2007 *Langmuir* **23** 8170

Journal of Physics: Conference Series **602** (2015) 012008 doi:10.1088/1742-6596/602/1/012008

# Rheology of drilling muds

**G Kudaikulova**[1]

[1] Kazakh National Technical University, Almaty, KazakhstanRiga,

kgulzhanabd@mail.ru

**Abstract.** Therheological properties of drilling muds were studied. It is shown that for receiving pseudo-plastic liquids with an indicator of nonlinearity of N<0.3 use of polymers with a high molecular weight or a combination of polymers to various influence on water structure is most expedient.

## 1. Introduction

In the process of drilling wells drilling muds prevent possible technological complications, minimize layer pollution by components of drilling muds, provide the fullest carrying out of drilling cutting and let us achieve the highest the technician – economic indicators.

The hydraulic program of washing of wells, which is realized by regulation of rheological indicators of drilling muds plays an important role in the solution of most of these problems.

Rheological properties of drilling muds have considerable impact on carrying out of drilling cutting on a day surface, creation of hydrodynamic pressure in well. The major factor in defining the efficiency of carrying out of cutting is a ratio of plastic viscosity of drilling mud with a dynamic resistance to shift in the course of a current. Hydrodynamic pressure, in turn, defines possibility of complications in the course of construction of wells: manifestation or absorption of drilling mud, hydraulic fracturing of rocks, and also layer pollution [1, 2]. Therefore there is a problem of management of rheological characteristics of drilling muds in a well occurs which can be solved by chemical processing of drilling mudsby various polymers.

## 2. Theory

All existing rheological models reflect the ideal behavior of real bodies. A number of complex models which are used in various rheological systems were introduced by a combination of three main models.

Drillers are interested inthose models, received by an experimental or semi-empirical way, only if they are applicable for the liquids with non-Newtonian behavior.

In order to describe the behavior of drilling muds with the low maintenance of the solid phase, processed by polymeric reagents (developed by us and considered in present work), Gershel – Balkli's model is commonly used, whichin its turn was achieved by a combination of viscoplastic model of Bingham with Ostvald – de Vaale's model.Bingam'sviscoplastic model describes substances which have the yield point, below which substances aren't deformed, whereas above – they flow like viscous liquids. There are two rheological parameters at this model – yield point ($\tau_0$) and plastic viscosity ($\eta_p$):

$$\tau = \tau_0 + \eta_p \gamma, \quad (1)$$

where$\tau$– stress of shear, $\gamma$ – rate of shear.

In Ostvald– de Vaale's model there are two rheological parameters – experimental constants K and N; K – consistence and N – nonlinearity indicator:

$$\tau = K\gamma^N \quad (2)$$

It is possible to calculate apparent viscositycomparing this equation to Newton's law:

$$\eta = K\gamma^{N-1} \tag{3}$$

As at the drilling muds developed by us N <1, that points to degree of non-Newtonian behavior of liquid, so η = K. Thus, Gershel – Balkli's model will look as follows:

$$\tau = \tau_0 + K\gamma^N \tag{4}$$

Researches of the processes proceeding in a well, show that as drilling muds, it is most expedient to use the pseudo-plastic liquids possessing size of an indicator of nonlinearity of N<0.3 [2 – 4]. Such liquids provide effective cleaning a hole of well of drilling cutting and minimum possible losses of pressure in a well.

## 3. Experimental

Unfortunately, provisions on management of a nonlinearity indicator at drilling muds aren't developed yet. It is caused by that studies of properties of drilling mudsand their analytical description in the course of movement carry out from a position of classical mechanics of continuous mediums[5, 6]. Thus it isn't considered that drilling mud is the continuous mediumat rest which properties are defined by properties of bonds in it.

We suggest considering of drilling mudas the discrete system consisting of the discrete dispersive medium and a disperse phase. Thus it is necessary to consider that in drilling mud on an interface between two phases of disperse system there is a double electric layer (DEL) which charge on the separate centers is presented by discrete ions.

The indicator of nonlinearity N will depend on ability of hydrogen bonds to form new bonds in mud and is an indicator estimating its internal continuity, and a consistence indicator K – the size of durability of hydrogen bonds. According to it for pseudo-plastic liquid it will be fair that than more size N especially solution is discrete, and than it is more K, that bigger durability hydrogen bonds in volume of mud possess.

From practice of drilling it is known that at drilling mud have size N<0.3 it keeps a continuity of the structure, and at values N>0.3 costs of energy of overcoming of internal resistance of liquid sharply increase that testifies about bigger discrete of drilling mud and fast destruction of bonds in its structure.

Research of rheological parameters of the drilling muds developed was made by FANN – viscometer, model 35 SA (USA) and the VSN-3 rotational viscometer (Russia).

In this research two types montmorilloniteclays – clay of the Taukentsky field of South Kazakhstan and the Tagansky field of East Kazakhstan are used. The polymers of French “Promafor” (Polycol 60S, Polycol 60SM, Polyfor 30, Polyfor 50), German “Bayer” (Antisol 100, Antisol 30000), USA “M-I Swaco” (EZ MUD, Barazan, Modivis)and Kazakh polymer SOP – threefold copolymer of an acrylic row are used. For each type of drilling mud, polymers are added in different weight concentrations ranging between 0.03 % to 0.7 %.

Consistence (K) and an indicator of nonlinearity (N) were calculated on formulas (5) and (6) [3]:

$$K = \tau_{300} / \gamma_{300}^{\ N} = \gamma_{300}\, 0.511/511^N \tag{5}$$

$$N = 3.32\ lg \gamma_{600}/\gamma_{300} \tag{6}$$

In present work we discuss results of the received values of rheological parameters of drilling muds, as major practical parameters connected with speed of a stream of drilling mud in a well and sharp fluctuations of pressure at lowering-hoisting operations and also at start-up or a stop of drilling pumps.

## 4. Results and Discussion

Rheograms of the developed drilling muds testify that they aren't ideal viscoplastic liquids as shear stress at zero stress above viscometer indications at low speeds of rotation. In Figure 1 are presented developed rheograms such drilling muds. Rheological parameters of drilling muds processed by various reagents are specified in Table 1.

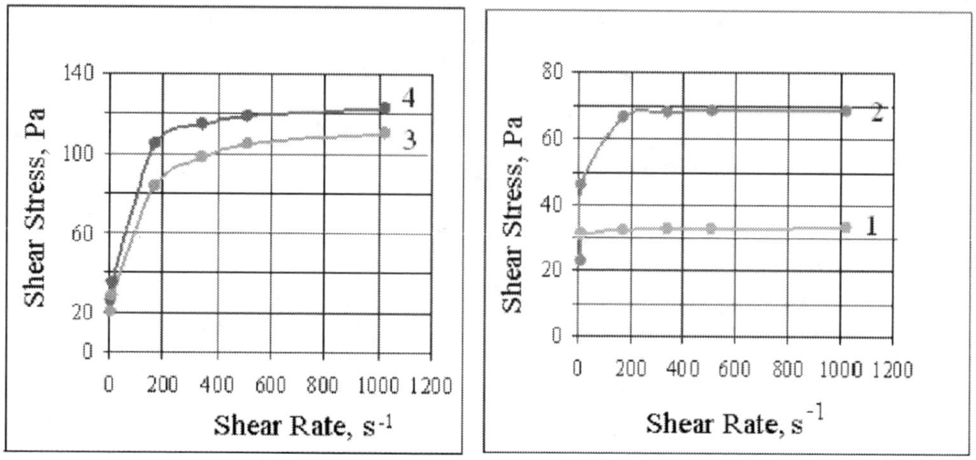


1 – Taganskyclay suspension (CS), 2 – CS + 0.04 % SOP, 3 – CS + 0.04 % Polyfor 30, 4 – CS + 0.04 % SOP + 0.04 % Polyfor 30

Figure 1. – Rheograms of various drilling muds

In the course of a current of drilling mud owing to deformation of DEL there is an effect of braking of particles. The more concentration of clay particles (in a case with Taukentsky clay), the more electrically loaded covers of particles interact by means of attraction or pushing away forces.

Addition of polymers in clay suspension leads to destabilization and decrease in degree of dispersion of system at the expense of aggregation of particles. This process happens owing to reduction of thickness of DEL and increase in forces of an attraction of Van der Waals and adsorption of chains of polymer by surfaces of clay particles. These processes considerably influence viscosity of system, as is observed at increase in concentration of polymers.

The yield point ($\tau_0$)characterizes resistance at a current, caused by the electric forces of an attraction pushing away existing between solvated particles. Value $\tau_0$ depends on concentration of a disperse phase, extent of ionization, thickness of DEL, the nature of ions, concentration and the nature of ions in the dispersive medium and concentration of polymer in system. Change of $\tau_0$ is caused mainly by change of concentration of polymers and their adsorption, displacing balance of forces of an attraction and pushing away.

Apparently from Table 1 values of an indicator of nonlinearity are influenced by structure, concentration and the nature of reagents. Big values N as it was already specified, testify about bigger discrete of drilling mud because bonds between reagents and particles of clay are less strong, than their bonds with water molecules. The hydrated compounds in the course of movement of liquid will represent a separate phase, similar to a firm phase of mud. Such mud is more similar to viscoplastic liquid and have nonlinearity indicator more than 0.3. In this way reagents with the increased concentration in mud – Polycol 60S, Polycol 60SM, Polyfor 50 proved.

Other option when durability of hydrogen bonds in volume of mud is distributed evenly and as a whole increases not at so considerable value. It conducts to deficient rigidity and the increased plasticity of communications. According to it solution in the course of mechanical movement doesn't break the microstructure at the expense of fast restoration of the destroyed bonds. Such reagents give to mud of property of pseudo-plastic liquids. In this way showed reagents with higher molecular weight such, as Antizol 30000, EZ MUD, Barazan and Modivis. It is connected with that molecules with a high weight owing to the size make more uniform changes to water structure in comparison with molecules of smaller weight and respectively provide more uniform distribution of durability of hydrogen bonds in volume of mud. Because the number of single contacts at molecules with high weight above, than at molecules with a low weight, and liquid durability on destruction of structure of solution is higher also, as well as the speed of formation of single contacts.

**Table 1.**Rheological parameters of drilling muds

| № | Compositionofmud(claysuspension CS +polymer) | Plastic Viscosity mPa·s | YieldPoint, $\tau_0$ Pa | Consistence K Pa·s | Nonlinearity indicatorN |
|---|---|---|---|---|---|
| 1. | CS 1 (Taukentskyclay 10 % + $H_2O$) | 6.0 | 28.73 | 9.34 | 0.22 |
| 2. | CS 2 (Taganskyclay5 % + $H_2O$) | 1.0 | 30.24 | 28.51 | 0.022 |
| 3. | CS 2 + 0.03 % SOP | 0.5 | 53.52 | 37.1 | 0.006 |
| 4. | CS 2+ 0.04% SOP | 0.0 | 64.32 | 68.47 | 0 |
| 5. | CS 2+ 0.05 % SOP | 2.0 | 56.64 | 53.14 | 0.023 |
| 6. | CS2+ 0.03 % Polyfor 30 | 6.0 | 100.60 | 75.96 | 0.063 |
| 7. | CS 2+0.04 % Polyfor 30 | 7.5 | 107.75 | 88.2 | 0.048 |
| 8. | CS 2+0.05 %Polyfor 30 | 8.5 | 129.54 | 111.86 | 0.038 |
| 9. | CS 2+0.06 % Polyfor 30 | 17.0 | 115.65 | 63.28 | 0.122 |
| 10. | CS 2+0.07 % Polyfor 30 | 23.0 | 115.89 | 58.5 | 0.137 |
| 11. | CS 1+ 0.1 % Polycol60S | 6.0 | 28.73 | 9.33 | 0.22 |
| 12. | CS 1+0.3 % Polycol60S | 8.0 | 16.28 | 5.35 | 0.30 |
| 13. | CS 1+ 0.5 % Polycol60S | 8.5 | 22.51 | 3.92 | 0.34 |
| 14. | CS1+0.7 % Polycol60S | 13.0 | 16.76 | 1.08 | 0.53 |
| 15. | CS 1 + 0.1 % Pol. 60SM | 6.5 | 29.21 | 9.00 | 0.23 |
| 16. | CS 1 + 0.3 % Pol.60SM | 9.0 | 18.67 | 4.11 | 0.34 |
| 17. | CS 1 + 0.5 % Pol.60SM | 10.0 | 22.03 | 3.57 | 0.38 |
| 18. | CS 1 + 0.7 % Pol.60SM | 15.0 | 12.93 | 0.65 | 0.61 |
| 19. | CS 1+ 0.1 % Polyfor 50 | 3.5 | 28.25 | 17.03 | 0.13 |
| 20. | CS 1 + 0.2 % Polyfor 50 | 7.0 | 26.34 | 6.24 | 0.28 |
| 21. | CS 1 + 0.3 % Polyfor 50 | 7.5 | 27.30 | 6.84 | 0.27 |
| 22. | CS 1 + 0.5 % Polyfor 50 | 9.5 | 22.03 | 3.30 | 0.37 |
| 23. | CS 2 +0.04%Antisol100 | 1.9 | 31.13 | 12.48 | 0.18 |
| 24. | CS 2 +0.04% Antisol30000 | 4.8 | 35.92 | 23.36 | 0.09 |
| 25. | CS 2 + 0.04 % EZ MUD | 6.0 | 31.60 | 10.76 | 0.21 |
| 26. | CS 2 + 0.04 % Barazan | 7.0 | 34.48 | 19.35 | 0.12 |
| 27. | CS 2 + 0.04 % Modivis | 3.5 | 34.96 | 19.35 | 0.12 |

In order toreceive the drilling muds being pseudo-plastic liquid with value N less than 0.3 was of interest to find combinations of the reagents, capable to provide uniform distribution of durability of hydrogen bonds in structure of water of a drilling mud.For this purpose we developed recipesof the drilling muds containing mutually compatible reagents, allowing to operate durability of hydrogen bonds in volume of mud. Drilling muds were prepared with Tagansky clay. Results of the carried-out researches are given in Table 2.

The mechanism of action of binary composition of reagents such is. At introduction of composition of reagents in clay suspension one reagents will create an initial grid on the basis of hydrogen bonds, to set thereby mud structure, others at the expense of the hydration will redistribute it among themselves, with simultaneous alignment of durability of hydrogen bonds on mud volume. Therefore, for example, the combination in drilling mud of reagents with the various nature of dissolution, or strongly differing on molecular weight, promotes obtaining optimum properties of bonds. Example of similar combinations are combinations Antisol(improved cellulose) and the SOP acrylic copolymer, or Antisol with the Polycol 60S and Polycol 60SM reagents representing a mix of polyacrylamides and sodium acrylates.

Combinations Antisol with Barazan or Modivis biopolymers, and also Polyfor 30 reagents with Barazan or Modivis allow to receive muds with a nonlinearity indicator N <0.3 where biopolymer plays the priming role, focusing original structure of mud.

In case of combination use reagent Polyfor 30 with Antisol arises strong synergetic increase of durability of bonds in mud (high value K), but without change of an indicator of nonlinearity in comparison with the mud processed only by Polyfor 30 reagent.

Journal of Physics: Conference Series **602** (2015) 012008 doi:10.1088/1742-6596/602/1/012008

**Table 2.** Influence binary composition of reagents on rheological parameters drilling muds

| № | Compositionofmud(claysuspension CS +polymer 1 + polymer 2) | Plastic Viscosity mPa·s | Shear Strength1 | Shear Strength10 | Consistence K Pa·s | Nonlinearity indicator N |
|---|---|---|---|---|---|---|
| 1. | CS + 0.04 % SOP + 0.04 % Antisol | 30.0 | 19.68 | 20.16 | 50.32 | 0.154 |
| 2. | CS+ 0.04 % SOP + 0.05 % Antisol | 33.5 | 21.60 | 22.56 | 44.58 | 0.175 |
| 3. | CS+ 0.04 % SOP + 0.06 % Antisol | 26.5 | 22.08 | 23.04 | 55.50 | 0.140 |
| 4. | CS+ 0.04 % SOP + 0.07 % Antisol | 22.0 | 33.12 | 33.60 | 50.45 | 0.129 |
| 5. | CS + 0.04 % SOP + 0.08 % Antisol | 22.5 | 32.64 | 36.00 | 58.13 | 0.125 |
| 6. | CS + 0.04 % SOP + 0.04 % Polyfor30 | 12.0 | 17.28 | 19.44 | 63.11 | 0.081 |
| 7. | CS + 0.04 % SOP + 0.05 % Polyfor30 | 14.5 | 17.76 | 20.40 | 58.96 | 0.096 |
| 8. | CS + 0.04 % SOP + 0.06 % Polyfor30 | 22.0 | 18.48 | 19.44 | 46.98 | 0.138 |
| 9. | CS + 0.04 % SOP + 0.07 % Polyfor30 | 26.8 | 17.28 | 20.88 | 45.47 | 0.155 |
| 10. | CS + 0.04 % Polyfor30 + 0.08 % Antisol | 10.5 | 41.76 | 45.60 | 107.42 | 0.051 |
| 11. | CS + 0.04 % Polycol 60S + 0.04 % Antisol 30000 | 10.0 | 2.85 | 9.56 | 13.66 | 0.17 |
| 12. | CS + 0.04 % Polycol60SM + 0.04 % Antisol100 | 8.0 | 4.86 | 6.04 | 5.83 | 0.30 |
| 13. | CS + 0.04 % Antisol 30000+ 0.04 % Barazan | 5.5 | 4.2 | 4.7 | 11.01 | 0.20 |
| 14. | CS + 0.04 % Antisol 30000+ 0.04 % Modivis | 4.5 | 7.05 | 7.55 | 14.7 | 0.16 |
| 15. | CS + 0.04 %Polyfor30 + + 0.04 % Barazan | 5.5 | 1.01 | 5.03 | 12.03 | 0.19 |
| 16. | CS + 0.04 %Polyfor30 + + 0.04 %Modivis | 4.5 | 4.86 | 5.03 | 16.03 | 0.15 |

## 5. Conclusion

Thus, proceeding from specification of ideas of interaction of reagents in volume of drilling muds and the given experimental results, follows that for receiving pseudo-plastic liquids with an indicator of nonlinearity of N<0.3 use of polymers with a high molecular weight or a combination of reagents to various influence on water structure is most expedient.

## References

[1] Kister E.G. 1972 Chemical processing of drilling muds. M: "Nedra", 392

[2] Gray G.R., Darley H.C.H. 1985 Composition and Properties of Oil Well Drilling Fluids Transl. Engl. M: "Nedra", 509

[3] Makovey N. 1986 Hydraulics of drilling. M:"Nedra", 536

[4] Ogibalov P. M., Mirzadzhanzade A. Kh. 1970 Non-stationary movements viscously-plastic mediums. M: Moscow State University Publishing, 415

[5] Krasnov K.S. 1984 Molecules and chemical bond. M: "Vyssh. Sh.", 295

[6] Rabinovich N.R. 1989 Engineering problems of mechanics of the continuous medium in drilling.M:"Nedra", 270

# Influence of mineral aggregates on the rheological properties of concrete mixture

**A Klovas**[1,2], **M Daukšys**[1]
[1] Department of Civil Engineering Technologies, Kaunas University of Technology, Studentu str. 48, LT-51367 Kaunas, Lithuania

**Abstract.** The aim of this research was to determine how the change of concrete mixture constituents: concentration of fine and coarse aggregate as well as the amount of fine particles, not exceeding 0.25 mm, influence concrete mixture`s rheological properties. Firstly, inner-concentration of fine aggregate (sand, fraction of 0/1 and 0/4) was changed. Secondly, coarse aggregate (gravel, fraction of 4/16) concentration was changed and finally, the amount of fine particles was changed. Results have shown that with the increase of sand (fraction of 0/1) quantity, the plastic viscosity also increased. On the other hand, yield stress, at the beginning decreased, but eventually – increased. The increase of coarse aggregate quantity acted differently: plastic viscosity and yield stress decreased. Finally, the increase of fine particles quantity decreased the plastic viscosity as well as yield stress of concrete mixture.

## 1. Introduction

Aggregate characteristics: size, gradation, shape and surface texture have significant effects on concrete rheology. The different rheology behaviour is the result of the aggregate inter-particle forces and the particle movement in the liquid phases of fresh concrete [1-4]. The investigations indicate that the aspect ratio, angularity, and surface texture of aggregates affect the viscosity and yield stress of concrete mixture. Yield stress and viscosity of concrete mixture significantly increase with increased coarse aggregate volume fraction [5]. The test of concrete mixture rheological properties was conducted replacing the ordinary Portland cement with some mineral additives: finely ground blast furnace slag, fly ash and silica fume. Results revealed yield stress and plastic viscosity decreased with replacing ordinary Portland cement with blast furnace slag and fly ash [6]. The water requirement for concrete decreases with increased aggregate particle size. Finer the aggregate is more water for a given consistency it demands [7, 8]. Friction among aggregate has a big contribution to concrete rheology. Less the friction between the aggregate particles is more workable concrete it is possible to obtain. Particles with a nearly spherical shape and a smooth surface texture provide more workable concrete [9]. An optimal aggregate gradation provides a higher degree of packing and requires less paste to reach a given consistency since less cement paste is needed to fill the space among the aggregate [1].

It was determined that concrete mixtures with ultrafine fly ash represented the best rheological improvement while silica fume represented the worst. It was shown that the replacement of cement by silica fume results in an increase in the water demand and super-plasticizer dosage to maintain the rheological properties of the control. In contrast, the replacement of cement by ultrafine fly ash

[2] Corresponding author. Tel.: +370-619-13266; fax.: +370-37-438744.
E-mail address: *Albertas.Klovas@ktu.lt* (A. Klovas)

resulted in a reduction of the water demand and super-plasticizer dosage to maintain the same rheological properties of the control [10].

It is known that yield stress and viscosity of concrete mixture typically increase with the sand content in the mortar [1].

The main aim of this research was to obtain the limits of rheological properties by changing aggregate proportions and fine particles quantity that the concrete mixture would possess the best workability and at the same time would be possible to avoid its segregation. The reference concrete mixture composition BA1-0 was recommended by native concrete mixture producers which they use for surface appealing concrete.

## 2. Materials and methodology

JSC "Akmenes cementas" (Lithuania) Portland cement CEM II/A-LL 42.5 R was used. Physical and mechanical properties of cement are given in table 1.

**Table 1.** Physical and mechanical properties of Portland cement CEM II/A-LL 42.5 R (MA)(A)

| Property, units | Value |
|---|---|
| Specific surface area, $m^2/kg$ | 410 |
| Normal consistency of cement paste, % | 26.5 |
| Initial setting time, min. | 195 |
| Compressive strength after 2 days / after 28 days, MPa | 27.1/54.0 |

Kvesu quarry sand with the fraction of 0/4, bulk density of 1550 $kg/m^3$ and fineness module of 1.67 was used as fine aggregate for concrete mixtures. 0/1 sand fraction (density 1460 $kg/m^3$, fineness module 2.37) was also used as fine aggregate. Gravel with the fraction of 4/16 and bulk density of 1327 kg/m3 was used as the coarse aggregate. Granulometric composition of aggregates was conducted according to LST EN 12620:2003+A1:2008 and presented in table 2.

**Table 2.** Granuliometric composition of aggregate

| Sieve's mesh, mm | 0/1 fr. sand | 0/4 fr. sand | 4/16 fr. gravel |
|---|---|---|---|
| 32.0 | - | - | 100.0 |
| 16.0 | - | - | 95.6 |
| 8.0 | 100.0 | 100.0 | 34.9 |
| 4.0 | 100.0 | 97.8 | 2.9 |
| 2.0 | 99.9 | 86.3 | 0.7 |
| 1.0 | 94.8 | 68.5 | 0.7 |
| 0.500 | 39.1 | 37.9 | 0.7 |
| 0.250 | 3.0 | 4.9 | 0.7 |
| 0.125 | 0.3 | 1.2 | 0.7 |
| 0 | 0.2 | 0.1 | 0.1 |

Plasticizing admixture based on policarboxile polymers Glenium SKY 628 (BASF Construction Chemicals) was used with density of solution 1.06 g/ml. The total dosage of admixture was 1.2% of cement. During the research, dry aggregates were used for concrete mixtures. Cement and aggregates were dosed by weight while water and chemical admixture were dosed by volume. Chemical admixtures in the form of solutions were mixed with water and used in preparation of concrete mixtures. Concrete mixtures were mixed for 3 minutes in the laboratory in forced type concrete mixers.

The calculation of concrete mixture`s rheological properties was based according to formulas 1 and 2 [8, 11]. The yield stress of concrete mixture was calculated according to 1 formula.

$$\tau_0^{SL} = \frac{260 \cdot \rho_{mi\check{s}}}{(\sqrt{\frac{21000}{30-SL}} - 75 - 5)^2} \tag{1}$$

Here:

$\tau_0^{SL}$ Yield stress of concrete mixture calculated by its slump, kPa;

$\rho_{mi\check{s}}$ Density of concrete mixture, g/cm$^3$;

$SL$ Slump of concrete mixture, cm.

The plastic viscosity of concrete mixtures was calculated according to formula 2.

$$\eta_b = \eta_v \cdot \exp\left( \frac{a_c \cdot \rho_v}{\rho_v + \frac{V}{C} \cdot \rho_c - b_c \cdot \rho_v} + \frac{a_{sm}\left(1 - \varphi_{st} - \varphi_o - \frac{V}{\rho_v} - \frac{C}{\rho_c}\right)}{1 - \varphi_{st} - b_{sm}\left(1 - \varphi_{st} - \varphi_o - \frac{V}{\rho_v} - \frac{C}{\rho_c}\right)} + \frac{a_{st} \cdot \varphi_{st}}{1 - b_{st} \cdot \varphi_{st}} \right) \tag{2}$$

Here:

$\varphi_o$ Air content in concrete mixture, %;

$V$ Water amount in concrete mixture per 1m$^3$, l;

$C$ Cement content in concrete mixture per 1m$^3$, kg.

$\rho_v$, $\rho_c$ Density of water and cement, kg/m$^3$;

$\varphi_{sm}$ Volume concentration of fine aggregate in mortar;

$\varphi_{st}$ Volume concentration of coarse aggregate in concrete mixture;

$a_c$, $a_{sm}$, $a_{st}$ Coefficients concerning the surface texture of cement, fine and coarse aggregates. ($a_c$ = 2.6; $a_{sm}$ = 2.5; $a_{st}$ = 2.6);

$b_c$, $b_{sm}$, $b_{st}$ Coefficients concerning the placement density of cement, fine and coarse aggregates particles in mortar and concrete mixture.

## 3. Results

### *3.1. The influence of fine aggregate quantity on the rheological properties of concrete mixture*

Concrete mixture composition combined two fractions of fine aggregate: 0/1 and 0/4. In order to evaluate the influence of fine aggregate (sand, fraction 0/1) quantity to the rheological properties of concrete mixture (figure 1), the quantity of sand (fraction 0/1) was increased while the quantity of sand (fraction 0/4) was decreased. The quantity of sand (fraction 0/1) was increased from 11 to 36 % according to mass in respect to total fine aggregate`s quantity. The compositions of concrete mixtures (BA1) used in this research are presented in Table 3.

**Table 3.** Composition of concrete mixtures BA1 (a)

| Materials | The amount of materials for 1m$^3$ of concrete mixture, kg | | | | | |
|---|---|---|---|---|---|---|
| | BA1-0 | BA1-1 | BA1-2 | BA1-3 | BA1-4 | BA1-5 |
| CEM II/A-LL 42.5 R (MA)(A) | 330 | 330 | 330 | 330 | 330 | 330 |
| Water | 178 | 178 | 178 | 178 | 178 | 178 |
| W/C | 0.54 | 0.54 | 0.54 | 0.54 | 0.54 | 0.54 |
| Sand fraction 0/1 | 208<br>11% | 303<br>16% | 398<br>21% | 493<br>26% | 588<br>31% | 682<br>36% |

(b)

| Materials | The amount of materials for $1m^3$ of concrete mixture, kg | | | | | |
|---|---|---|---|---|---|---|
| | BA1-0 | BA1-1 | BA1-2 | BA1-3 | BA1-4 | BA1-5 |
| Sand fraction 0/4 | 701<br>37% | 607<br>32% | 512<br>27% | 417<br>22% | 322<br>17% | 227<br>12% |
| Gravel fraction 4/16 | 986<br>52% | 986<br>52% | 986<br>52% | 986<br>52% | 986<br>52% | 986<br>52% |
| Glenium SKY 628 | 3.96 | 3.96 | 3.96 | 3.96 | 3.96 | 3.96 |

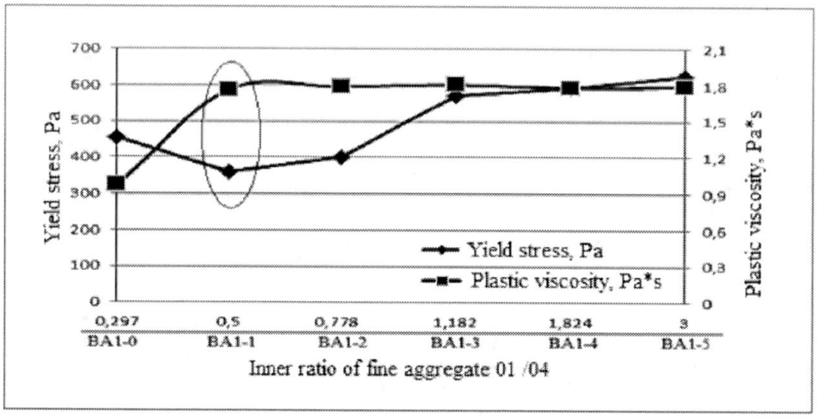


**Figure 1.** The influence of inner ratio of fine aggregate on the rheological properties of concrete mixture

The optimal concrete mixture composition was chosen according to research results shown at figure 1. The inner concentration of fine aggregates was 16/32 (0.5) (BA1-1). The values of yield stress and plastic viscosity of concrete mixture were as follow: 362.0 Pa and 1.764 Pa*s.

### *3.2. The influence of coarse aggregate quantity on the rheological properties of concrete mixture*

This part of the research was based on changing ratio between fine (sand with fractions 0/1 and 0/4) and coarse (gravel, fraction 4/16) aggregates. The concentration of coarse aggregate was increased from 22 % to 84 % in respect to total aggregate's quantity, while at the same time fine aggregate's concentration was decreased. It must be noted that the inner ratio between fine aggregate (fraction 0/1 and 0/4) was always at the level of 0.5. Concrete mixtures compositions (BA2) used in this research are presented in Table 4.

**Table 4.** Composition of concrete mixtures BA2

| Materials | The amount of materials for $1m^3$ of concrete mixture, kg | | | | | | |
|---|---|---|---|---|---|---|---|
| | BA2-0 | BA2-1 | BA2-2 | BA2-3 | BA2-4 | BA2-5 | BA2-6 |
| CEM II/A-LL 42.5 R (MA)(A) | 330 | 330 | 330 | 330 | 330 | 330 | 330 |
| Water | 178 | 178 | 178 | 178 | 178 | 178 | 178 |
| W/C | 0.54 | 0.54 | 0.54 | 0.54 | 0.54 | 0.54 | 0.54 |
| Sand fraction 0/1 | 492 | 429 | 366 | 303 | 240 | 177 | 114 |
| Sand fraction 0/4 | 986 | 860 | 733 | 607 | 480 | 354 | 228 |
| Gravel fraction 4/16 | 417<br>22 % | 607<br>32 % | 796<br>42 % | 986<br>52% | 1175<br>62 % | 1365<br>72 % | 1554<br>82 % |
| Glenium SKY 628 | 3.96 | 3.96 | 3.96 | 3.96 | 3.96 | 3.96 | 3.96 |

The influence of ratio between coarse and fine aggregates on the rheological properties of concrete mixture is presented at figure 2.

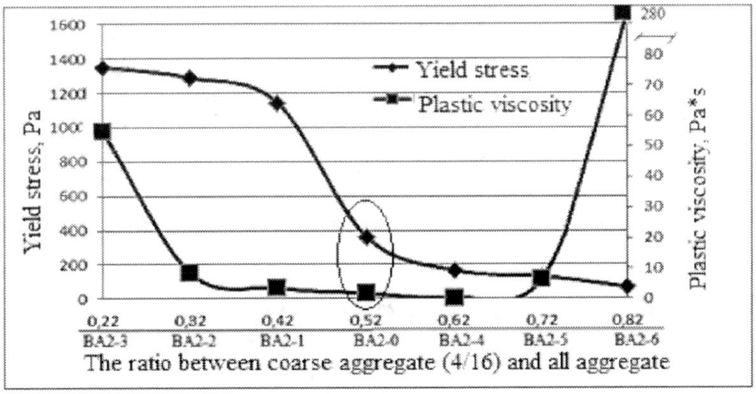


**Figure 2.** The influence of ratio between coarse aggregate and total aggregates on the rheological properties of concrete mixture

BA2-0 concrete mixture composition was chosen as an optimal one which yield stress was 361.95 Pa and plastic viscosity was 1.764 Pa*s. Further decrease of yield stress (increase of coarse aggregate ratio) resulted in segregation therefore this point was chosen as optimal.

### *3.3. The influence of fine particles quantity on the rheological properties of concrete mixture*

The total quantity of fine particles combined particles from sand and cement with the certain diameter. In order to investigate the influence of fine particles (not exceeding 0.25 mm) quantity on the rheological properties of concrete mixture, the amount of cement was increased (table 5). The quantity of fine aggregate (sand, fraction 0/1 and 0/4) was decreased respectively. It must be noted that the inner ratio between fine aggregate (fraction 0/1 and 0/4) was kept at the level of 0.5. The ratio of coarse aggregate in respect to total quantity of aggregates was kept at 0.52. Concrete mixture compositions (BA3) used in this research are presented in Table 5.

**Table 5.** Composition of concrete mixtures BA3

| Materials | The amount of materials for $1 m^3$ of concrete mixture, kg | | | | | | |
|---|---|---|---|---|---|---|---|
| | BA3-0 | BA3-1 | BA3-2 | BA3-3 | BA3-4 | BA3-5 | BA3-6 |
| CEM II/A-LL 42.5 R (MA)(A) | 180 | 230 | 280 | 330 | 380 | 430 | 480 |
| Water | 178 | 178 | 178 | 178 | 178 | 178 | 178 |
| W/C | 0.99 | 0.77 | 0.64 | 0.54 | 0.47 | 0.41 | 0.37 |
| Sand fraction 0/1 | 353 | 336 | 320 | 303 | 286 | 270 | 253 |
| Sand fraction 0/4 | 707 | 674 | 640 | 607 | 574 | 540 | 507 |
| Gravel fraction 4/16 | 985 | 985 | 985 | 985 | 985 | 985 | 985 |
| Glenium SKY 628 | 3.96 | 3.96 | 3.96 | 3.96 | 3.96 | 3.96 | 3.96 |

The influence of fine particles (not exceeding 0.25 mm) quantity on the rheological properties of concrete mixture is presented at figure 3. The base concrete mixture composition was chosen BA2-0 (figure 2) when conducting tests with BA3 mixtures.

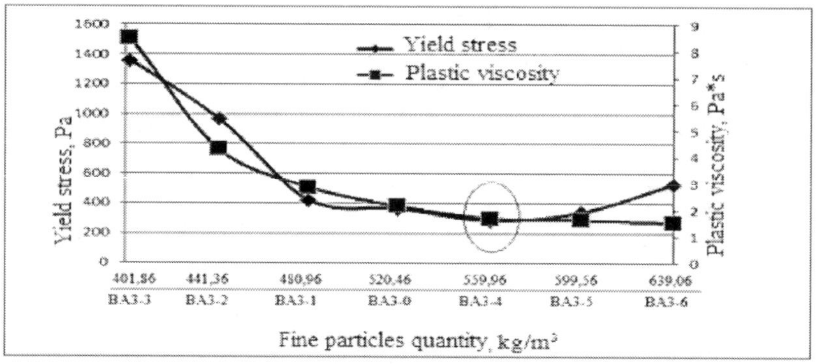


**Figure 3.** The influence of fine particles quantity on the rheological properties of concrete mixture

BA3-4 concrete mixture composition was chosen as an optimal one which yield stress was 286.0 Pa and plastic viscosity was 1.683 Pa*s.

## 4. Conclusions

The optimal inner ratio between fine aggregate (fraction of 0/1 and 0/4) was 0.5. The increase of ratio from 0.297 to 0.778 resulted in decrease of yield stress. Within the ratio of 0.297 and 0.5, plastic viscosity increased and further stayed at the steady value.

The optimal ratio between quantities of coarse aggregates and total aggregates was 0.52 (mixture without segregation). The increase of ratio resulted in the decrease of yield stress and plastic viscosity. Over the ratio limit of 0.72, plastic viscosity increased dramatically and mixture became very stiff.

The optimal quantity of fine particles was 559.96 kg/m$^3$. The increase (up to 559.96 kg/m$^3$) of fine particles resulted in decrease of yield stress and plastic viscosity. Further increase of the fine particles quantity resulted in increase of yield stress and concrete mixture became too stiff to work with.

## 5. References

[1] Hu J and Wang K 2011 Effect of coarse aggregate characteristics on concrete rheology *Constr Build Mater* **25** 1196-204
[2] Mehta P K and Monteiro P J M 1993 *Concrete: structure, properties, and materials*: Prentice Hall)
[3] Kosmatka S H, Panarese W C, Allen G E and Cumming S 2002 *Design and control of concrete mixtures* vol 5420: Portland Cement Association Skokie, IL)
[4] Tattersall G H 2003 *Workability and Quality Control of Concrete*: Taylor & Francis)
[5] Geiker M R, Brandl M, Thrane L N and Nielsen L F 2002 On the effect of coarse aggregate fraction and shape on the rheological properties of self-compacting concrete *Cement, concrete and aggregates* **24** 3-6
[6] Park C, Noh M and Park T 2005 Rheological properties of cementitious materials containing mineral admixtures *Cement Concrete Res* **35** 842-9
[7] Jamkar S S and Rao C B K 2004 Index of Aggregate Particle Shape and Texture of coarse aggregate as a parameter for concrete mix proportioning *Cement Concrete Res* **34** 2021-7
[8] Skripkiūnas G 2007 *Statybinių konglomeratų struktūra ir savybės* (Kaunas: Vitae Litera)
[9] Kurokawa Y, Tanigawa Y, Mori H and Nishinosono K 1996 Analytical study on effect of volume fraction of coarse aggregate on Bingham's constants of fresh concrete *T Jpn Concr I* **18** 37-44
[10] Ferraris C F, Obla K H and Hill R 2001 The influence of mineral admixtures on the rheology of cement paste and concrete *Cement Concrete Res* **31** 245-55
[11] Gumuliauskas A A G 2000 Betono mišinio ir jo struktūros dalių ribinių šlyties įtempimų tyrimas *Betonas ir gelžbetonis* Konferencijų pranešimų medžiaga. Kaunas: Technologija 47-52

# Monitoring of vulcanization process using measurement of electrical properties during linear increasing temperature

**E Seliga[1], O Bošák[1], P Koštial[2], Z Dvořák[3], M Kubliha[1,4], S Minárik[1] and V Labaš[1,5]**

[1] Department of Physics, Institute of Materials, Faculty of Materials Science and Technology SUT, J. Bottu 25, 917 24 Trnava, Slovakia
[2] Faculty of Metallurgy and Materials Engineering, VŠB – Technical University of Ostrava, 17.listopadu 15, 708 33 Ostrava – Poruba
[3] Department of production engineering, Faculty of technology, Thomas Bata University in Zlin, Nad Stráněmi 4511, 760 05 Zlín, Czech Republic
[4] Department of Physics, Faculty of Civil Engineering SUT, Radlinského 11, 813 68 Bratislava, Slovakia
[5] Department of Physics, Faculty of Education, Catholic University, Hrabovská cesta 1, 034 01 Ružomberok, Slovakia

E-mail: emil.seliga@stuba.sk, ondrej.bosak@stuba.sk, pavel.kostial@vsb.cz, zdvorak@ft.utb.cz, marian.kubliha@stuba.sk, stanislav.minarik@stuba.sk, vladimir.labas@stuba.sk

**Abstract**. The article presents the possibilities of diagnostics of irreversible chemical reaction - vulcanization in case of laboratory prepared rubber mixture based on styrene - butadiene (SBR) using measurements of selected physical parameters. Our work is focused on the measurement of current rheologic parameters (torque at defined shear deformation) and selected electrical parameters (DC conductivity) during linear increasing temperature. The individual steps of vulcanization are well identified by means of measurements of rheologic parameters, while significantly affecting the value of the electrical conductivity. The value of the electrical conductivity increases with the increasing of rate of the crossbridging reactions during vulcanization. The rate of the heating affects both types of measurements. When the rate of the heating is increasing the temperature of the beginning of networking step of reactions and also the rate of vulcanization grow. The sensitivity of the both types of measurements allows a good mathematical description of the temperature dependence of the torque and the electric conductivity during the vulcanization of rubber mixtures based on SBR.

## 1. Introduction

Most important process in elastomers production is vulcanization process. Three-dimensional network of polymer chains creates during vulcanisation. This network is generated by cross-bonds among linear rubber chains [1]. Significant increase of resistance against external forces caused by the origin and the increasing of number of cross-linking bonds can be observed. There is correlation between the number of cross-linking bonds and values of some mechanical properties manly Young modulus. It makes possible to observe and analyse the creation of cross-linking bonds indirectly by means of measurement of amounts of mechanical units required for realization of the deformation. Generally used method is measurement of the time dependence of torque generating constant shear deformation at constant temperature (figure 1).

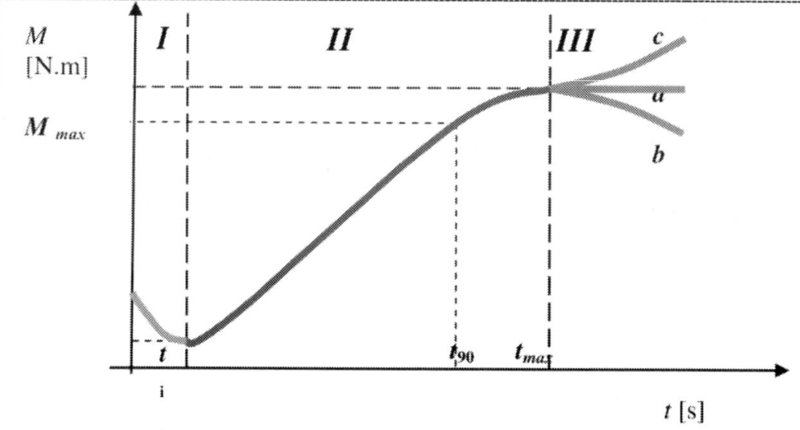


**Figure 1**. Scheme of time dependence of torque - vulcanization curve.

Whole vulcanization process can be divided on three stages which determine changes in material structure. The first of them is induction stage (I) during which preliminary chemical reactions in material are in progress. Next cross-linking bonds among linear chains are generated during second stage (II). The increasing of their number causes growth of value of measured torque. Depending on used kind of rubber the number of bonds can be slightly changed in third stage (III) (cases a, b, c) [2]. Requirement of defined samples preparation and impossibility of practical realization of these methods directly in technological equipment determined for real production are disadvantages of analysis of vulcanization process by means of measurement of chosen mechanical units. Content of this contribution is oriented on presentation of possibilities of the using of electrical methods for observation of vulcanization process in case of laboratory prepared rubber mixture based on styrene-butadiene (SBR) at linear increasing of temperature.

## 2. Experiment

Experimental measurements were carried out on laboratory prepared rubber mixture based on SBR with simple vulcanizing reagent based on sulphur. Mixing of the mixture shown in table 1 was realized in double-roller.

**Table 1.** Composition of the tested rubber mixture

| Components | pphr |
|---|---|
| SBR Rubber<br>Buna VSL 5025-0 | 100.0 |
| Vulcan C-72 R | 8.0 |
| ZnO | 3.0 |
| Sterin III | 1.0 |
| Dusantox IPPD | 1.0 |
| Sulfenax CBS | 1.5 |
| Sulphur | 1.5 |
| Total | 116 |

In order to the fact that the number of cross-linking bonds generated during vulcanization process increases values of mechanical units, rheological measurements were used as a basis for evaluation.

Values of torque needed for creation of constant value of shear modulus 0.5° [3] were measured. Measurements were realized by means of equipment D-MDR 3000 MonTech. Value of torque was observed for more values of heating rates (0.2°C/min; 0.4°C/min; 0.6°C/min; 0.8°C/min; 1.0°C/min; 2.0°C/min; 3.0°C/min; 4.0°C/min; 5.0°C/min; 7.0°C/min; 10°C/min). Mentioned measurements were realized repeatedly at increasing temperature ranged from 30 °C up to 200 °C. Heating rates were determined on the basis of possibilities of the measuring equipment and mathematical processing of results. Measurements of torque carried out during linear increasing temperature were supplemented by measurement of DC electrical conductivity realized by means of equipment Novocontrol concept 90. Measurements of DC electrical conductivity were realized by direct method in constant voltage 10V at the same heating rates as measurements of vulcanization curves during linear increasing temperature.

## 3. Results

Example of measured temperature dependencies of torque during heating rate 2 °C.min$^{-1}$ is shown in figure 2. The first measured curve can be divided on two areas while transition between them corresponds to the temperature ~155 °C. The change of torque is influenced by two factors in this case. The first one is the decrease of torque values of rubber mixture caused by the change of viscosity during the increasing of temperature (see area I in figure 2). The second one is the beginning and progress of cross-linking reaction – networking with relatively abrupt increase of torque values (see area II in figure 2). Results of repeated measurement are not already influenced by potential changes resulted from the subsequent cross-linking of rubber chains.

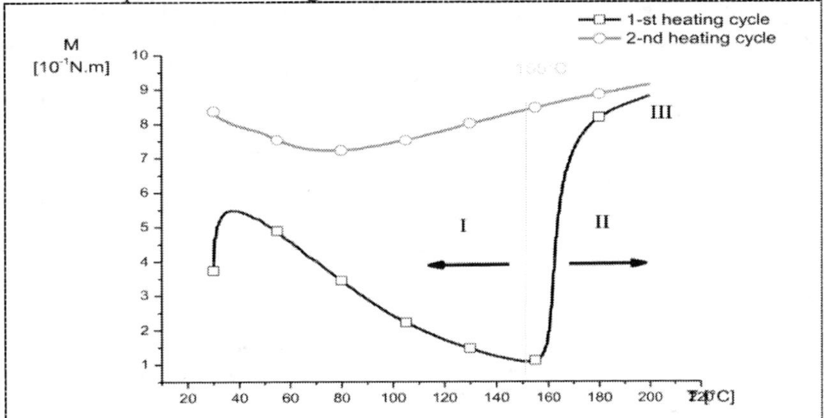


**Figure 2.** Dependence of torque $M$ of rubber compound at a temperature $T$, by heating rate $\vartheta$=2 °C/min, at the 1-st heating cycle (▪):area represents the induction period, II - the area represents the crosslinking reaction.

There is observable influence of heating rate when temperature dependencies of torque are compared (see figure 3).

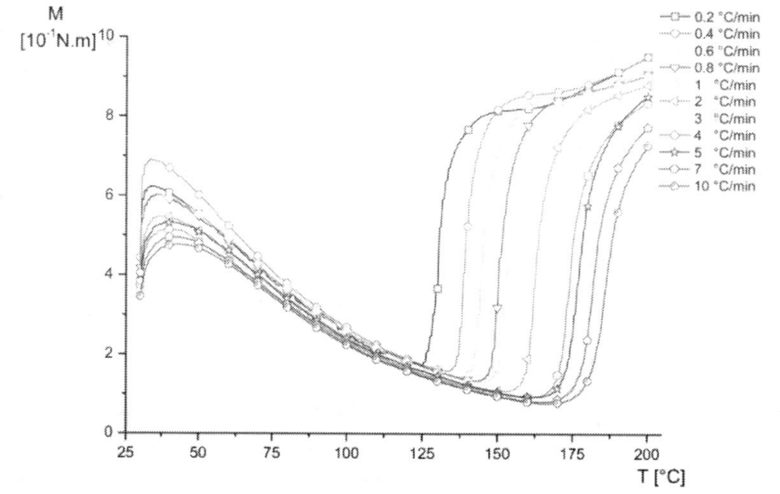


**Figure 3.** Dependence of the torque $M$ on the temperature of measured rubber mixture in different heating rates at 1-st heating cycle.

With regard to the fact that the increase of torque is in correlation with the increasing of number of generated cross-linking bonds [4], in case of linear increasing of temperature the temperature derivation of torque represents value correlating with the rate of cross-links creation $\frac{dM}{dT} \approx \frac{dN}{dt}$. Temperature dependence of the torque derivation (rate of cross-links creation) was determined numerically from measured data and it is shown in figure 4.

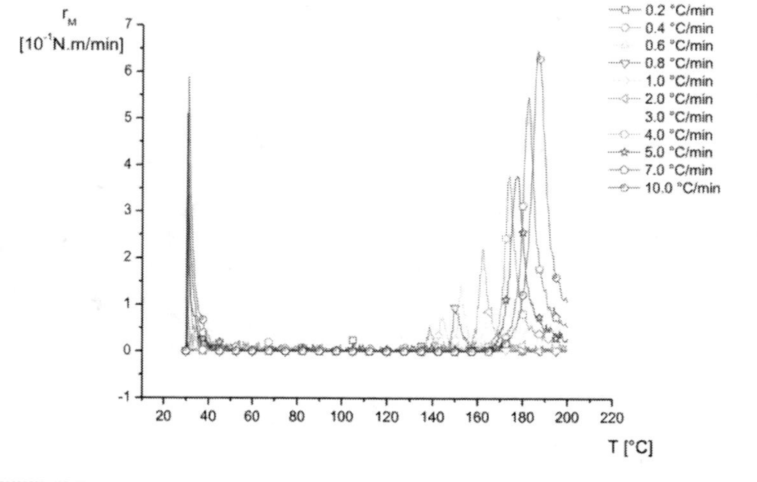


**Figure 4.** Temperature dependence of rates of torque change $r_M$ (numerically determined from temperature dependence of torque values) of rubber mixture at temperature by different heating rates $\vartheta$.

Journal of Physics: Conference Series **602** (2015) 012010 doi:10.1088/1742-6596/602/1/012010

Rubber mixture based on SBR at which cross-linking bonds during vulcanization process are created via sulphur ions was used for experimental measurements. There is possibility of the increasing of the electrical charge transport during vulcanization process caused by mentioned ions occurrence. Influence of vulcanization reaction – crosslinking, in accordance with prediction, can be identified on measured curves of DC electrical conductivity as an increase of DC electrical conductivity during 1st. heating cycle which is observable at divers values of heating rate (see figure 5). Influence of another cross-linking of rubber mixture was not already observed during second heating cycle.

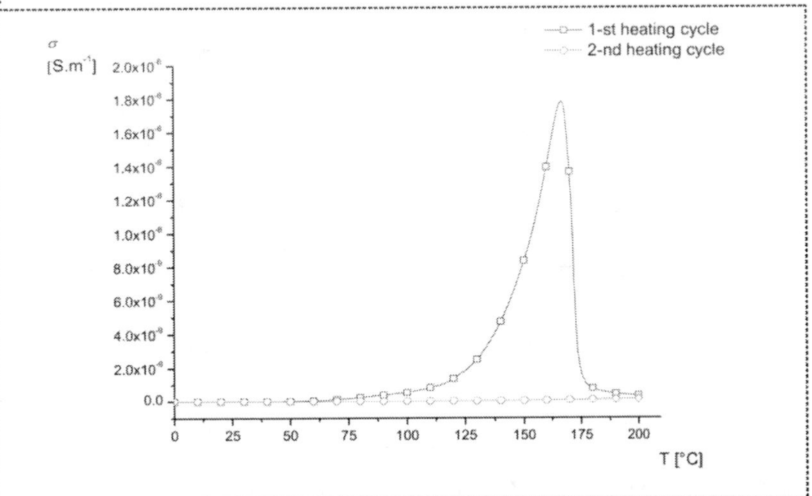


**Figure 5.** Dependence of DC conductivity $\sigma$ at temperature $T$ SBR8 rubber mixture at heating rate $\vartheta$=2°C/min, 1-st heating cycle (▪) a 2-nd heating cycle (∗) by intermediate cooling to 0°C.

In case of these measurements can be observed influence of heating rate similarly as in the case of torque measurements.

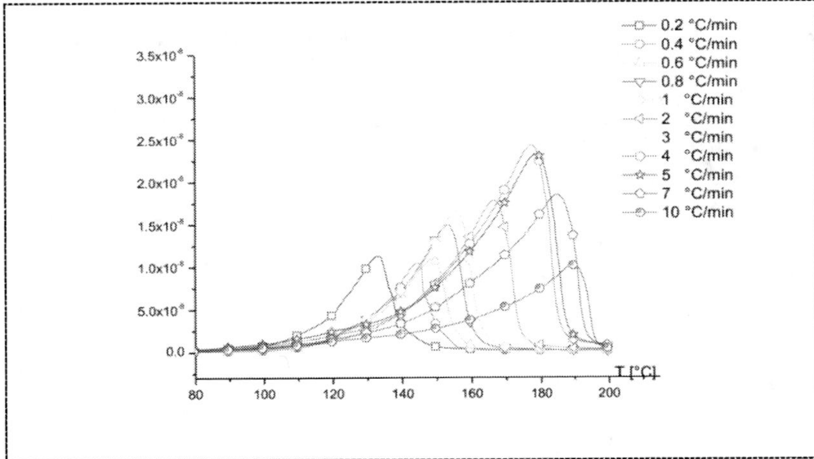


**Figure 6.** Temperature dependencies of DC electrical conductivity $\sigma$ of rubber mixture for different heating rate $\vartheta$ at 1-st heating cycle.

There is observable the significant correlation between temperature at which the torque value is maximal $T_{rM\max}$ and temperature corresponding with maximal value of DC electrical conductivity during 1-st heating cycle $T_{\sigma\max}$ (see figure 5 and figure 6).

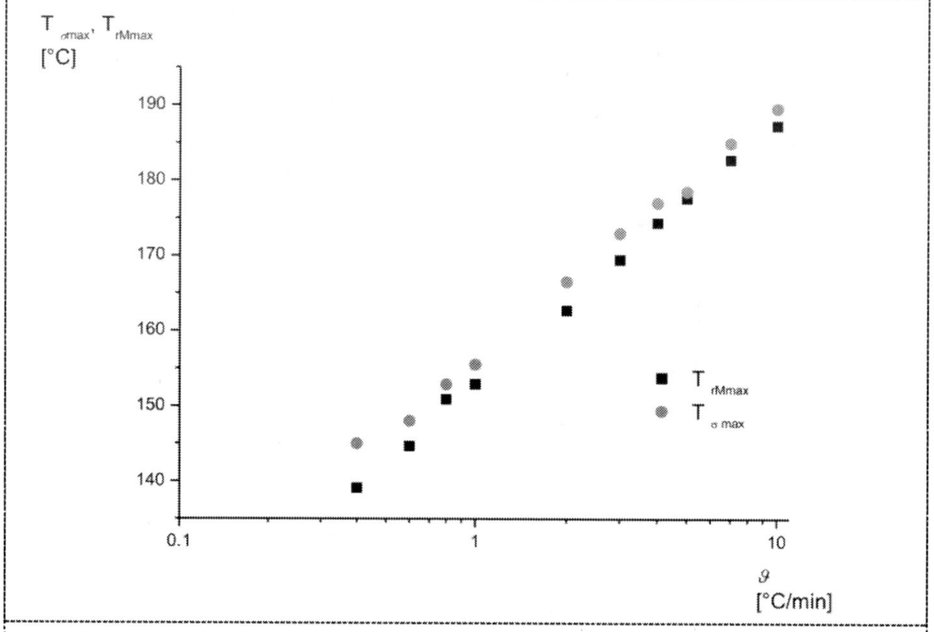


**Figure 7.** Dependence $T_{\sigma\max}$ and $T_{rM\max}$ on heating rate $\vartheta$ for SBR8 rubber mixture.

Difference between $T_{\sigma\max}$ a $T_{rM\max}$ can be caused by different dimensions of samples as well as their temperature inhomogeinity. However, values of DC electrical conductivity increase also in temperature area of induction period (figure 5 up to 150°C) in which there is observed no increase of number of cross-linking bonds (figure 4 up to 150°C).

## 4. Conclusion

Observation of changes during cross-linking reaction of rubber mixtures and their subsequent quantification is one of the criteria for evaluation of final mechanical properties of rubber products.

Measurement of vulcanization characteristics at linear increasing temperature is of great importance for determination of transition temperature from induction period to area of cross-linking reaction. This transition is manifested by the increasing of mechanical properties when cross-linking bonds among rubber macromolecules are generated in the mixture. Measurement of vulcanization characteristics by means of rheometer is typical for evaluation of mechanical properties in technical practice.

Measurement of DC electrical conductivity at linear increasing temperature is non-standard measurement in technical practice. Value of temperature $T_{rM\max}$, at which largest speed of reaction is observed (largest value of derivation of torque) correlates with temperature $T_{\sigma\max}$, at which the maximal value of DC electrical conductivity is detected. We observed similar dependencies of temperatures $T_{rM\max}$ a $T_{\sigma\max}$ on used heating rate $\vartheta$.

Results presented in this work create good basis for development of material science in the area of rubber mixtures. Correlation between mechanical and electrical properties has been proven. The application of electrical methods at diagnosis of process of vulcanization appears as a possible alternative to non-destructive observation of cross-linking process.

This research was supported by APVV-SK-CZ-0168-11, VEGA 1/0356/13 and ITMS:26220120014

**References**

[1] Young Kee Chae, Woo Young Kang, Joong-Hee Jang and Sung-Seen Choi 2010 *Polymer Testing* **29** 953
[2] Stark W, Jaunich M, Bohmeyer W and Lange K 2012 *Polymer Testing* **31** 904
[3] Bošák O, Tóth M, Minárik S, Psota J 2011 Equipments for investigation of non-metallic materials. *XXIV Didmattech 2011 : Problems in teachers education. - Kraków : Instytut Techniki UP* ISBN 978-83-7271-679-8. 63-68
[4] Salgueiroa W, Somozaa A, Marzocca A J, Consolati G and Quasso F 2007 *Radiation Physics and Chemistry* **76** 142
[5] Milani G, Leroy E, Milani F and Deterre R 2013 *Polymer Testing* **32** 1052
[6] Zanchet A, Carli L N, Giovanela M, Brandalise R N and Crespo J S 2012 *Materials and Design* **39** 437
[7] Johns J and Nakason Ch 2011 *Journal of Non-Crystalline Solids* **357** 1816
[8] Likozar B and Krajnc M 2008 *Chemical Engineering Science* **63** 3181
[9] Mitra S, Ghanbari-Siahkali1 A and Almdal K 2006 *Polymer Degradation and Stability* **91** 2520
[10] Bishay I K, Abd-El-Messieh S L and Mansour S H 2011 *Materials and Design* **32** 62
[11] Reffaee S A, El Nashar D E, Abd-El-Messieh S L and Abd-El Nour K N 2009 *Materials and Design* **30** 3760

Journal of Physics: Conference Series **602** (2015) 012011 doi:10.1088/1742-6596/602/1/012011

# Mathematical characterization of values of rheological variables during the networking reaction of rubber mixtures based on SBR

**E Seliga[1], O Bošák[1], S Rusnáková[2], S Minárik[1] and M Tóth[3]**

[1] Department of Physics, Institute of Materials, Faculty of Materials Science and Technology SUT, J. Bottu 25, 917 24 Trnava, Slovakia
[2] Department of production engineering, Faculty of technology, Thomas Bata University in Zlin, Nad Stráněmi 4511, 760 05 Zlín, Czech Republic
[3] FLEXOPRINT, s. r. o., Bystrická cesta 69, 034 01 Ružomberok, Slovakia

E-mail: emil.seliga@stuba.sk, ondrej.bosak@stuba.sk, rusnakova@ft.utb.cz, stanislav.minarik@stuba.sk, toth@flexoprint.sk

**Abstract**. The article presents an analysis of cross-linking reaction using measurements of time dependencies of torque at constant temperature from the range 100 up to 200 °C. The measured results obtained on sample based on styrene - butadiene (SBS) prepared in the laboratory show the behaviour which can be well described by equations of chemical reactions of first-order kinetics. It is possible mathematically describe significant constants of the kinetics of networking reaction (induction period, reaction rate coefficient) by the solution of differential equations and by mathematical approximation. Constants are exponentially dependent on the temperature of vulcanization, while dependencies are Arrhenius like. Math description allows describe the progress of the vulcanization reaction also in the temperature range outside of the monitoring interval, i.e. in the area of extremely long times required for the realization of the cross-linking reaction.

## 1. Introduction

Elastomers used in production of divers gaskets, tires, insulant claddings etc. belong among meaning group of technical materials. These materials gain their properties in vulcanization process. Fundamentals of vulcanization process are based on generation of three-dimensional network by origin of cross-linking bonds among linear chains of rubber [1]. Generally, the process of vulcanization predominantly influences the final properties of rubber and course of this process can be influenced by compounds in vulcanization system (kind of rubber, vulcanization reagent insuring creation of cross-linking bonds, accelerator for acceleration and shortening of process of cross-linking bonds creation and activator for initialization of cross-linking bonds creation). Others compounds insuring properties of incoming vulcanizers can be divided on fillers, antidegradation reagents and additional compounds (colorants, retarders of combustion and etc.). Production of vulcanizers of rubber mixtures is most often realized at constant increased temperature. Mentioned procedure of cross-linking creation is well described mathematically [1,2]. In technical practice number of crosslinking bonds can be observed by means of values of mechanical variables increasing [2].

Presented contribution is oriented on determination and experimental approval of mathematical formula describing vulcanization process at linear increasing temperature. Laboratory prepared mixture based on styrene-butadiene rubber (SBR) was prepared for experimental approval.

## 2. Experiment

Experimental work has been carried out on laboratory prepared rubber mixture based on SBR with a simple vulcanization system with vulcanization reagent based of sulphur. The blending of the mixture composition showed in table 1 was realized on double- roller.

**Table 1.** Composition of the tested rubber mixture

| Components | pphr |
|---|---|
| SBR Rubber Buna VSL 5025-0 | 100.0 |
| Vulcan C-72 R | 8.0 |
| ZnO | 3.0 |
| Sterin III | 1.0 |
| Dusantox IPPD | 1.0 |
| Sulfenax CBS | 1.5 |
| Sulphur | 1.5 |
| Total | 116 |

In view of the fact that the number of cross-links resulting from the process of vulcanization increases growth of mechanical properties, rheological measurements have been used in the case of the mixture. Torque values required to create a constant value of shear deformation 0,5° were measured [3]. Measurements were carried out by means of the device D-MDR 3000 MonTech. Two types of measurements have been realized. The first type of measurement was carried out at a constant temperature from the range 100°C to 200°C. The temperature interval was chosen with a view to a typical vulcanization temperature of mixture based on SBR rubbers. Measuring time intervals were 4 hours for temperatures (100°C – 120°C), 3 hours for temperatures (130°C – 150°C) and 2 hours for temperatures (150°C – 200°C). Time dependences of the torque at a constant temperature was the result of the experiment. Measurements of temperature dependence of torque at linear increasing temperature with constant heating rate 0.2°C/min; 0.4°C/min; 0.6°C/min; 0.8°C/min; 1.0°C/min; 2.0°C/min; 3.0°C/min; 4.0°C/min; 5.0°C/min; 7.0°C/min; 10°C/min were the second type of experiment [4].

## 3. Determination of the formula describing the vulcanization process at linear increasing temperature

In General, the process of vulcanization realized on a rubber mixture is observable at a constant temperature by a characteristic time dependence of torque. This dependence can be divided into three parts: I - induction stage II- cross-linking stage and III-stage of additional reactions. In the induction stage the chemical preparation of the mixture on the networking process, which follows in the cross-linking stage is realized. In the third stage, depending on the type of rubber, it can be realized further networking (a), the number of links may not be changed (b) or cross-links reduction can be occurred (c).

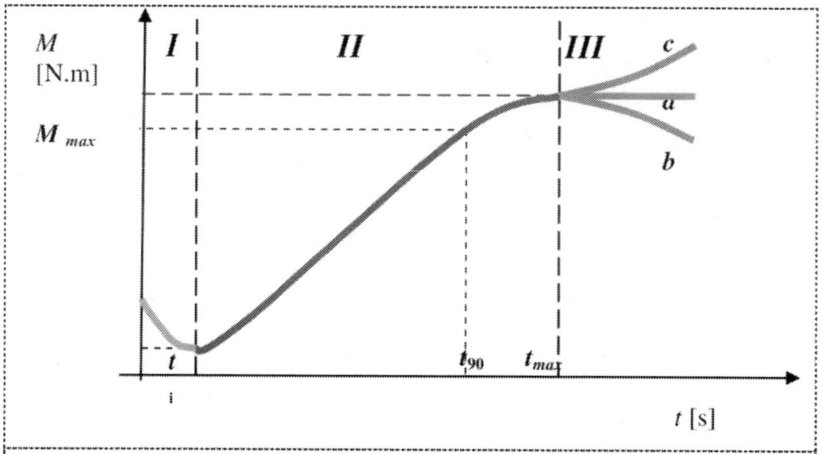


**Figure 1.** Schematic view of the vulcanization curve (time dependence of the torque).

For a mathematical description of networking (stage II in Figure 1), we can use next equation [5]:

$$M = M_{\max} - \left(M_{\max} - M_{\min}\right) \cdot e^{-k(t-t_i)}, \quad (1)$$

where $M_{\max}$ represents maximum value of torque, $M_{\min}$, minimum value of torque, $k$ vulcanization rate constant, $t_i$ time of induction period.

Equation (1) is the solution to the differential equation that explains the proportionality between time change of torque and instant value of torque [6]. The vulcanization rate constant $k$ and the time of the induction period $t_i$ are constants in case of measurements at a specific temperature. For this reason, they are not suitable for description of the behaviour of rubber mixtures during networking at linearly increasing temperature. Thermal dependence of induction period can be described as:

$$t_i = e^{-S_2 + \frac{E_2}{RT}}, \quad (2)$$

where $S_2$ is kinetic constant independent on temperature and $E_2$ is activation energy of initiation [7]. Thermal dependence of "vulcanization rate" can be described as:

$$k = e^{S_1 - \frac{E_1}{RT}}, \quad (3)$$

where $S_1$ is kinetic constant independent on temperature and $E_1$ is activation energy of vulcanisation.

Mentioned kinetic constants ($S_1$, $S_2$, $E_1$, $E_2$) are independent from the temperature and they allow to describe the kinetics of the cross-linking process during vulcanization. Taking account of both relationships (2), (3) in (1) may be obtained formula:

$$M = M_{\max} - \left(M_{\max} - M_{\min}\right) \cdot \exp\left[ -e^{S_1 - \frac{E_1}{RT}} \left( t - e^{-S_2 + \frac{E_2}{RT}} \right) \right] \quad (4)$$

and also the rate of the cross-linking reaction in the next form:

$$r_M = \frac{dM}{dt} = e^{S_1 - \frac{E_1}{RT}} \cdot \left(M_{\max} - M_{\min}\right) \cdot \exp\left[ -e^{S_1 - \frac{E_1}{RT}} \left( t - e^{-S_2 + \frac{E_2}{RT}} \right) \right]. \quad (5)$$

Relationships (4), (5) provide a description of the vulcanization also in case of temperatures, which cannot be used for the experimental observations (too long or short periods of vulcanization).

Formula for determination of the rate of change of torque was used in the next form:

$$r_M = \dot{M} = \frac{dM}{dt}. \tag{6}$$

Considering the time and temperature dependence of torque $M = M(t;T)$ and time dependence of temperature $T = T(t)$ it can be found:

$$\dot{M} = \frac{dM}{dt} = \frac{\partial M}{\partial t} + \frac{\partial M}{\partial T}\left(\frac{dT}{dt}\right) = \frac{\partial M}{\partial t} + \frac{\partial M}{\partial T}\vartheta, \tag{7}$$

where $\vartheta$ is heating rate, which corresponds to time derivation of temperature:

$$\vartheta = \frac{dT}{dt}. \tag{8}$$

Using the relations (5), (7), (8) the temperature dependence of the rate of change of the torque can be written in the form:

$$M = \int_{T_0}^{T} (M_{max} - M_{min}) k \left\{1 + \frac{1}{RT^2}\left[E_1(T - T_0) + \vartheta(E_2 - E_1)t_i\right]\right\} e^{-k\left(\frac{T-T_0}{\vartheta} - t_i\right)} dT + M_{min}, \tag{9}$$

where $k$ and $t_i$ are values of rate constants and time of the vulcanization period, writable by temperature dependencies (relations 0 and 0) by means of constants $S_2$, $E_2$ and $S_1$, $E_1$**Hiba! A hivatkozási forrás nem található.**. The constant $T_0$ corresponds to the initial temperature of cross-linking reaction at the linear increasing temperature.

## 4. Results and discussion

Vulcanization curves i.e. time dependencies of torque in determined temperature range are shown in figure 2. It can be recognized from the figure 2, that only the induction period area and the beginning of cross-linking reaction can be observed at low temperatures (the temperature range 100°C – 130°C). For this reason, we chose the longer period of measurements (4 hours). So it is already possible to observe the whole vulcanization curve at the increasing of temperature.

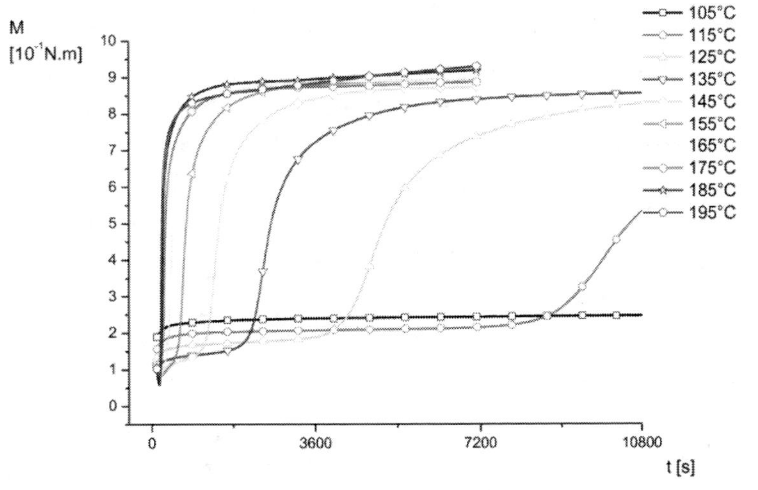


**Figure 2.** Time dependences of the torque $M$ measured at the constant temperature from temperature range 100°C – 200°C.

Time dependencies shown in the figure 2 were approximated by non-linear least squares method using formula (1). Values of four constants describing the shape of the vulcanization curve at constant temperature were determined by approximation (table 2)

**Table 2**. Approximated parameters $M_{max}$, $M_{min}$, $k$, $t_i$ at selected temperatures obtained by nonlinear least-squares method at the formula for time dependence of torque

| $T$ [°C] | $M_{max}$ [$10^{-1}$ Nm] | $M_{min}$ [$10^{-1}$ Nm] | $k$ [$s^{-1}$] | $t_i$ [s] |
|---|---|---|---|---|
| 100 | 3.99704 | 0.49189 | 1.3E-4 | 18779.08 |
| 105 | 2.68953 | 2.49577 | 2.1E-4 | 12941.58 |
| 110 | 6.9127 | 1.75399 | 4.4E-4 | 14557.63094 |
| 115 | 7.35884 | 3.22511 | 4.9E-4 | 9427.539 |
| 120 | 7.74198 | 0.31809 | 6E-4 | 6662.87541 |
| 125 | 8.39588 | 3.20519 | 6.5E-4 | 4525.968 |
| 130 | 8.02582 | 6.01818 | 9.7E-4 | 4958.79029 |
| 135 | 8.38727 | 1.35968 | 0.00116 | 2049.691 |
| 140 | 8.48268 | 7.95749 | 0.0014 | 3663.10373 |
| 145 | 8.59956 | 1.89259 | 0.00194 | 1206.522 |
| 150 | 8.35191 | 7.81586 | 0.002 | 2091.17697 |
| 155 | 8.80918 | 1.63488 | 0.00266 | 552.5776 |
| 160 | 8.5945 | 4.80595 | 0.0025 | 652.4464 |
| 165 | 8.93491 | 6.17877 | 0.00185 | 661.0386 |
| 170 | 8.76327 | 5.73496 | 0.00219 | 498.7025 |
| 175 | 8.77729 | 5.61095 | 0.00262 | 387.2245 |
| 180 | 8.90283 | 6.02321 | 0.00288 | 383.2369 |
| 185 | 9.0135 | 6.21982 | 0.00308 | 369.0329 |
| 190 | 9.07853 | 6.25042 | 0.00343 | 331.6193 |
| 195 | 8.9259 | 5.77486 | 0.00334 | 299.905 |
| 200 | 9.43829 | 5.76528 | 0.00352 | 263.9531 |

Rate constant values $k$ and values of time of induction period listed in table 2 are exponentially dependent on the temperature (figure 3, figure 4). Their dependencies are well aproximable by relationships (2) and (3). Values of constants $S_1$, $E_1$, $S_2$, $E_2$ were determined after the approximation. Values of constants $S_1$ = 1,68 [-] and $E_1$ = 28487 [J.mol$^{-1}$] were obtained by approximation of the temperature dependence of the rate constant $k$ (3). Values of constants $S_2$ = 10,31 [-] and $E_2$ = 62581 [J.mol$^{-1}$] were obtained by approximation of the temperature dependence of the time of induction period $t_i$ (2).

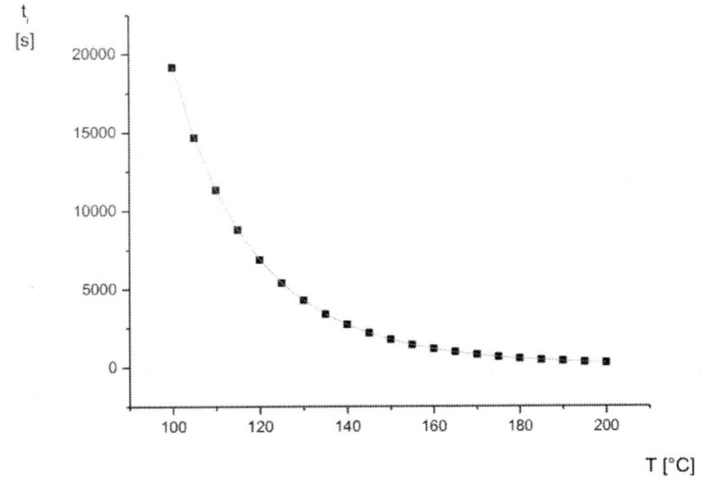


**Figure 3.** The temperature dependence of the induction period with the displaying the of the approximation using the formula (2).

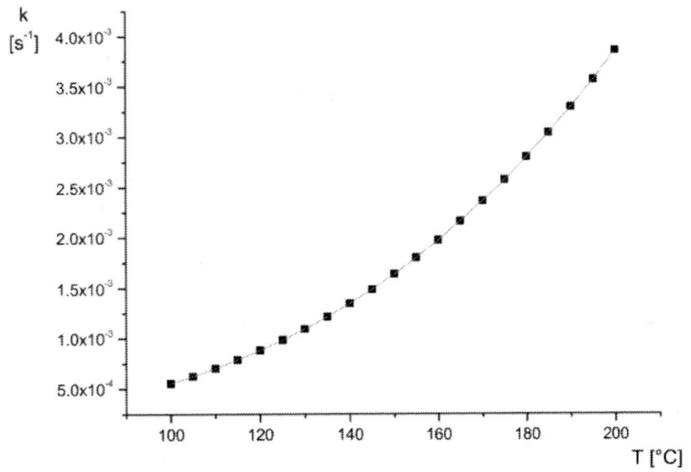


**Figure 4.** The temperature dependence of the values of the vulcanization rate constant of the rubber mixture SBR8 with the displaying of the approximation using the formula (3).

Non-linear least squares approximation by means of formula (9) was realized using mentioned constants. Example of the measured temperature dependence of torque and also approximation of measured values in the cross-linking area at the temperature increasing 0.8°C/min are shown in figure 5. Decrease of measured values of the torque associated with the change of viscosity at the mixture heating is observable up to temperature 140°C. Significant increase of torque values from the temperature approx. 150°C can be associated with the cross-linking reaction (II. stages of vulcanization). The same character of the measured dependencies was recorded at all speeds of the heating used in the experiment, 0.2°C/min; 0.4°C/min; 0.6°C/min; 0.8°C/min; 1.0°C/min; 2.0°C/min; 3.0°C/min; 4.0°C/min; 5.0°C/min; 7.0°C/min; 10°C/min. Good agreement between measured data and values obtained by approximation by means of formulas (9) (table 3) can be seen at comparison of values of the cross-linking beginning at linear increasing temperature.

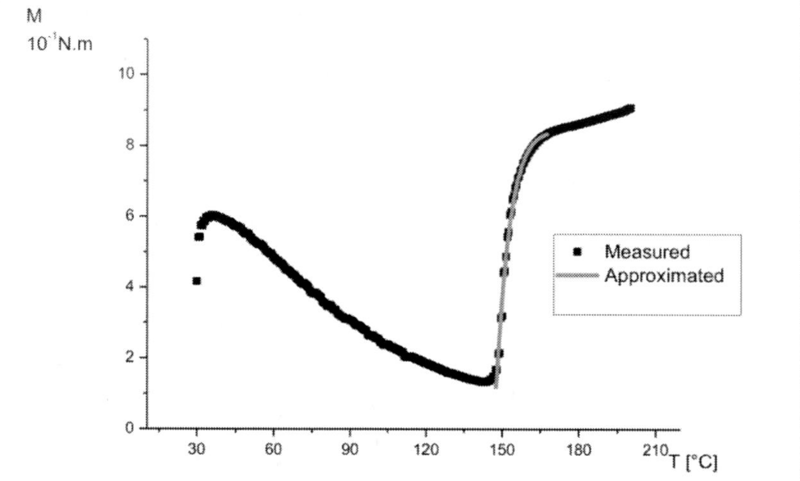


**Figure 5.** Temperature dependence of torque at the speed of heat 0,8°C/min.

**Table 3.** Table of values of specific constants $M_{min}$, a $T_0$.

| | Values obtained from the measured dependence | | Values obtained by approximation measured dependencies using equation (9) | |
|---|---|---|---|---|
| Heating rate $\vartheta$ [°C.min$^{-1}$] | $T_0$ [°C] | $M_{min}$ [10$^{-1}$ N.m] | $T_0$ [°C] | $M_{min}$ [10$^{-1}$ N.m] |
| 10.0 | 180.5 | 0.558797 | 179.709 | 0.559 |
| 7.0 | 177.5 | 0.556334 | 177.004 | 0.556 |
| 5.0 | 172.5 | 0.67664 | 171.807 | 0.674 |
| 4.0 | 170 | 0.680469 | 169.271 | 0.673 |
| 3.0 | 165.5 | 0.714617 | 165 | 0.697 |
| 2.0 | 159 | 0.83019329 | 158.626 | 0.820 |
| 1.0 | 150 | 1.0354992 | 149.751 | 1.135 |
| 0.8 | 148 | 1.14451 | 147.731 | 1.140 |
| 0.6 | 142 | 1.2401 | 142.155 | 1.24 |
| 0.4 | 136.5 | 1.386896 | 136.998 | 1.392 |
| 0.2 | 126.5 | 1.567274 | 128.345 | 1.567 |

## 5. Conclusion

Time dependencies of torque during vulcanization can be described by formula (4) in case of laboratory prepared rubber mixture based on SBR with simple vulcanization system. Temperature dependence of the torque at the linear temperature increasing can be found by means of this formula. This dependence is expressed by (9). Formula (9) well describes the actual measured curves. Constants specifying the beginning of cross-linking during the linear increasing temperature ($M_{min}$, $T_0$) obtained from the measured data and from the relation (9) derived by approximation are in good agreement.

This research was supported by APVV-SK-CZ-0168-11, VEGA 1/0356/13 and ITMS:26220120014

**References**

[1] Stark W, Jaunich M, Bohmeyer W and Lange K 2012 *Polymer Testing* **31** 904

[2] Bielinski D M, Stepkowska A 2012 Mechanical properties and friction of rubber vulcanizates: Aspects of crosslink structure *Archives of civil and mechanical engineering* 192-198

[3] Bošák O, Tóth M, Minárik S and Psota J 2011 Equipments for investigation of non-metallic materials. *XXIV Didmattech 2011 : Problems in teachers education. - Kraków : Instytut Techniki UP* ISBN 978-83-7271-679-8. 63-68

[4] Milani G, Leroy E, Milani F and Deterre R 2013 *Polymer Testing* **32** 1052

[5] Goyanes S, Lopez C C, Rubiolo G H, Quasso F and Marzocca A J 2008 *European Polymer Journal* **44** 1525

[6] Zanchet A, Carli L N, Giovanela M, Brandalise R N and Crespo J S 2012 *Materials and Design* **39** 437

[7] Leroy E, Souid A and Deterre R 2013 *Polymer Testing* **32** 575

[8] Young Kee Chae, Woo Young Kang, Joong-Hee Jang and Sung-Seen Choi 2010 *Polymer Testing* **29** 953

[9] Johns J and Nakason Ch 2011 *Journal of Non-Crystalline Solids* **357** 1816

[10] Salgueiroa W, Somozaa A, Marzocca A J, Consolati G and Quasso F 2007 *Radiation Physics and Chemistry* **76** 142

[11] Bošák O, Kalužný J, Preto J, Vacval J, Kubliha M and Hronkovič J 2007 *Polymers for Advanced Technologies* **18** 141

# Correlation between Rheotens measurements and reinforcement of polymer nanocomposites in the injection molding compounder

**Markus G. Battisti**[a, 1], **Walter Friesenbichler**[1], **Ivica Duretek**[2] **and Peter Guttmann**[3]
[1] Department Polymer Engineering and Science, Chair of Injection Molding of Polymer, Montanuniversitaet Leoben, A-8700 Leoben, Austria
[2] Department Polymer Engineering and Science, Chair of Polymer Processing, Montanuniversitaet Leoben, A-8700 Leoben, Austria
[3] Department Polymer Engineering and Science, Chair of Materials Science and Testing of Polymers, Montanuniversitaet Leoben, A-8700 Leoben, Austria

markus.battisti@unileoben.ac.at, and walter.friesenbichler@unileoben.ac.at

**Abstract**. The evaluation of the effectiveness of reinforcement of polymers and polymer nanocomposites (PNCs), in particular the improvement of Young's modulus, is made by performing standardized tensile tests. Structural and morphological characterizations typically are investigated using expensive techniques like transmission electron microscopy (TEM), X-ray scattering and sometimes also rheological analyses (rotational rheometry). The objective of this study is to generate faster and economically advantageous data to verify the quality of the produced PNC-compound in an on-line measurement system. Subsequently injection molded parts are processed by using the Injection Molding Compounder (PNC-IMC) "by only one plasticizing process". In comparison to the conventional compounding process, where the compound has to be pelletized and fed into the injection molding machine for the second plasticizing process, injection molding compounding combines these two processing steps. This paper shows first results and problems with the implementation of the Rheotens equipment into the concept of the IMC. Different processing techniques and various processing conditions were compared and the occurring effects were detected both with tensile testing and extensional melt rheology. Both, the increase of the Young's modulus by using layered silicates as nanofillers is compared to the virgin polypropylene and the correlation of the level of melt strength with Rheotens measurements is shown. These results give a good overview on both the possibilities and the limitations of the material pre-tests by the use of extensional rheology in the concept of the IMC for producing PNCs. Further studies to enable a fast and efficient way of estimating the level of reinforcement in PNCs by means of rheotens measurements will be carried out towards industrial usability. Furthermore the verification of exfoliation and intercalation of the layered silicates in the polymer matrix using small angle X-ray scattering is planned.

**Keywords:** Injection molding compounding, polymer nanocomposites, layered silicates, extensional rheology

## 1. Introduction

[a] **Corresponding author:** Markus G. Battisti, research field: injection molding compounding. E-mail: markus.battisti@unileoben.ac.at

### *1.1. Polymer nanocomposites*

Polymer nanocomposites represent that class of polymer materials, which are currently investigated with greatest interest. They consist of the base polymer and organic or inorganic fillers where at minimum one dimension of the length is in the nanometer range. These nanofillers are dispersed during processing in a way that the different layers have a distance of a few nanometers to each other. The special feature of nanofillers in comparison to conventional fillers is, that already low volume fractions (< 10 wt%) in the polymer matrix lead to remarkable property improvements. For example, mechanical, optical and electrical properties are positively influenced. The reason for this improvement is due to the fact that the magnitudes of the nanofillers and the macromolecules of the polymer matrix are similar and thus even at low filler content, the distance between the particles are in the nanometer range. Furthermore, nanofillers have large specific surfaces and thus allow a very good coupling to the polymer matrix. Another criterion which especially affects the tensile strength is the form factor (length-thickness-ratio). This factor is closely related to the specific surface [1].

### *1.2. Injection molding compounder*

The injection-molding compounding process combines two processing steps: the material compounding, which normally takes place at the raw material manufacturer and the injection-molding process which usually is done at the injection molder. The final part is processed directly with only one plasticizing process, and the material used does not have to be granulated and plasticized again in the plasticizing unit of an injection-molding machine after the compounding step. The molten material exits the extruder and is directly fed into a melt accumulator and further into the injection unit. After the shot volume is reached, a conventional injection-molding process starts.

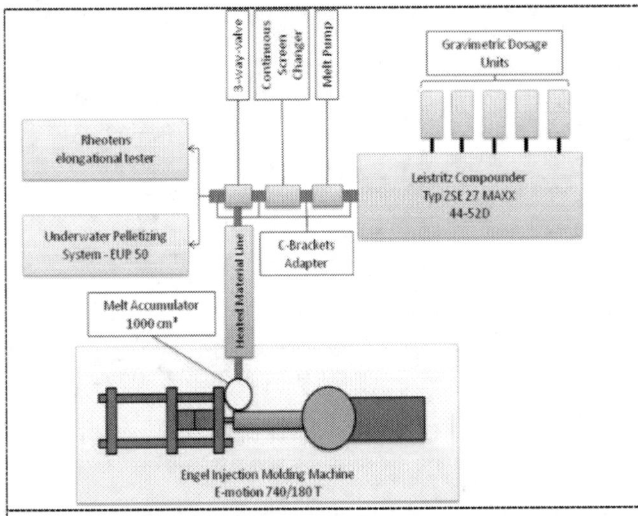


**Figure 1.** System schematic of the MUL Polymer Nano-Composite Injection Molding Compounder (PNC-IMC)

The accumulator serves as a buffer element between the continuously-operating extruder and the discontinuously-operating injection unit [2-3]. The MUL (Montanuniversitaet Leoben) polymer nanocomposite Injection Molding Compounder (PNC-IMC) described in figure 1 is the world's first specially adapted IMC for the manufacturing and injection molding of PNCs. It consists of a ZSE 27 MAXX Leistritz-Compounder and an 1800 kN fully-electric injection molding machine by Engel. A 3-way valve and its particular arrangement allows to use the IMC as three-in-one production line either for injection molding compounding, for injection molding or as a separately acting twin-screw compounder line.

### *1.3. Extensional rheometry (Rheotens)*

The drawability of polymer melts in many processes is of crucial importance. The elongational tester Rheotens allows determining the drawing force of a melt strand extruded through a round die as a function of the drawing speed. Based on the gained technical strain curves the polymer melt can be characterized [4-5].

The extrudate is pulled off from the upper draw down wheels, while the lower wheels are used to stabilize the strand. Both pairs of wheels are mounted on a weighing beam which is connected to a displacement transducer. Withdrawing of the melt strand forms a movement of the weighing beam which is detected by the transducer and converted into a pull-off force. It is important that at the beginning of the measurement, the exit speed of the strand corresponds to the drawing speed.

Therefore drawing speed is set by the wheel speed so that no force acts on the strand. During the measurement the drawing speed is constantly increased until the strand breaks. The pull-off force is recorded as a function of the drawing speed.

## 2. Materials and methods

### *2.1. Materials*

An isotactic PP homopolymer Bormed DM55pharm (MFR 2.8 g/10 min; 230 °C/2.16 kg), supplied by Borealis, was used for the preparation of nanocomposites in combination with the nanofiller Nanofil 5 (montmorillonite intercalated with dimethyl distearyl ammonium chloride) and the compatibilizer Scona TPPP 2112 GA (PP grafted with 1.2 wt% of maleic acid anhydride, PP-g-MA, MFR 6 g/10 min; 190 °C/2.16 kg), supplied by Kometra.

### *2.2. Preparation of polypropylene nanocomposites*

The preparation of PNC was carried out with the intermeshing, co-rotating twin-screw extruder of the PNC-IMC. As first step a compatibilizer/nanofiller masterbatch (MB) was manufactured using a standard screw configuration by feeding the compatibilizer through the main hopper and the nanofiller downstream at the 4th extruder barrel. The MB concentration was 50 wt% organoclay and 50 wt% compatibilizer.

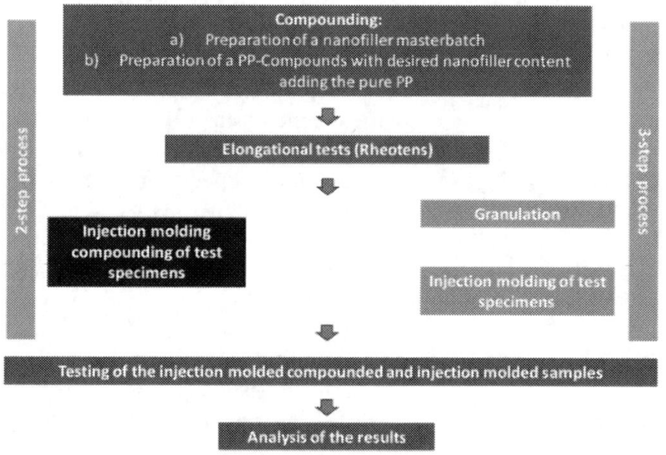


**Figure 2.** Experimental procedure

The MB was produced with an overall throughput rate of 5 kg/h and then pelletized with the underwater pelletizing system. As second step the MB pellets were diluted with pure PP in the twin-screw extruder to the final PNC formulation (90 wt% PP, 5 wt% nanofiller and 5 wt% compatibilizer) and either directly processed via the PNC-IMC to specimens (two-step process) or in a third step injection molded (three-step process). The whole procedure is schematically presented in figure 2.

For compounding, injection molding compounding and the Rheotens tests according to a Shainin DoE four process parameters were identified. Their different values are listed in table 1.

**Table 1.** Process parameters and their different values for the compounding of the PNCs.

| | Value -1 | Value +1 |
|---|---|---|
| Screw speed (ss) [rpm] | 50 | 200 |
| Screw geometry (sg) | standard | optimized |
| Temperature profile extruder (T) [°C] | 180-190-200-200-... | 210-220-230-230-... |
| Screen changer mesh (sc) | no | yes |

*2.3. Determination of the mechanical properties (tensile test)*

The tensile specimens according to ISO 3167 (multipurpose specimens type A) for the investigation of the mechanical properties were produced both with the two-step PNC-IMC process and with the three-step process using the injection molding machine of the PNC-IMC.

A Zwick Z250 universal tensile testing machine was used to carry out the tensile tests according to ISO 527. All tests were done at standardized conditions (23 ± 2°C/50 ± 5% r.H.). For evaluation the Young's modulus $E_t$, tensile strength $\sigma_M$ and strain at break $\varepsilon_B$ were determined (according to ISO 527).

*2.4.* Determination of the melt strength (elongational test - Rheotens)

For all four process parameters (see table 2) two factor levels (low -1 and high +1) were used. At first one series with the pure PP and all four process parameters at factor -1 were processed. Then the nanofiller reinforced material was examined by the same factor settings (all -1). The remaining series of each material group were produced by changing one factor at a time to the high level +1 and setting it back to the low level -1 (see table 2). The purpose of these series was to compare the elongational tests with the tensile tests and to compare the results within the series. It is important to mention that these elongational tests were made at a throughput rate of 1 kg/h due to the fact that at higher throughput rates the extruded strands did not break in the Rheotens test. Therefore the elongational test results cannot be compared exactly with the tensile tests.

**Table 2.** Overview of the elongational tests settings at a throughput rate of 1 kg/h.

| Series | Material | Settings | Temperature die | Wheel gap | | Starting speed wheels |
|---|---|---|---|---|---|---|
| | | | | upper | lower | |
| [/] | [/] | [/] | [°C] | [/] | [/] | [mm/s] |
| 1 | DM55 | pure, all -1 | 210 | 0.85 | 1.3 | 55 |
| 2 | | filled, all -1 | | | | 65 |
| 3 | | filled, ss +1 | | | | 80 |
| 4 | | filled, sg +1 | | | | 80 |
| 5 | | filled, t +1 | | | | 80 |
| 6 | | filled, sc +1 | | | | 65 |

In all experimental series, grooved wheels were used and the drawing acceleration was 60 mm/sec$^2$. Further the same die with a diameter of 2 mm and a length of 40 mm was used for all measurements maintaining the die temperature of 210 °C. The starting speed of the draw down wheels had to be varied due to flow inhomogeneity in order to be able to ensure a pull-off force of zero at the beginning of each measurement.

## 3. Results and Discussion

*3.1. Tensile tests*

Each experimental series consisted of two tensile test series. On the one hand specimens were produced by means of injection molding compounding (IMC) and on the other hand by conventional injection molding (IM). 15 test specimens were tested and the arithmetic averages x as well as the standard deviations σ are listed in table 3.

Table 3 shows high scattering of elongation at break for PP DM55. As compared to pure PP DM55 the average increase of all compounded materials shows approx. 25% increase in Young's modulus. The tensile strength and the elongation at break increased by approx. 10 % compared to the unfilled material. The highest increase was found for both IMC and conventional IM process at a high screw speed.

**Table 3.** Arithmetic average (x) and standard deviation (σ) for the tensile tests with PP DM55

| | | Young's modulus [MPa] | | Tensile strength [MPa] | | Elongation at break [%] | |
|---|---|---|---|---|---|---|---|
| | | x | σ | x | σ | x | σ |
| pure, all -1 | IMC | 1678 | 46 | 35.7 | 0.3 | 55 | 18 |
| | IM | 1668 | 66 | 35.8 | 0.4 | 32 | 13 |
| filled, all -1 | IMC | 2155 | 28 | 38.8 | 0.2 | 58 | 11 |
| | IM | 2137 | 29 | 39.0 | 0.2 | 74 | 17 |
| filled, ss +1 | IMC | 2201 | 22 | 39.0 | 0.2 | 76 | 24 |
| | IM | 2155 | 26 | 39.0 | 0.2 | 118 | 41 |
| filled, sg +1 | IMC | 2121 | 23 | 38.6 | 0.2 | 51 | 25 |
| | IM | 2135 | 21 | 38.9 | 0.3 | 66 | 10 |
| filled, t +1 | IMC | 2105 | 26 | 38.5 | 0.2 | 68 | 19 |
| | IM | 2089 | 33 | 38.5 | 0.3 | 68 | 9 |
| filled, sc +1 | IMC | 2065 | 102 | 38.9 | 0.3 | 53 | 21 |
| | IM | 2146 | 54 | 39.2 | 0.3 | 59 | 16 |

To verify these results statistically, an analysis of variance (ANOVA) was made. ANOVA is a collection of statistical models used to analyze the differences between means of groups of results and their associated process variants and process parameters. Three important values which indicate the significance of a result will be shown. The first one is the F-value which examines the likelihood of a difference between two different observations (the higher the better). In order to use the F-value for statements about the processing influence, the values of 5 groups and 15 measured values ($F_{4,10}$-distribution) must be looked up in the so-called Fisher-table. The F-value is 3.48 which means that each value greater than 3.48 suggests significance. The second one is the post hoc Tukey range test, which compares every group mean with every other group mean to reveal if a difference is significant. The third one is the coefficient of determination $R^2$. The confidence levels of all analyses with ANOVA were set to 95%, which corresponds to an α-value of 0.05 (see table 4).

**Table 4.** Overview of the significance of the processing parameters on Young´s modulus, tensile strength and elongation at break for PP DM55

| | F-value [/] | | | Tukey test [/] | | | $R^2$ [%] | | |
|---|---|---|---|---|---|---|---|---|---|
| | Y | $\sigma_m$ | $\varepsilon_b$ | Y | $\sigma_m$ | $\varepsilon_b$ | Y | $\sigma_m$ | $\varepsilon_b$ |
| filled, all -1 | | | | AB | AB | B | | | |
| filled, ss +1 | | | | A | A | A | | | |
| filled, sg +1 | 14.16 | 18.24 | 14.05 | BC | B | B | 28.37 | 33.78 | 28.21 |
| filled, t +1 | | | | C | C | B | | | |
| filled, sc +1 | | | | C | A | B | | | |

Considering the significance analysis of processing parameters in table 4, it turns out that like already assumed significance is visible. The Tukey method is a method of multiple comparisons. If identical letters are assigned to different groups, the corresponding mean values with a defined likelihood do not differ significantly. If two groups do not have letters in common, there is a significant difference in the mean values. Looking at the individual parameters of the Tukey method it is obvious that two blocks can be arranged in which no significance difference occurs (AB/A/BC/C/C and AB/A/B/C/A). However these blocks are different for Young's modulus and tensile strength although one would rather expect a consistent trend. In addition, the coefficients of determination have only a maximum of around 30 % indicating that 70 % of the variation is caused by other factors than those listed.

### *3.2. Elongational tests*

In order to compare the results of the elongational tests the average curves of each series were used. Since the average value curves of the various series have different starting speeds and therefore also the draw down force is different, the curves were shifted to a start speed of 70 mm/s. This means all average curves have a drawn down speed of 70 mm/s at a pull-off force of 0 N. A representative value for each curve was determined. In this case the pull-off force at a draw down speed of 400 mm/s was chosen. The average curves are represented in figure 3 with their characteristic values marked with a red vertical line at 400 mm/s.

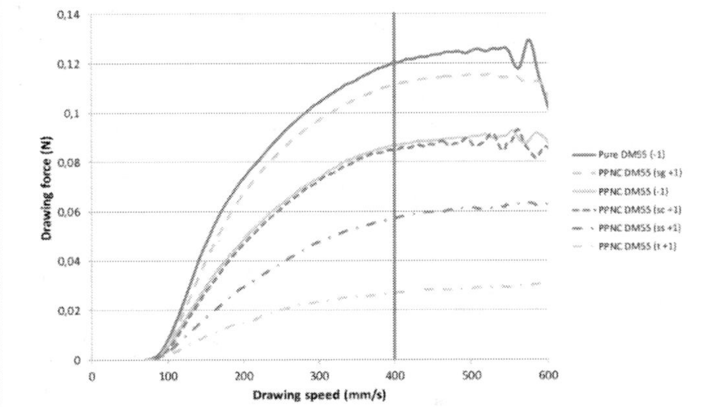


It is completely unexpected that the pull-off force of the nanofilled PP DM55 (see table 5) is approx. 25 % lower compared to the pure PP DM55. The processing of PP DM55 with an optimized screw tends to have better results of the nanofilled material, whereas a decrease in the pull-off force for higher temperature and higher screw speed is observed.

**Figure 3.** Rheotens curves at different process parameters

**Table 5.** Pull-off force (n) at an drawing speed of 400 m/s.

| **Process parameters** | **PP DM55** |
|---|---|
| pure, all -1 | 0.12 ± 0.01 |
| filled, all -1 | 0.09 ± 0.01 |
| filled, ss +1 | 0.06 ± 0.02 |
| filled, sg +1 | 0.11 ± 0.01 |
| filled, t +1 | 0.03 ± 0.01 |
| filled, sc +1 | 0.09 ± 0.01 |

### *3.3. Comparison tensile tests and elongational tests*

Elongational tests and tensile tests were performed to get an overview of the trends and the possible correlation between these test methods. Since the trends of the tensile strength results are very similar to the Young's modulus results and the results of elongation at break scattered strongly, only the Young's modulus values are shown in Figure 4.

Journal of Physics: Conference Series **602** (2015) 012012 doi:10.1088/1742-6596/602/1/012012

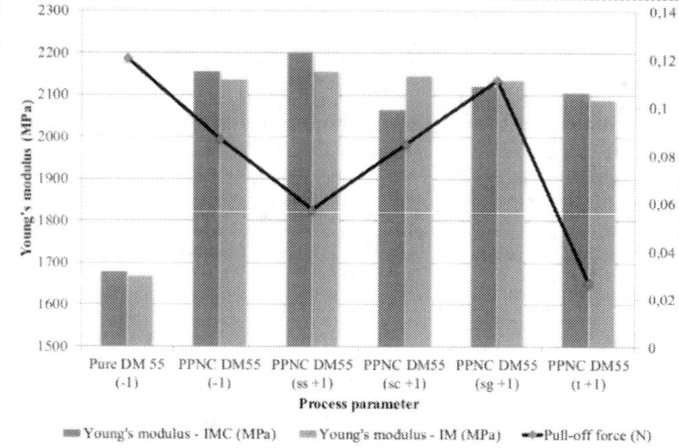


**Figure 4.** Comparison of Young's modulus and pull-off force for PP DM55

For PP DM55 results of conventionally injection molded (IM) and injection molding compounded (IMC) specimens are shown in comparison to the pull-off force at a drawing speed of 400 mm/s. Based on tensile tests the positive influence of the layered silicates on the mechanical properties is detectable. While the significance of the pull-off force for the different settings is obvious, the Young's modulus does not show corresponding changes in the mean values. This could mean that the factor levels of the process parameters were chosen too small due to the restriction by the elongation tests. Thus a comparison between the elongational test results and the Young's modulus values seems to be questionable. Furthermore figure 4 shows an opposite trend of pull-off force and Young's modulus (usually the pull-off force decreases with increasing Young's modulus and vice versa). A possible reason for this effect is the considerable lower throughput rate (1 kg/h) which had to be implemented for performing the Rheotens elongational tests, while the production of the test specimens had been carried out with a throughput rate of 5kg/h.

## 4. Conclusion

The aim of this study was to find a correlation between elongational test measurements (Rheotens) and layered silicate reinforcement of polymer nanocomposites for the injection molding compounding process. Another objective was to determine, if the variation of selected process parameters during compounding, which results in different mechanical properties, is detectable by using the Rheotens elongational tests.

The preparation of the polymer nanocomposites had been carried out with the intermeshing, co-rotating twin-screw extruder of the PNC-IMC. At first a masterbatch (50 wt% organoclay and 50 wt% compatibilizer) was manufactured with the standard screw configuration at an overall throughput rate of 5 kg/h and subsequently pelletized with the underwater pelletizing system. In the second step the masterbatch pellets were diluted with pure PP DM55 in the twin-screw extruder. The final formulation of the compounded PNCs was constant at 90 wt% polypropylene, 5 wt% nanofiller and 5 wt% compatibilizer. The tensile bars for the investigation of the mechanical properties were produced both, with the two-step PNC-IMC process and the three-step process using the injection molding machine of the PNC-IMC. The calculated pull-off forces of the elongational tests (throughput rate 1 kg/h) and the values of the tensile tests (throughput rate 5 kg/h) were compared. It was found that the Young's modulus and the tensile strength often showed no significant changes. Although there were significant differences in the pull-off forces, changes in the process parameters could not be detected by using Rheotens elongational tests.

The comparison of the pull-off forces and tensile test results with pure and reinforced PP showed an unexpected trend. While the Young's modulus of the filled material was significantly higher, the PP DM55 showed lower pull-off forces when nanofilled. This shows, that no correlation could be found between the pull-off forces and the tensile test values under varying processing conditions.

In conclusion, it is crucial to say that the used concept of elongational tests for monitoring material property changes were feasible only for very low throughput rates. I must be pointed out again, that the elongational tests could only be performed at a throughput rate of 1 kg/h, depending on limitations of the Rheotens tester, while compounding was carried out at a throughput rate of 5 kg/h.

Furthermore, even for extremely careful experimental work, slightly different experimental settings (draw down wheel, weighing beam) can influence the results. These factors strongly restrict the applicability of the presented on-line extensional test in the laboratory and industrial application.

Encouraging approaches for the on-line determination of the melt quality would be either the use of a by-pass elongational testing equipment, which would enable measurements independent of the throughput rate, or a FT-NIR measurement system.

**Acknowledgements**
The authors would like to thank Borealis, Linz for the provided PP and Prof. Dr. Gerald Pinter for the provided testing equipment.

**References**

*[1]* Kim G-M 2007 Verstärkungsmechanismen auf Makro-, Mikro- und Nano-Längenskalen in heterogenen Polymerwerkstoffen, *Habilitation an der Martin-Luther-Universität Halle-Wittenberg*

[2] Jensen R 2001 Making intelligent use of synergies. IMC injection moulding compounder creates extra added value **91** 40

[3] Sieverding M, Bürkle E, Zimmet R 2005 *Kunststoffe Plast Europe* **8** 34

[4] Aglassinger S. 2011 Calculation of the elongational viscosity of a LDPE using Rheotens-mastercurves, *Bachelor thesis, Montanuniversitaet Leoben*

[5] Laske S, Kracalik M, Gschweitl M, Feuchter M, Maier G, Pinter G, Thomann R, Friesenbichler W, Langecker G R 2008 *Journal of Applied Polymer Science* **111** 2253

# Rheological Characterization of Polyoxyethylene (POE) and Carboxymethyl Cellulose (CMC) Suspensions with Added Solids

**A. Mora[a y b], O. Skurtys[a]*, F. Osorio[b]**
[a]Department of Mechanical Engineering, Universidad Técnica Federico Santa María, Av. Vicuña Mackenna 3939, Santiago, Chile
[b]Department of Food Science and Technology, Universidad de Santiago de Chile, Avenida Ecuador 3769, Santiago, Chile

fernando.osorio@usach.cl, olivier.skurtys@usm.cl, alfredo.mora.miles@gmail.com

**Abstract** The rheological properties of high molecular weight POE and CMC suspensions by adding micro-metric solid particles such as fibers or spheres were studied. The volume fraction, $\Phi$, was varied between 0 and 0.4. Their rheological properties were obtained after fitting a Cross model. For POE suspending fluid with spherical particle, the behavior of the normalized steady shear viscosity, $\mu/\mu_0$, as function of the fraction volume followed a Thomas model. However, for CMC suspensions, $\mu/\mu_0$ seems to be lineal with $\Phi$. For a pure fluid or a suspension with $\Phi = 0; 2$, the suspension presented an elastic behavior whereas it was observed a viscous behavior when the volume fraction was increased.

## 1. Introduction

Suspensions are a class of complex fluids (and dispersed system) in which a finely divided solid is dispersed uniformly in a liquid dispersion medium [1]. Suspensions of solid particles in polymeric fluids are present in many industrial applications, including: food engineering, pharmaceutical production, material engineering [1-4]. Knowledge of the rheological behavior of these viscoelastic fluids is fundamental to understand their processing behavior and relations between structure and property. Indeed, the addition of solid particles to polymeric fluids does not simply change the magnitude of the viscosity, it can modify strongly the rheological properties of the fluid. Rheological properties of polymeric fluids with solid particles are strongly related to chemical formulation, molecular weight and polymer concentration, solvent properties, volume fraction and particle geometry. Moreover, external variables such as temperature and pressure can also influenced the rheological behavior. Thus, understanding the relationship between the microstructure of the polymer solutions and their macroscopic properties is of fundamental importance.

Considerable attention in the literature is given to the rheology of suspensions of particles in Newtonian fluids. Due to their technological importance, most studies are focused on highly filled systems (typically, volume fractions greater than 10%), with polydisperse particles of irregular shapes. In contrast, even if the rheology of suspensions in non Newtonian media is, from a technological point of view, more important than its counterpart in Newtonian media, relatively few studies are conducted on the rheology of dilute or semi-dilute suspensions of solid particles in no-newtonian fluids. If both theoretical and experimental studies for dilute and semi-dilute suspensions in Newtonian matrices are relatively scarce and somewhat contradictory, the situation is even worse for suspensions in viscoelastic fluids.

In engineering, pseudo-plasticity or shear-thinning fluids as Polyoxyethylene (POE) or Carboxymethyl cellulose (CMC) are the most common type of time-independent non-Newtonian fluid

Journal of Physics: Conference Series **602** (2015) 012013 doi:10.1088/1742-6596/602/1/012013

[5]. These fluids are characterized by an apparent viscosity η (defined as shear stress ɵ / shear rate ɤ), which decreases with increasing shear rate. The rheological properties of polymer solutions are strongly related to the chemical formulation, the molecular weight and the concentration of the polymer and external variables such as temperature and pressure [6].

Polyoxyethylene (POE) is a polymer of ethylene oxide, flexible and non-ionic water-soluble used in many applications from industrial manufacturing to medicine [7]. The shear rheology of POE solutions has been characterized by several authors. Effects of concentrations, molecular weights, salt and anionic surfactants on the viscoelastic charter of POE in aqueous solution were commented by [8]. Carboxymethyl cellulose (CMC) is derived from cellulose, it is used as a viscosity modifier or thickener, and to stabilize emulsions in various products including ice cream; it is known for its excellent water retaining capacity [7]. Moreover, this is the sodium salt of CMC (CH2COONa) that promotes water solubility, which is not affected by the temperature of the water itself. The rheological properties of CMC solutions at high concentrations have both non-Newtonian and viscoelastic properties. The aim of this article is to present the changes in apparent viscosity of high molecular weight POE and CMC suspensions by adding micro-metric solid particles such as fibers or spheres. Steady shear flow tests were done to measure the steady-state viscosity of the suspension. Linear oscillatory tests (upon verification of the linearity limit) were also performed to determine the viscoelastic frequency response.

## 2. Materials and methods

### *2.1. Suspending fluids*

Two different types of suspending fluids were used. Poly(ethylene oxide) (POE) is used as shear-thinning fluid whereas sodium carboxymethyl cellulose (CMC) are, instead, used as viscoelastic carrier fluid. Both fluids have a density $\rho$ = 1000 kg*m$^{-3}$ at 20° C. POE and CMC were provided by Sigma-Aldrich. POE has a nominal molecular weight $M_w$ of 4 x 10$^6$ g*mol$^{-1}$ whereas CMC had $M_w \approx$ 7 x 10$^{-5}$ g*mol $^{-1}$ with a degree of substitution of 0.80 – 0.95. Aqueous solutions of POE and CMC were prepared by dissolving the appropriate amount of POE and CMC in distilled water at room temperature. Sufficient time (> 24h) of continuous magnetic stirring was allowed to achieve complete homogenization. POE concentration varied from 1% to 1.5% w/w whereas CMC varied from 1.5% to 2.2% w/w.

### *2.2. Solid particles*

Glass bead spheres (Sigmund Lindner, Germany) with a particle diameter of 0 - 50 µm and 40 - 70 µm and carbon fibers (Panex 35, Zoltek, USA) were used as solid suspensions. In Table 1, physical and geometry properties of the solid particles are presented. Solid particle concentrations varied

Table 1: Properties of solid particles

| Particle | $\rho$ (kg.m$^{-3}$) | $l_p(\mu m)$ | $d_p(\mu m)$ | $r_p = l_p/d_p$ |
|---|---|---|---|---|
| sphere $(0-50\mu m)$ | 2550 | | | 1 |
| sphere $(40-70\mu m)$ | 2550 | | | 1 |
| fiber | 1810 | 100 | 7.2 | 13.88 |

between 0 and 40% for each range and geometry.

### *2.3. Preparation of the suspensions*

Preparation of the suspensions was done with care as homogeneous dispersion of non agglomerated particles are difficult to obtain especially for low solid concentrations. The solid particles were dried to remove moisture. The polymer solution and solid particles were then weighted and gently stirred (to not introduce air) in a beaker for XXX minutes in the desired proportion. Optical observation

$$t_{settling} = \frac{0.45\mu_0 h}{R^2 g(\rho_p - \rho_f)} \quad (1)$$

confirmed that the particles were well dispersed and did not form aggregates. It was checked that in all cases the time scale associated with sedimentation was much larger than the time scale of both sample preparation and experiment. For a sphere, the settling time required to migrate 10% of the rheometer gap, h, under the influence of gravity, g, acting perpendicular to the gap is:
From the settling vertical velocity of the fiber:

$$V_z = \frac{d_p^2 g(\rho_p - \rho_l)}{16\mu}\left[2\ln 2r_p - 1.614 - 0.355(\ln 2r_p)^{-1} + O(\ln 2r_p)^{-2}\right] \quad (2)$$

it was also able to calculate a settling time.

*2.4. Rheological measurements*
All the rheological measurements were performed on a rheometer (Carri Med, CSL2 100, TA Instruments, England) using a 40 mm diameter parallel plate fixture with a gap h = 400 µm. The lower plate is equipped with a Peltier temperature control system; all tests were conducted at 20°C with 3 repetitions. The samples were carefully loaded to the measuring plate of the rheometer using a spatula and then the measuring plate was lowered at a very slow speed, in order to prevent the disruption of the solution structure. The flow curves were obtained by applying an increasing shear stress ramp at a constant stress rate of XXXPa*$s^{-1}$. Storage G′(ω) and loss G′′(ω) moduli were measured at frequencies ranging from 0.01 to 20Hz and, by applying a chosen stress value, allowing measurements within the linear viscoelastic region. Rheological behavior of the suspensions was modeled using the Cross model:

$$\frac{\mu - \mu_\infty}{\mu_0 - \mu_\infty} = \frac{1}{1 + (\lambda\dot{\gamma})^m} \quad (3)$$

where $\mu$ is the dynamic viscosity at any given shear rate $\dot{\gamma}$, $\mu_0$, and $\mu_\infty$ are the asymptotic values of viscosity at zero-shear rate and infinity shear rate, respectively. $\lambda$ is a time constant with the dimensions of time and m is a dimensionless rate constant indicating the degree of dependence of the viscosity on shear rate in the shear-thinning region. This model was chosen because it can describe both monotonic and non-monotonic flow curves.

## 3. Results and discussion

For four suspending fluids, the viscosity as function of the shear rate is presented in Figure 1. Their rheological properties obtained after fitting the Eq. 3 are reported in Table 2. All pure solutions are shear-thinning since m > 1; however, when the polymer concentration increase m was close to 1. For POE and CMC polymers, the viscous interactions between polymers chains are higher when the concentration increase since the viscosity at zero shear rate and infinity shear rate increase. When the concentration was fixed to 1.5%, the viscosity at zero shear rate of CMC is higher than POE. The time constant, λ, also increases with the polymer concentration therefore the entanglement-disentanglement process with shear rate is longer.

In Figure 2 and 3, the normalized steady shear viscosity as function of volume fraction is presented for POE and CMC solutions. As it is generally observed, the additive of solid particles increased the viscosity over those of the pure polymeric solutions. Moreover, it is possible to express the normalized viscosity in terms of third order polynomials in the volume fraction. Indeed, for volume fraction slightly higher than 30%, for Newtonian fluid with

Table 2: Cross model constants of pure suspending solutions

| Fluid | $\mu_\infty$ (Pa.s) | $\mu_0$ (Pa.s) | $\lambda$ (s) | m |
|---|---|---|---|---|
| POE 1% | 0.04 | 1.49 | 0.439 | 0.741 |
| POE 1.5% | 0.11 | 11.25 | 2.187 | 0.757 |
| CMC 1.5% | 0.24 | 14.25 | 0.318 | 0.773 |
| CMC 2.2% | 0.58 | 62.69 | 0.514 | 0.827 |

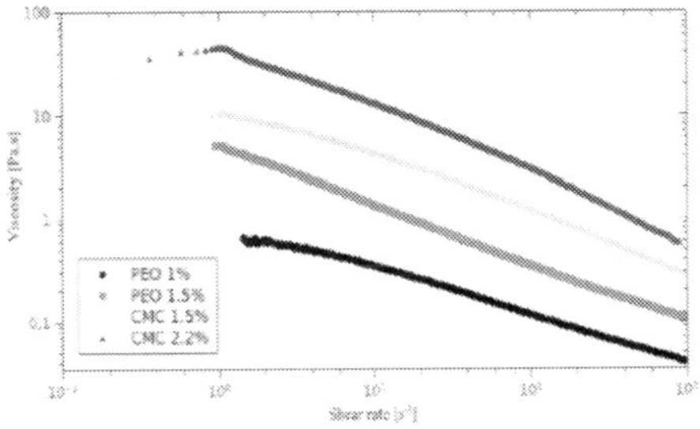


Figure 1: Viscosity as a function of shear rate of suspending fluids

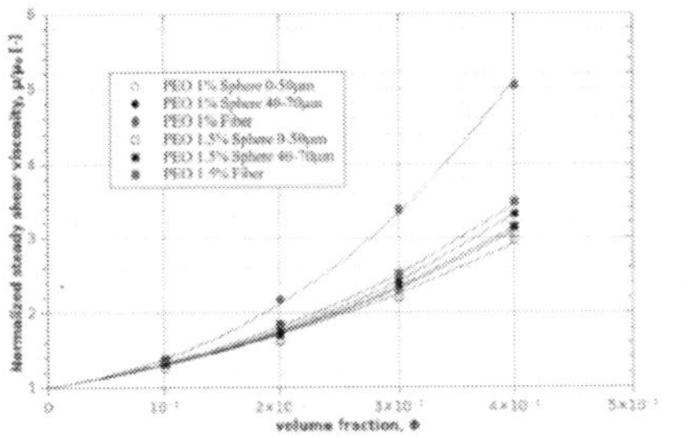


Figure 2: For PEO suspensions normalized steady shear viscosity as function of volume fraction.

spherical particles [9], considered a third term in the polynomial expansion that take into account hydrodynamic interactions of an arbitrary number of spherical spheres:

The polynomial fits of the experimental data are shown as solid lines in Figure 2 and 3. For POE

$$\frac{\mu}{\mu_0} = 1 + 2.5\Phi + 4.83\Phi^2 + 6.4\Phi^3 \tag{4}$$

solutions with particles (spheres or fibers), it can be noticed that the fit on the first order coefficient is quite robust and always in good agreement with Einstein's prediction since $a_1 \approx 0.25$. The maximum deviation from the theoretical value of 2.5 is 3%. The 2.5 value confirmed for non Newtonian suspensions by our results, as theoretical reported by [10]. When spherical particles were added to POE solutions, the coefficient $a_2$ is close to 4.83 and the hydrodynamic interaction between particles was well described by the Thomas's model. However, as expected, in case of fiber particles, the model (Ec. 4) was not correct. The normalized steady shear viscosity as function of volume fraction seems to be linear. Four linear lines (guides to the eyes) are shown as dot lines in Figure 3. More experimental results are necessary to confirm these results but clearly the Thomas model was not correct to model the rheological behavior. In particular, it will be interesting to work with low concentration of solid particles. However, for CMC and POE suspensions, it can be observed that the addition of fibers in the suspending fluid allow to increase more the viscosity of the solution than the addition of the spherical particles.

Table 3: Fit values of the polynomial coefficients of $\frac{\mu}{\mu_0} = 1 + a_1\Phi + a_2\Phi$. Standard deviations are also shown.

| Fluid | Suspensions | $a_1$ | $a_2$ | $a_3$ |
|---|---|---|---|---|
| POE 1% | sphere $(0 - 50\mu m)$ | $2.47 \pm 0.27$ | $4.79 \pm 0.31$ | $5.34 \pm 0.20$ |
| | sphere $(40 - 70\mu m)$ | $2.53 \pm 0.41$ | $4.83 \pm 0.23$ | $6.87 \pm 0.017$ |
| | Fiber | $2.48 \pm 0.09$ | $11.83 \pm 0.94$ | $19.29 \pm 0.20$ |
| POE 1.5% | sphere $(0 - 50\mu m)$ | $2.49 \pm 0.17$ | $4.90 \pm 0.31$ | $2.32 \pm 0.01$ |
| | sphere $(40 - 70\mu m)$ | $2.51 \pm 0.14$ | $4.82 \pm 0.72$ | $5.84 \pm 0.74$ |
| | Fiber | $2.53 \pm 0.12$ | $6.37 \pm 0.58$ | $6.64 \pm 1.03$ |

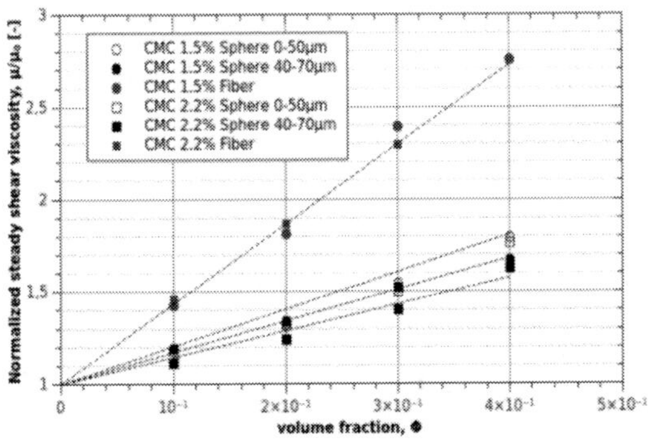


Figure 3: For CMC suspensions normalized steady shear viscosity as function of volume fraction.

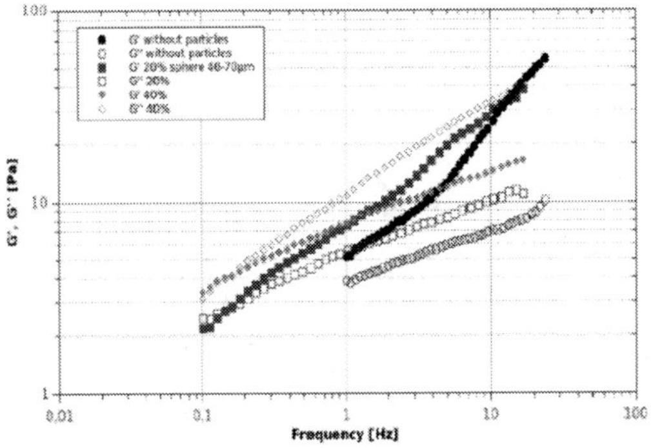


Figure 4: For 1.5% of PEO, elastic modulus, G', and loss modules, G'', as function as function of volume fraction.

To observe and explain more the influence of the spherical particles concentrations on the rheological behavior, for POE at 1.5%, the linear viscoelastic response of the pure fluids and fluids with added solid particles are presented in Figure 4. The elastic modulus, G', of the pure suspending fluid shows an erratic behavior at frequencies lower than 1 Hz; in particular, values for G' were negative, indicating that the elastic modulus was non measurable, within the sensitivity limits of the rheometer. For a pure fluid or a suspension with $\Phi = 0.2$, G" was smaller than G' indicating an elastic behavior. However, when the volume fraction was increased it was observed a viscous behavior since G" > G'. It would be very interesting to study the transition between both behaviors. This phenomenon was also observed for POE at 1% is (results not show).

## 4. Conclusion

As it is generally observed, the addition of solid particles increased the viscosity over those of the pure polymeric solutions. The increase of viscosity was more notorious when using fiber. For POE suspending fluid with spherical particles, the behavior of the normalized steady shear viscosity, $\mu/\mu_0$, as function of the fraction volume followed a Thomas model. These results are consistent with those reported theoretically by [10]. For CMC suspensions, $\mu/\mu_0$ seems to be lineal with $\Phi$. However, more experimental are necessary especially for low volume fraction. For a pure fluid or a suspension with $\Phi = 0{,}2$, G" < G' whereas G" > G' when the volume fraction was increased.

## Acknowledgment

O.S wishes to thanks the financial support provided by Fondecyt (project 1120661).

## References

[1] Tadros T. F. 2010 Rheology of Dispersions: Principles and Applications, *Wiley-VCH Verlag*
[2] Kulshreshtha A K, Singh O N and Wall G M 2010 Pharmaceutical Suspensions, *Springer*
[3] Saeseaw S, Shiowatana J and Siripinyanond A 2005 *Food Research International* **38** 777
[4] Yao X, Tan S, Huang Z and Jiang D 2005 *Materials Letters* **59** 100
[5] Chhabra R. P. 2010 Rheology of Complex Fluids, chapter Non-Newtonian Fluids: an Introduction, p 3–34, *Springer-Verlag,*
[6] Doi M and Edwards S F 1994 The Theory of Polymer Dynamics, *Oxford University Press*
[7] Phillips G O and Williams P A 2000 Handbook of hydrocolloids, *CRC - Woodhead Publishing*
[8] Lance-Gomez E and Ward T 1986 *Journal of Applied Polymer Science* **31** 333
[9] Thomas C and Muthukumar M 1991 *The Journal of Chemical Physics* **94** 5180
[10] Palierne J 1990 *The Journal of Chemical Physics* **29** 204

# Rheological properties of wood polymer composites and their role in extrusion

**I. Duretek[a], S. Schuschnigg, A. Gooneie, G.R. Langecker, C. Holzer**

Department of Polymer Engineering and Science, Chair of Polymer Processing, Montanuniversitaet Leoben, 8700 Leoben, Austria

Ivica.Duretek@unileoben.ac.at

**Abstract.** The influence of the rheological behaviour of PP based wood plastic composites (WPC) has been investigated in this research by means of a high pressure capillary rheometer incorporating dies having different geometries. The rheological experiments were performed using slit and round dies. The influence of moisture content on the flow properties of the WPC has been investigated as well. It was observed that higher moisture contents lead to wall slippage effect. Furthermore, measured viscosity data have been used in flow simulation of an extrusion profile die. Also, the influence of different rheological models on the simulation results is demonstrated. This research work presents a theoretical and experimental study on the measurement and prediction of the die pressure in the extrusion process of wood-plastic composite (WPC).

**Keywords**: Wood Plastic Composites, Rheology, Slit Die, Wall Slippage, Extrusion Profile Die

## 1. Introduction

Wood plastic composites (WPC) are an important and growing part of the plastic industry. In recent years WPC have evolved the considerable possibilities of design and functionality of the end product. The most common method for manufacturing WPC is either profile or sheet extrusion.

WPC products cover a wide range of composites including polyethylene, polypropylene, and polyvinyl chloride (PVC) as the polymeric matrix and sawdust or wood fibres as the filler. Most of the commercial WPC products have a wood content in the range of 40 – 60 wt%. Although WPC is not yet a highly common material, its usage is intensively growing [2].

Predicting the required die profile to achieve good products is a very complex task and requires detailed knowledge of material properties and extensive experience with extrusion processing. Design of the dies for extrusion of products having complicated cross sectional shapes is still an art rather than a science. In many cases costly experiments and in-plant trials could be replaced by numerical simulation [16]. The flow behaviour of WPCs is strongly dependent on the wood filler content and on the moisture content of the wood particles. However, due to different flow behaviour of filler and polymer and the complex interaction between the two phases, the rheological behaviour of the WPC is

[a] **Corresponding author:** Ivica.Duretek, Department of Polymer Engineering and Science, Chair of Polymer Processing, Montanuniversitaet Leoben, 8700 Leoben, Austria , Ivica.Duretek@unileoben.ac.at

complex. Several researchers have investigated the rheological behaviour of WPC [e.g. 4, 5, 6, 9, 11, 12, 13, 14, 15]. A better understanding of the rheological behaviour of the WPC and accurate rheological data are the prerequisite for the process optimization and the accuracy of simulation of extrusion of a WPC. These prerequisites for WPC are mostly not available.

## 2. Materials and methods

### *2.1. Material*

The rheological experiments were performed using two PP based WPC, 70 wt% and 50 wt% soft wood particles (pine and firewood). Wood particles have an average length of approximately 500 µm. The measurements were done on undried material (relative moisture 3.1 - 3.8 %) and vacuum dried material (relative moisture 0.5 %).

### *2.2. Rheometry*

The shear viscosity was measured at 190 °C on a commercial high pressure capillary rheometer Rheograph 2002 (GÖTTFERT Werkstoff-Prüfmaschinen GmbH) using round dies with a diameter of 2 mm, L/D 10, 20 and 30 and 180° entry angle and slit dies with different gap heights (1 mm, 1.5 mm, 2 mm and 2.5 mm). The width of the flow channel is 10 mm. The pressure gradient in the slit die is measured by using flush mounted pressure transducers along the length of the slit (figure 2). The figures 1 and 2 show a schematic view of measurement configurations.

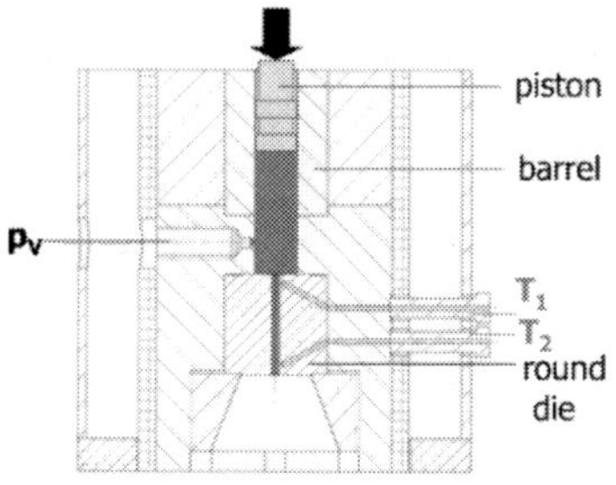


**Figure 1.** Schematic cross sectional view of the high pressure capillary rheometer with round die; $p_v$ – pressure transducer, $T_1$, $T_2$ – thermocouples.

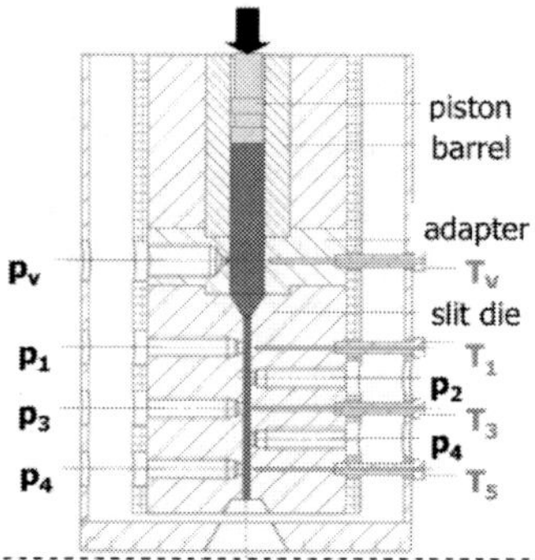


**Figure 2.** Schematic cross sectional view of the high pressure capillary rheometer with slit die; $p_v$, $p_1$ $p_2$, $p_3$, $p_4$, $p_5$ – pressure transducers, $T_V$, $T_1$, $T_3$, $T_5$ – thermocouples.

The volumetric flow rate $\dot{V}$ is calculated by using the following equation

$$\dot{V} = \frac{\pi \cdot D_p}{4} \cdot v_p \quad (1).$$

Where: $D_p$ - diameter of the cylinder/piston, $v_p$ – piston velocity.

For a capillary of radius R the volumetric flow rate causes an apparent (Newtonian) wall shear rate $\dot{\gamma}_{ap}$ of

$$\dot{\gamma}_{ap} = \frac{4 \cdot \dot{V}}{\pi \cdot R^3} \quad (2).$$

The wall shear stress $\tau_w$ follows from the measured extrusion pressure $p_V$, Bagley ends correction $p_C$ [1] and the length over radius ratio L/R

Journal of Physics: Conference Series **602** (2015) 012014 doi:10.1088/1742-6596/602/1/012014

$$\tau_w = \frac{p - p_C}{2 \cdot L/R} \quad (3).$$

The conversion from apparent to true wall shear rate $\dot{\gamma}_W$ is given by the Weissenberg-Rabinowitsch correction [8]

$$\dot{\gamma}_w = \frac{\dot{\gamma}_{ap}}{4} \cdot \left( 3 + \frac{d \log \tau_W}{d \log \dot{\gamma}_{ap}} \right) \quad (4).$$

Finally the true viscosity is calculated as

$$\eta = \frac{\tau_W}{\dot{\gamma}_W} \quad (5).$$

In the case of the slit die, with the help of shape factor $F_p$ [17] the corrected volumetric flow rate is given by the following equation

$$\dot{V}_{H,corr} = \dot{V} \cdot \frac{1}{F_p} \quad (6).$$

The shape factor accounts the width influence in the rectangular flow channel.

The apparent shear rate for the slit die is expressed as

$$\dot{\gamma}_{H,ap} = \frac{6 \cdot \dot{V}_{H,corr}}{B \cdot H^2} \quad (7).$$

Where: B - width of the slit, H - height of slit gap.

The true shear rate $\dot{\gamma}_w$ is determined by applying the Weissenberg-Rabinowitsch correction [8]

$$\dot{\gamma}_{H,w} = \frac{\dot{\gamma}_{H,ap}}{3} \cdot \left( 2 + \frac{d \log \tau_W}{d \log \dot{\gamma}_{ap}} \right) \quad (8).$$

The pressure drop $\Delta p$ can be determined by taking the difference of two pressures measured along the flow length and the corresponding distance $\Delta L$ between the pressure transducers. When the ratio of width to thickness of the flow channel is smaller than 20 the wall shear stress is given by

$$\tau_{H,W} = \frac{\Delta p}{2 \cdot \Delta L} \cdot \left( \frac{B \cdot H}{B + H} \right) \quad (9).$$

After correction, the true viscosity can be obtained by dividing the wall shear stress by true shear rate. This calculated true viscosity is used for further analysis.

In case of wall adhesion (shear flow) the velocity in the slit die ( $v_S$ ) has a parabolic profile and the velocity is maximum in the middle and at the walls zero (figure 3, left). According to Mooney [10] and Knappe [7] the reduced volumetric flow rate $\dot{V}_{red}$ vs. wall shear stress curve for different slit heights H should be identical and lie over each other (figure 3, right).

$$\dot{V}_{red} = \frac{\dot{V}}{B \cdot H^2} \quad (10).$$

In case of wall slippage (ideal plug flow) the velocity profile of the flowing fluid is constant across the channel height (figure 4, left). The reduced volumetric flow rate curves lie apart from each other and cannot be represented as geometry independent parameter (figure 4, middle). In such case the slip velocity curves ( $v_G$ vs. $\tau_W$ ) for different slit heights lie over each other forming a master curve (figure 4, right). The slip velocity $v_G$ is a function of wall shear stress and can be well described as

$$v_G = \frac{\dot{V}}{B \cdot H} \quad (11).$$

Journal of Physics: Conference Series **602** (2015) 012014 doi:10.1088/1742-6596/602/1/012014

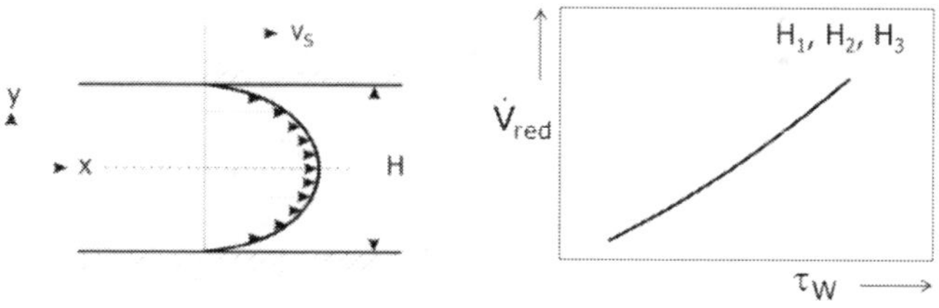


**Figure 3.** Left: velocity profile of shear flow; right: reduced volumetric flow rate at the existence of wall adhesion.

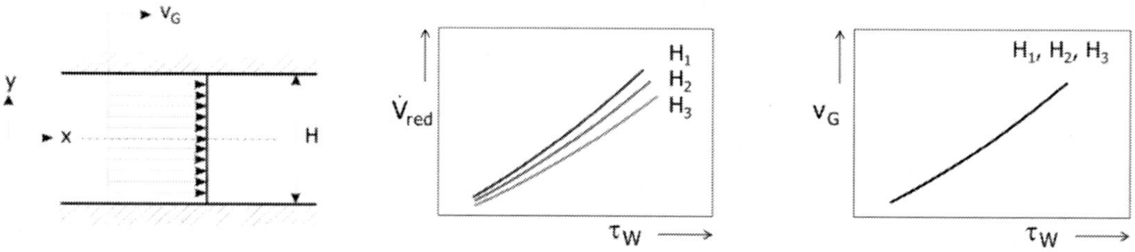


**Figure 4.** Left: velocity profile for pure wall slip; middle: reduced volumetric flow rate at the existence of wall slip; right: slip velocity as a function of wall shear stress.

### *2.3. Profile extrusion*

In the present work, extrusion trials were carried out on the twin screw extruder Konos 38 (battenfeld-cincinnati Austria GmbH) using a rectangular hollow profile die (figure 5) in the melt throughput range between 10 and 50 kg/h. The profile die with installed measurement sensors is shown in figure 6.

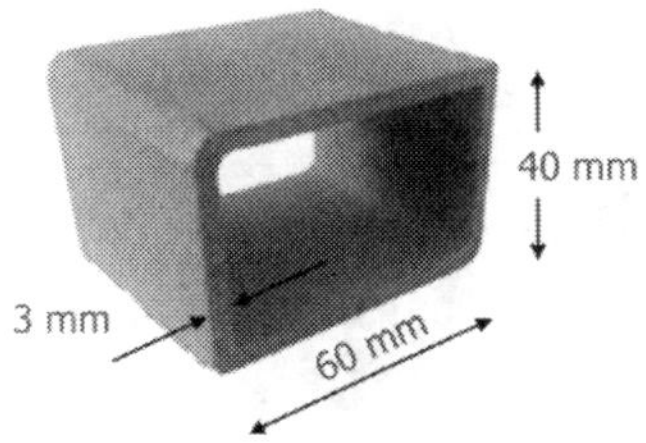


**Figure 5.** Rectangular hollow profile.

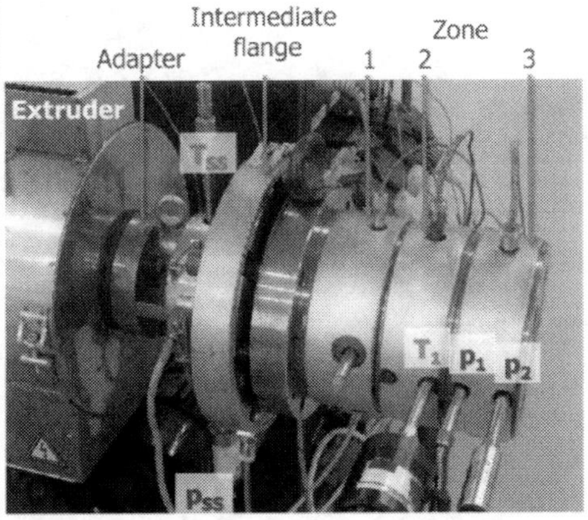


**Figure 6.** Profile die with measurement sensors (right); $T_{SS}$ – melt temperature at the screw tip, $p_{SS}$ – melt pressure at the screw tip, $T_1$ – melt temperature in the exit zone, $p_1$, $p_2$ – melt pressure in the exit zone.

## *2.4. Calculation of pressure drop*

In order to predict the pressure drop along the die, two approaches were adopted. As a first attempt, a 2-dimensional finite-difference method (FDM) was incorporated to capture the response of the material at two flow extremes, i.e. shear and plug flow. Later, 3-dimensional simulations of the die were performed with a finite-element method (FEM) scheme utilizing ANSYS Polyflow software (v. 14.0.0). This approach enables us to calculate the pressure drop by considering the flow as a combination of both shear and plug flows. The results are discussed in section 3.2.

For FDM calculations, the profile die was discretized into smaller parts consisting of 4 cylindrical and 5 rectangular sections, wherein each length of the sections is divided into thousand subdivisions. The pressure drops were calculated on the cylindrical sections. For the rectangular sections we had to balance the flow between the four sides of the profile so that the axial pressure drop is equal on each side. The sum of the individual pressure drops equals the needed pressure for the whole die.
These calculations scheme could be used either for a shear flow with a power-law viscosity model, or a plug flow with the Navier's slip law:

$$\tau = k \cdot v_G^n \quad (12).$$

In the case of FEM, a triangular meshing was applied with 77 094 nodes and 387 862 elements. Two approaches were adopted; first, considering the flow as a shear type with the power-law viscosity model and second, considering the flow as the combination of shear and plug flow.

While the slippage coefficients are derived according to the well-known indirect gap-dependency method, one cannot apply the traditional viscosity measurement methods in the presence of wall slippage phenomenon. Therefore, the viscosity value was optimized to obtain valid material data for the simulation of the profile die. It was found that a very high viscosity value in the order of 1e6 Pa·s could best represent the material behaviour in the simulations. Such a high viscosity value induces a higher portion of plug flow in combination with the shear flow which can describe the pressure drop with a reasonable accuracy.

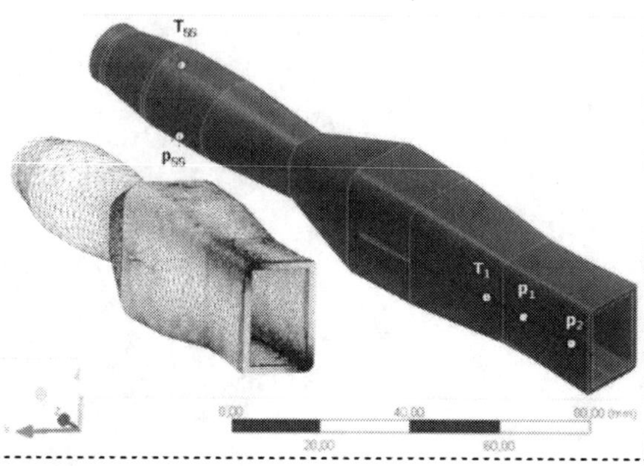


**Figure 7.** Geometry and mesh for FEM simulation.

## 3. Results and discussion

### *3.1. Flow behaviour*

Figure 8 shows measured viscosity curves of predried (0.5 % moisture) WPC with 70 wt% wood using the slit and the round die. It can be seen that the viscosity measured with the slit die is 20 % lower than the viscosity measured using the slit die. The main cause for this effect is that the pressure in the round die system is measured using the pressure hole. Due to poor pressure transmission through the pressure hole onto the pressure transducer diaphragm the measured pressure is lower. The wood fibres are accumulating in the pressure hole which has a diameter of 1 mm and constrict a good pressure measurement. This pressure hole effect can be avoided by using the slit die instead of round die. By implementing flush mounted pressure transducer, the pressure fluctuation can be minimized.

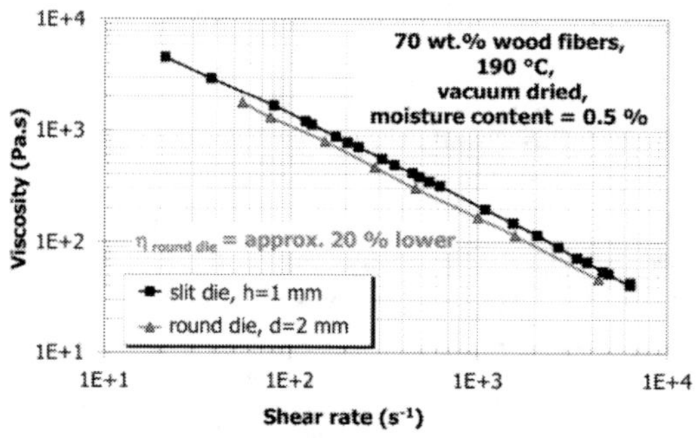


**Figure 8.** Comparison between viscosities of predried WPC measured with round and slit die.

Figure 9 illustrates the reduced volumetric flow rate as function of the wall shear stress of vacuum dried WPC with 70 wt% wood for different slit heights. It can be seen that the flow curves produce a master curve independent of the slit height, which is indicative of wall adhesion and therefore shear flow. In case of WPC with higher moisture content 3.8 %, the curves do not produce such a master curve, which is an indication of wall slip.

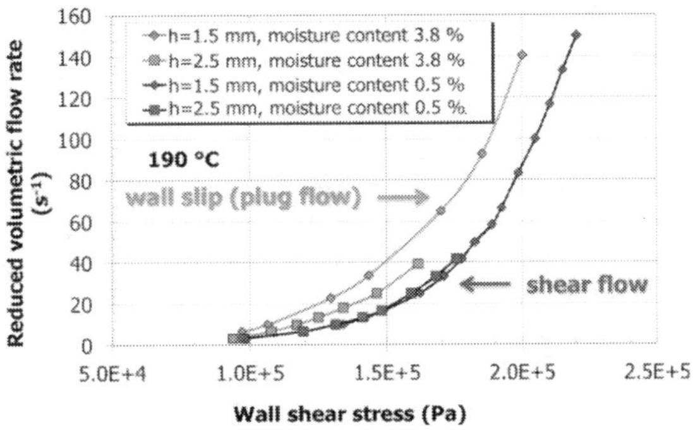


**Figure 9.** . Influence of moisture content on flow behaviour of WPC with 70 wt% wood particles; 3.8 % moisture (undried), 0.6 % moisture (vacuum dried), measured slit with slit die heights of 1.5 and 2.5 mm.

WPC with 50 wt% and 3.1 % moisture content provide the similar results. As can be seen in figure 10, the flow curves measured using different slit heights show the dependence upon the slit height which proves the existence of wall slip.

Figure 11 shows the wall slip velocity curves. As explained before, in case of pure wall slip, the slip velocity curves for different slit die heights coincide to a single curve.

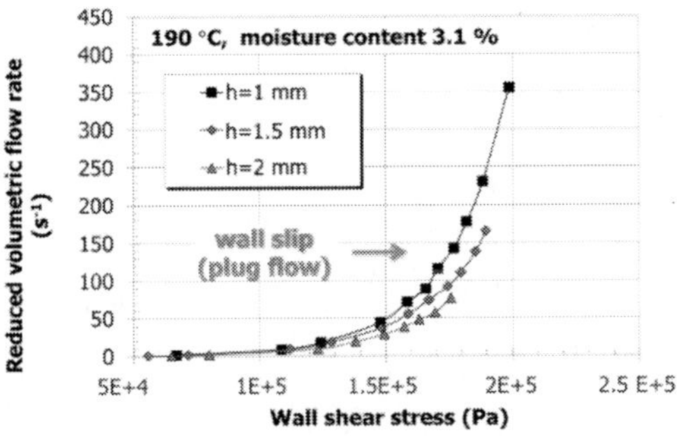


**Figure 10.** Reduced volumetric flow curves of WPC with 50 wt% wood particles and moisture content of 3.1 %; measured with slit die heights of 1, 1.5, and 2 mm.

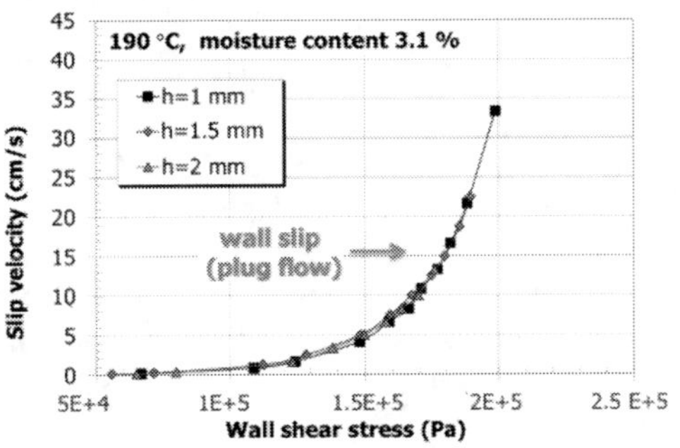


**Figure 11.** Slip velocity curves of WPC with 50 wt% wood particles and moisture content of 3.1 %; measured with slit die heights of 1, 1.5, and 2 mm.

### *3.2. Simulation results and their analyses*

Pressure drop along the profile die for the WPC with 50 wt% wood and a moisture content of 3.1 % was measured at different flow rates (figure 12). A 2D-FDM calculation was performed in order to calculate the pressure drop for either a complete wall adhering (shear) or a complete wall slipping (plug) flow type. According to figure 12, the pressure drops for the fully shear flow (2D-FDM shear flow) is observed to be quite lower than the measurement data while the predictions for the fully plug flow (2D-FDM, wall slip) are higher than the measurements especially at lower flow rates. This suggests that a flow type comprised of a combination of shear and plug flows should be dominant in the WPC extrusion through the profile die.

Using the viscosity model (figure 8) for wall adhesion (shear flow), in the 3D-FEM calculation, as material data, one can clearly observe that the calculated values are lower than the measurements and very close to the 2D FDM which was expected as it is indeed a shear flow simulation. The deviation between the FDM and the FEM calculations can be explained by the used methods, as in the 2D calculations the radii in the corners and other features as the spider legs are neglected.

By introducing Navier's slip law into the calculations, it was possible to add the plug flow in combination with the shear flow (figure 8). The data are found to be well in tally with the measurements. From the results presented here, one can safely conclude that not only is it necessary to incorporate the wall slippage coefficients in the simulations, but also the shear flow component is important to be included to give best results.

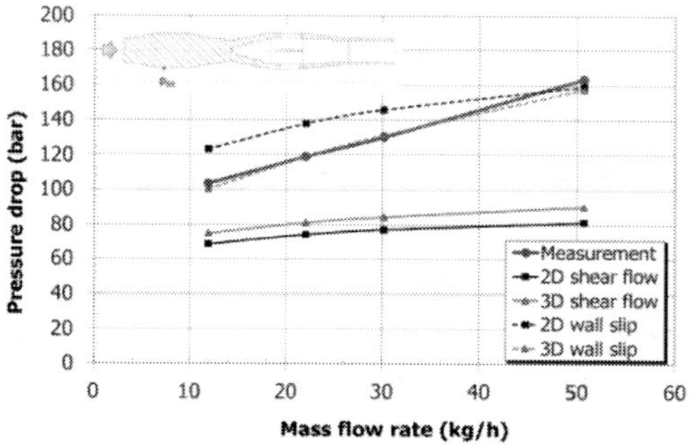


**Figure 12.** Resulting pressure drops against mass flow rate measured values, 3D simulation and 2D FDM results.

## 4. Conclusion

In the present work, the flow behaviour of two WPC with different contents of wood particles has been investigated utilizing high pressure capillary rheometer with two different die geometries. It was proven that an increase in moisture content of WPC leads to wall slippage effect. It was founded that the slit die system equipped with the flush mounted pressure transducers along the flow channel is the most appropriate choice for the viscosity measurements of WPCs and the characterization of wall slippage behaviour.

Simulation results show that a combination of shear and plug flow should be used to describe the material behaviour in the presence of wall slippage. An optimized viscosity value was found to best describe the WPC flow in addition to the wall slippage coefficients of the Navier's slip law determined from the gap-dependency experiments of the slit die setup. The coefficients for the simulations of the profile die where approximated using the experimental data and simulating the slit die. Based on the results presented here, it can be stated that a careful interpretation of flow behaviour is crucial to achieve good prediction results independent of the numerical methods.

**Acknowledgements**
The authors would like to thank battenfeld-cincinnati Austria GmbH for supplying the material used in this work and especially Ms. Sonja Kahr from battenfeld-cincinnati Austria GmbH for her valuable discussions.

**References**

[1] Bagley E B 1956 *Journal of Applied Physics* **28** 624-627
[2] Behravesh A H, Shakouri E, Zolfaghari A, Golzar M 2010 *J. of Comp. Mat.* **44** 11 1293–1304
[3] Braun J, Duretek I, Müller U, Friesenbichler W, Endesfelder A 2007 *Monatshefte für Chemie /*

*Chemical Monthly*, **138** 4 337–341
[4] Carrino L, Ciliberto S, Giorleo G, Prisco P 2011 *Polymer Composites* **32** 5 796–809
[5] Chastagner M W 2005 master thesis, *Washington State University, Pullman WA*
[6] Georg J, Janardhan R, Anand J S, Bhagawan S S Thomas S 1996 *Polymer* **37** 24 5421–5431
[7] Knappe W, Krumböck E 1986 *Rheologica Acta* **25** 296-307
[8] Macosko C W 1994 Rheology, Principles, Measurements and Applications, *Wiley-VCH*
[9] Markov A V 2008 *Mat.-wiss. und Werkstofftech.* **39** 3 227–233
[10] Mooney M 1931 *Journal of Rheology* **2** 210-222
[11] Liew K C, Harum J, Tahir P M, Yusoff M N M, Dahlan K Z M 2000 *Journal of Tropical Forest Products* **6** 1 21–27
[12] Li T Q, Wolcott M P 2004 *Composites Part A* **35** 3 303–311
[13] Li T Q, Wolcott M P 2005 *Polymer Engineering and Science* **45** 4 549–559
[14] Li T Q, Wolcott M P 2006 *Polymer Engineering and Science* **46** 4 464–473
[15] Li T Q, Wolcott M P 2006 *Polymer Engineering and Science* **46** 1 114–121
[16] Pepliński K, Mozer A 2011 Journal of Polish CIMAC **6** 3 221–226
[17] Tadmor Z, Gogos C G 2006 Principles of polymer processing. *John Willey & Sons Inc.*

# Rheological parameters of dough with inulin addition and its effect on bread quality

**T Bojnanska[1], M Tokar[1] and A Vollmannova[2]**
[1]Department of Storing and Processing Plant Products - Faculty of Biotechnology and Food Sciences - Slovak University of Agriculture in Nitra - 94976 Nitra - Slovak Republic
[2]Department of Chemistry - Faculty of Biotechnology and Food Sciences - Slovak University of Agriculture in Nitra - 94976 Nitra - Slovak Republic

tatiana.bojnanska@uniag.sk, alena.vollmannova@uniaf.sk, marian.tokar@uniag.sk

**Abstract**. The rheological properties of enriched flour prepared with an addition of inulin were studied. The addition of inulin caused changes of the rheological parameters of the recorder curve. 10% and more addition significantly extended development time and on the farinogram were two peaks of consistency, what is a non-standard shape. With increasing addition of inulin resistance to deformation grows and dough is difficult to process, over 15% addition make dough short and unsuitable for making bread. Bread volume, the most important parameter, significantly decreased with inulin addition. Our results suggest a level of 5% inulin to produce a functional bread of high sensory acceptance and a level of 10% inulin produce a bread of satisfactory sensory acceptance. Bread with a level over 10% of inulin was unsatisfactory.

## 1. Introduction

Food industry aims at the development of new products towards functional foods and ingredients with regard to the consumer´s demands on healthy nutrition. Being a source of proteins, dietary fibres, vitamin and micronutrients, bread is considered to be of global importance in nutritional equilibrium. As a response to public health problems innovative bread formulations have been in progress [1].

Wheat bread is a major component of people´s diet all over the world. Its consumption in Western Europe is stable, although it varies greatly between states. The Germans, Austrians and Scandinavians eat the most bread at about 70 – 80 kg per person per year while the UK, Ireland, France and Benelux are at the bottom of the list with an annual consumption of less than 60 kg [2]. In Slovakia consumption of bread (including bread under 400 g) has been recently about 70 kg per person per year. There has been increasing demand for food products with additional health benefits. In this respect, the enrichment of bread with the functional fibre (inulin) is of interest to the consumer as well as the cereal industry.

Contrary to whole bread, which is relatively high in fibre content (7 – 8 % of dry matter), white bread contains only 2 – 3 % fibre on a dry matter basis. Fibres, and more particularly the soluble ones, like inulin and oligofructose, might help to prevent diseases such as intestinal infection, colorectal cancers, obesity, cardiovascular diseases and type II diabetes [3], [4]. Bread represents a suitable food product for the addition of functional ingredients, such as the prebiotic inulin.

Water-soluble fibre predominantly consists of non-starch polysaccharides. Soluble fibre inhibits passing of the chime through the intestines, hampers the stomach discharge, reduces the absorption of glucose and sterols in the intestines, and decreases serum cholesterol, blood glucose, and the content of insulin in the human body [5], [6]. Inulin and oligofructose are not hydrolysed in the gastrointestinal tract, except in the colon where they have a specific effect on the microflora, more specifically by promoting very selectively the growth of bifidobacteria to the detriment of other (potentially pathogenic) microorganisms [7], [8].

Inulin is a storage carbohydrate found in many various plants like chicory and Jerusalem artichoke and contains fructan chains of different degree of polymerization. Inulin is a mixture of fructose polymers of various lengths with β(2-1) glycosidic bonds, terminated generally by a single glucose unit. Native inulin has the degree of polymerisation (DP) of 2 – 60 but high performance inulin has an average DP of about 23 – 25, depending on the manufacturer [9]. It is used to increase the amount of dietary fibre or as prebiotic ingredient in particular in dairy and bakery products [10]. Industrially it is nearly exclusively produced from chicory.

In food inulin presents a hydroscopic property and is able to reduce the available water contents during gelatinization of starch, causing carbohydrates to be less absorbed, leading to a lower glycemia rate. Besides the physiologic effect on glycemia outlined, the addition of these fructans is promising as they do not alter the appearance of flavour of foods [11].

Inulin influences bread quality characteristics too. Fibre addition increases mixing time and dough stability and decreases softening degree and farinograph water absorption [12], which depend on DP [13]. Inulin in bread produced a significant reduction in loaf and specific volume in relation to the standard sample and the most evident effect is the increase of crumb firmness and the dark crumb appearance [12].

Rheological measurements are often used to predict the behaviour of wheat dough during processing and the quality of the final product (bread). The aim of the present research was to incorporate this soluble fibre (inulin) into wheat and evaluate its effects on rheological properties of the dough and bread quality.

## 2. Materials and Methods

Mixes of wheat flour (T650) and inulin from chicory (Deracel CF 20, Dera Food Technology, Belgium) were prepared by substitution of wheat flour by 5 %, 10 %, 15 %, 20 % and 25 % of inulin.

### *2.1. Rheological measurements*

Farinograph – E, Brabender OhG, Duisburg, Germany (ICC-Standard 115/1, 1992, AACC Method 54T21, 1995) was used for the study of the rheological properties of dough. The mixing curve is characterised by an ascending part that indicated the changes during the dough development process, while the subsequent decline in the resistance is taken as a sign of a steady breakdown of the dough structure upon mixing beyond the point of optimum development. Optimum development from the standpoint of bread quality may occur slightly past "mixing peak". The following parameters were determined:

*Water absorption capacity – defined as the amount of water required to centre the highest part of the mixing curve on the arbitrary 500 FU (farinograph units) line.*

*Mixing time – dough development time – time interval in minutes from the first addition of water until the curve reaches its maximum height.*

*Stability – time in minutes for which the top of the curve remains over 400 line.*

*Degree of softening in FU – the distance between the centre of the mixing curve and the 500 line after 10 min mixing.*

*Elasticity in FU – defined as the width of the recorder curve after 10 min mixing.*

*2.2. Bread-making procedure*
Six different bread were formulated following a full experimental design. 1000 g of wheat flour (T650) or flours mixes were used for the preparation of test breads with water quantity determined by farinograph, and 4 % yeast, 1.8 % salt and 1 % sugar. Dough was mixed in an spiral mixer SP 12 Diosna (program: 10 sec. 20 Hz, 120 sec. 25 Hz, 300 sec. 50 Hz) and divided into 4 samples. For fermentation, samples were placed in a teflonplate iron form of 8.5 x 11 x 13 cm in a proofing cabinet for 40 min at 30°C, 95 % relative humidity. At the end of proofing the dough samples were baked in a static sole oven (MIWE Condo, Arnstein, Germany) 10 min at 240 °C (with 0,08 l of steam at the beginning of baking) and 25 min at 220 °C (without steam).

Bread quality was evaluated 24 hours after baking: weight (g), loaf volume ($cm^3$) - *was determined by the method of millet seeds displacement (OBK-03, Mezos)*, specific loaf volume ($cm^3$/100 g loaf), volume efficiency ($cm^3$/100 g flour), baking loss, height/width ratio (cambering), crumb acidity - determined *by titration method with sodium hydroxide and indicator phenolftalein.*

*2.3. Bread sensory analysis*
The sensory evaluation was conducted in line with ISO 8589 according to sensory descriptor. Following parameters were evaluated: general appearance and shape (significance coefficient 1), surface and properties of the crust (significance coefficient 2), fermentation and crumb appearance (significance coefficient 4), structure and flexibility of breadcrumb (significance coefficient 4), smell and taste (significance coefficient 9. The maximum possible score to achieve was 100. The sensory quality was analysed using sensory profiling and hedonic tests. Five experts have examined and tasted the samples and recorded their perceptions.

## 3. Results and Discussion

*3.1 Mixing properties*
Mixes of wheat and inulin were prepared by substitution of wheat flour by 5 %, 10 %, 15 %, 20 % and 25 % of inulin. The rheological properties of the doughs produced from these flour mixes were studied by Farinograph. Low addition of inulin (5 % and 10 %) decreased water absorption capacity of mixes in comparison to wheat flour but increasing of inulin addition (15 % - 25 %) proportionately increased water absorption capacity of mixes (Figure 1).

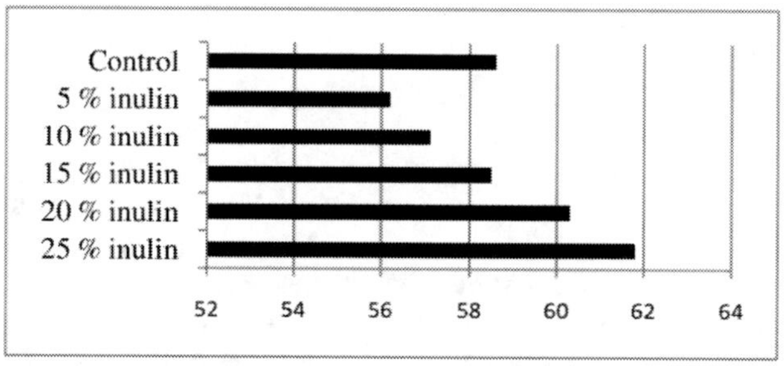


**Figure 1.** Effects the addition of 5 %, 10 %, 15 %, 20 % and 25 % inulin on Farinograph water absorption capacity

Dough is made from wheat flour to which an amount of water, based on the initial moisture content of the flour is added in order to reach a constant hydration level on a dry matter basis. During the kneading of this dough, the pressure on one side of the mixer is continuously monitored. The peak pressure recorded during kneading is used to calculate the water absorption of the flour at a given consistency (target pressure) [14].

The water absorption capacity is important parameter, which influences not only dough consistency but also economy of bread production. It is determined by the protein content of the flour, the amount of starch damaged during milling and the presence of non-starch carbohydrates. It is desirable that flours for bread-making possess a high water absorption capacity at normal working consistencies so that the yield of dough, and hence bread, will be relatively high [15]. Because the farinograph test measures a rheological quality of dough prepared from the tested flour, the relationship may be affected by any treatment or addition of material that alters the rheological character of dough [16].

**Table 1**. Rheological parameters of dough as affected by addition of inulin to wheat flour

| Parameter | Control | 5 % I | 10 % I | 15 % I | 20 % I | 25 % I |
|---|---|---|---|---|---|---|
| Water absorbing capacity | $58.6^{b}$ | $56.2^{b}$ | $57.1^{b}$ | $58.5^{b}$ | $60.3^{ab}$ | $61.8^{a}$ |
| Development time | $2.3^{d}$ | $2.4^{d}$ | $20.0^{a}$ | $16.5^{b}$ | $11.0^{c}$ | $10.5^{c}$ |
| Stability | $2.3^{c}$ | $8.9^{b}$ | $7.2^{b}$ | $8.0^{b}$ | $4.2^{b}$ | $4.4^{b}$ |
| Degree of softening | $48^{a}$ | $25^{b}$ | $0^{c}$ | $0^{c}$ | $0^{c}$ | $0^{c}$ |

The values of the parameters determined by farinograph for control wheat flour and flour mixes containing 5 %, 10 %, 15 %, 20 % and 25 % of inulin are given in Table 1. The addition of inulin caused significant changes of the rheological parameters of the recorder curve (statistical analysis was carried out using Kruskal-Wallis test). A higher change is observable in the case of more than 5 % inulin flour addition. The addition of 20 % of inulin caused important significant changes in all parameters measured, and consequently negatively affected the quality of the final product.

The addition of inulin causes extension of the dough development time. Dough stability is influenced as well by the degree of polymerization of inulin [12], [17], [18]. With the increased addition of inulin acceleration of hydration and decrease of consistency at the start of the kneading was recorded in comparison to control samples. On farinograms there are two peaks of consistency observed in samples with a portion of inulin. The first peak is at the beginning of processing of the mixture and it is caused by the primary hydration, the second peak represents the maximum consistency. Control samples showed the farinogram curve standard for wheat flour (Figure 2).

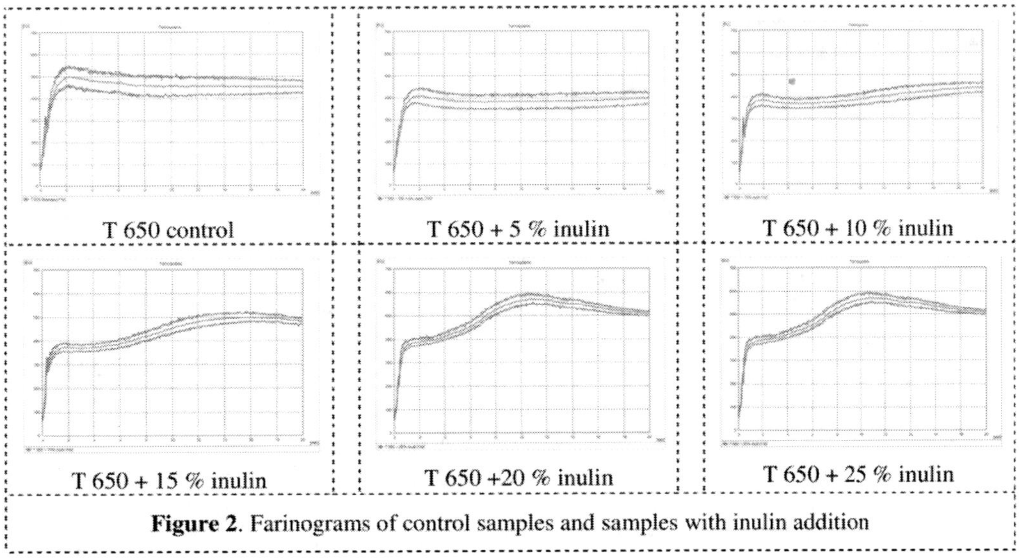


**Figure 2**. Farinograms of control samples and samples with inulin addition

Considering the curves of the samples with the addition of inulin, it is clear that there have been changes in the structure of the glycoprotein matrix. The intensity of these changes was affected by the increased portions of inulin in the flour. Inulin may have a potential to replace a portion of wheat flour without causing detrimental consequences for dough quality. The addition of inulin to wheat flour improved the strenght of the respective doughs, and the effect was especially prominent in the case of weaker flour [19].

Inulin might influence dough properties in various ways: it may affect dough consistency formation of elastic networks, which contribute to the overall dough elasticity and strength and interaction with gluten [19].

### *3.2 Bread quality*

The impact of different levels of inulin on bread properties is presented in Figure 3. The addition of inulin 10 % and more led to a decreased bread quality. Loaf volume and specific loaf volume decreased by the incorporation of inulin, which could be due to gelatinization delay that may promote higher dough expansion during the baking stage. Regarding the bake loss, the addition of inulin did not have a significant influence.

Hager et al. [20] state that addition of inulin darkened the crust of wheat loaves. The partial breakdown of the polysaccharides as well as the monosaccharides present in the inulin powder leads to a stronger Maillard reaction and, therefore, to a darker products [21]. However, in our experiments the additions of inulin in an amount of 5 % to 25 % did not cause darkening of crumb or crust.

Crumb hardness is correlated with the perception of bread freshness. Crumb hardening, which is one of the most obvious manifestations of bread staling, is caused by starch retrogradation as well as differences in vapour pressure between crumb and crust resulting in moisture migration. Additions of inulin did not significantly affect crumb hardness in comparison to control samples which is considered as positive [20].

Our results suggest a level of 5 % inulin to produce a functional bread of high sensory acceptance and a level of 10 % inulin to produce bread of satisfactory sensory acceptance. Bread with a level over 10 % of inulin was unsatisfactory (Figure 4). Similar results were reported by Brasil et al. [11]; bread with addition of inulin (6 %) was classified as good quality bread.

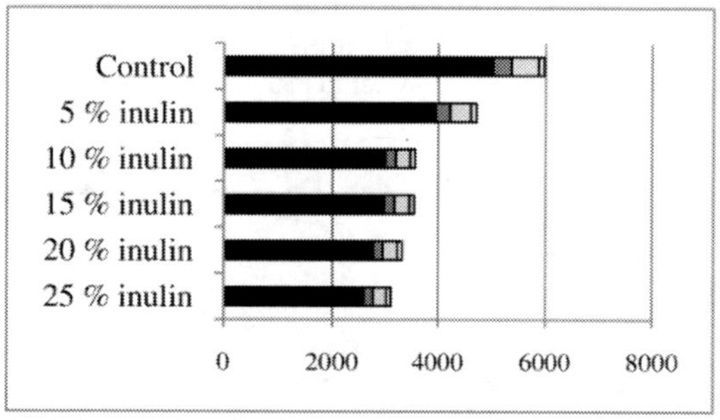


**Figure 3.** Bread properties
(**Bread properties**: loaf volume + specific loaf volume + volume efficiency + cambering x 100)

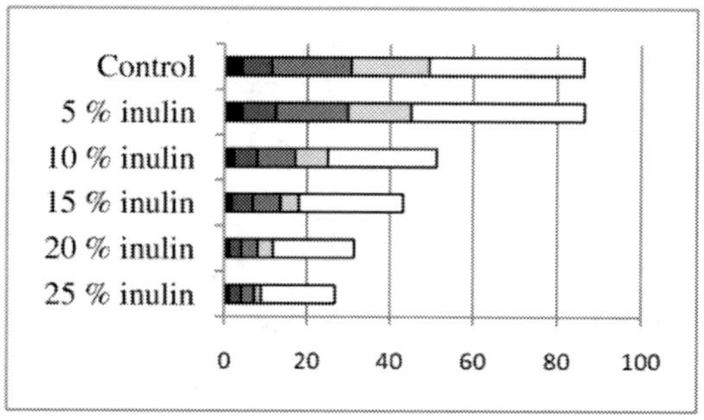


**Figure 4**. Sensory evaluation
(**Sensory evaluation**: general appearance and shape + surface and properties of the crust + fermentation and crumb appearance + structure and flexibility of breadcrumb + smell and taste)

## 4. Conclusion

Flour replacement at different levels (from 5 % up to 25 %) by inulin changed dough machinability, viscoelasticity and bread-making performances. Low addition of inulin (5 % and 10 %) decreased water absorption capacity of mixes in comparison to wheat flour but increasing of inulin addition (15 % - 25 %) proportionately increased water absorption capacity of mixes. The addition of inulin caused significant changes of the rheological parameters of the recorder curve, on farinograms there are two peaks of consistency observed in samples with a portion of inulin. The addition of 10 % and more of inulin led to a decreased bread quality, including the sensory quality. The present study has indicated that breads made with about 5 % inulin had high sensory acceptance.

Therefore, the addition of inulin could be an effective way to produce functional white flour bread without changing negatively its desirable physical properties. The incorporation of inulin has a positive effect on the nutritional value of wheat breads. European consumer prefers foods with properties strengthening human health. This study showed that several technological and textural properties are altered upon the inclusion of this dietary fibre.

**Acknowledgements**
The research leading to these results has received funding from the European Community under project no 26220220180: Building Research Centre „AgroBioTech", and the project: KEGA 015SPU-4/2011.

**References**
[1] Meyer D, de Wolf J 2008 Inulin in bread and other cereal-based products. Technology of functional cereal products. *Woodhead Publishing in Food Science, Technology and Nutrition* 409-427 (ISBN 978-1-84569-177-6)
[2] Berghofer E, Linsberger-Martin G 2009 Bread in Europe - an overview. *Deutsche Landwirtschafts-Gesellschaft e.V. The First Arab European Food Tec Days "Developing of Bread Making and Cereal Products"*, Cairo, 15 - 17 December 2008 (CD)
[3] Poinot P, Arvisenet G, Grua-Priol J, Fillonneua C, Le-Bail A, Prost C 2010 *Food Chemistry* **119** 1474
[4] Praznik W, Cieslik E, Filipiak-Florkiewicz A 2002 *Food/Nahrung* **46** 151
[5] Vanhoof K, De Schrijver R 1995 *Nutrition Research* **15** 1637
[6] Havrletová M, Petruláková Z, Burgárová A, Gago F, Hlinková A, Šturdík E 2011 *Czech Journal of Food Science* **29** 1
[7] Van Loo J, Coussement P, De Leenher L, Hoegregs H, Smits G 1995 *Critical Reviews in Food Science and Nutrition* **35** 525
[8] Castiglia-Delavaud C, Verdier E, Besle J M, Vernet J, Boirie Y, Beaufrere B, De Baynast R, Vermorel M 1998 *British Journal of Nutrition* **80** 343
[9] Franck A 2002 *British Journal of Nutrition* **87** 287
[10] Roberfroid M B 2007 *Journal of Nutrition* **137** 2493S
[11] Brasil J A, Correia da Silveira K, Magalhaes Salgado S, Souza Livera A V, Pinhero de Faro Z, Guerra N B 2011 *Brazilian Journal of Pharmaceutical Sciences* **47** 185
[12] Wang J, Rosell C M, Benedito de Barber C 2002 *Food Chemistry* **79** 221
[13] Peressini D, Gracco L, Sensidoni A 2006 *Technica Molitoria* **57** 855
[14] ICC No 171 Determination of the water absorption capacity of wheat flours and of physical properties of wheat flour dough using the Consistograph. Matrix: Wheat Flours and Wheat Flour Dough, 2004
[15] Ma W, Shuterland M W, Kammholz S, Banks P, Brennan P, Bowill W, Deggard G 2007 *Journal of Cereal Science* **45** **(3)** 302
[16] Rasper V F, DeMan J M 1980 *Cereal Chemistry* **57** **(1)** 27
[17] Wang J, Zhao M, Zhao Q 2007 *Journal of Cereal Science* **45** **(2)** 128
[18] Mayer D, Peters B 2009 *Agro Food Industry Hi-Tech* **20** **(3)** 48
[19] Peressini D, Sensidoni A 2009 *Journal of Cereal Science* **49** **(2)** pp 190
[20] Hager A S, Ryan L A M, Schwab C, Gänzle M G, O'Doherty J V, Arendt E K 2011 *Food Research and Technology* **232** 405
[21] Trabs K, Kasprick N, Henle T 2009 *Journal of Food Science* **27** 166

# Effects of ripening on rheological properties of avocado pulp (Persea americana mill. Cv. Hass)

**F Osorio[a, 1], A Roman[2], and J Ortiz[3]**
[1] Dpto. Ciencia y Tecnol. Alimentos. Fac. Tecnológica. Universidad de Santiago de Chile (USACH). Av. Ecuador 3769, Santiago, Chile.
[2]Dpto. Biotecnología – Escuela de Industria Alimentaria. Universidad Tecnológica Metropolitana. Las Palmeras 3360, Ñuñoa, Santiago, Chile.
[3]Dpto. Ciencia y Tecnol. Alimentos. Fac. Cs. Químicas y Farmacéuticas, U. de Chile. Vicuña Mackenna 20, Santiago, Chile.

fernando.osorio@usach.cl, alex.2.ro@gmail.com, jaortiz@uchile

**Abstract**. Avocado (Persea americana Mill) Hass variety is the most planted in Chile with a greater trade prospect. The aim of this study was to investigate the effect of maturity on rheological properties of Chilean Avocado Hass pulp. Fresh unripe avocados were washed and peeled, cut and stored at 3 different times; a portion was treated at 5°C and the other was treated at 20°C until it reached 2 lb puncture pressure. During maturation changes would develop due to temperature and time, with internal cellular structure changes. Preliminary results of the rheological characteristics of avocado puree show a Bingham plastic behavior.

## 1. Introduction

Avocado (Persea americana mill. c.v. Hass) industry is a very important export agribusiness in Chile, with sales over US$ 250 million and an important rural employment source. Exports represent 60% of the total production, mainly to the U.S. However, the increasing of plantations in Chile and from other producing countries, it is expected production increase and prices decrease, which will generate a high availability of raw material for domestic consumption and for the production of avocado oil which will result in an increase of their waste. For the latter, it is important for domestic producers to optimize the avocado industry and use their left overs.

Characterization of raw materials is of economic significance and has become very important for the development of innovative technological applications in various agro-productive areas. One of the most important factors that have a direct impact on the quality characteristics (color, texture, taste, palatability , etc.) is the avocado ripening process developed during postharvest storage, which indirectly depends on the harvesting season and biochemical changes that occur during fruit storage at a given temperature and time. It is interesting to characterize the intrinsic attributes of avocado, especially since the literature reports mainly chemical and textural characterization (Wang et al, 2010; Landahl et al, 2009). Related to flow and viscoelastic rheological properties, Tabilo-Munizaga (2005) characterized avocado pulp treated by high pressure (UHP) observing that did not show the behavior expected by the authors, defined by the Herschel-Bulkley model for flow properties and, on the other hand, it was determined that the UHP treatment resulted in a variation of the typical viscoelastic behavior of vegetable pulps obtained by monitoring storage (G') and loss (G") modules. Logaraj and Bhattacharya (2007) performed rheological analysis in model systems using avocado oil emulsions, determining a pseudoplastic behavior for emulsions . However, these authors did not consider the actual rheological behavior of avocado pulp which corresponds to complex matrix interactions of macronutrients influenced by the fruit maturity evolution. Industrially, green avocado is cooled at 5°C to minimize the effects of ripening and then after sale is kept at room temperature to reach the optimum ripening temperature. In this work, the rheological behavior of avocado pulp at different maturity stages during storage at 5°C and 20°C is characterized, which are actual conditions used by the industry.

## 2. Materials and methods

### *2.1. Samples preparation*

Raw materials. Fresh unripe avocados (Persea americana mill. cv. Hass) were obtained from INIA-La Cruz, Valparaíso- Chile. Avocados were harvested on November 26, 2012 and stored in 3 different times (T0 = 0 days; T1, T2 = 20 and 40 days respectively), a portion of the batch was treated at 5°C and the other batch portion was treated at 20°C until it reached 2 lb puncture pressure. For T0 it took 10 days to reach this pressure; it took four days for condition T1; and 5 days for condition T2.

### *2.2. Equipment.*

A Haake rheometer (Thermo Scientific™, RheoStress 1 Germany) which was operated from a computer loaded with the software package (HAAKE RheoWin version 4.30.0016), with cylindrical geometry (Ti DIN cup 40.1 ml) was used. Avocado puree temperature was controlled to 25 ± 0.1°C by a water bath (ThermoFisher SC 100). To determine mechanical ripeness a penetrometer (Durometer Gy-3 Fruit Scleromete) was used.

### *2.3. Preparation of avocado puree.*

Fresh unripe avocados were washed and peeled by hand. Avocado was sliced and placed in a food processor (Braun, 4162 model, 450 watts) using speed "Hi" to obtain a homogeneous paste or puree, using 5 minutes for non-mature or "hard" fruits, and 3 minutes for ripe fruit. See Figure 1.

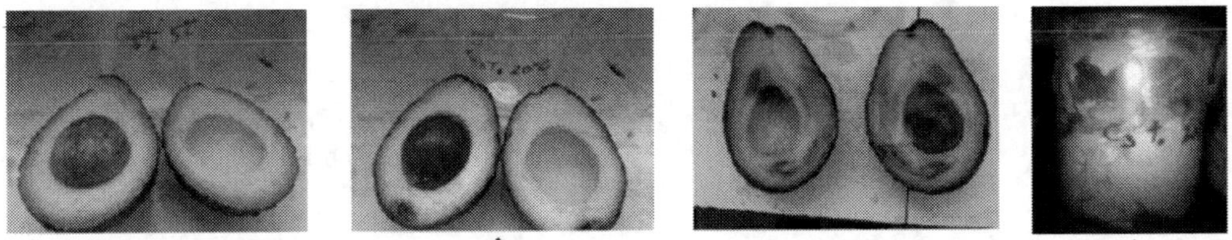

a b c d

Figure 1. Avocado a.- stored at T0; b.- stored at T1;c.- stored at T2; d.- pulp

### *2.4. Rheological measurements*

#### *2.4.1 Sample preparation.*

Pulp was mixed with 1.5% (w/w) sodium chloride and a citric acid: ascorbic acid ratio (3:1) to reduce the initial pH to 4.1. Approximately 150 g of avocado puree were poured into plastic whirl-pack bags, heat sealed and maintained under refrigeration no longer than 3 h until rheological tests were performed.

#### *2.4.2 Rheological tests.*

*2.4.2.1* Steady tests.

Tests were carried out using a cone and plate geometry. Six intervals spanning from 0.1 to 100/s, increasing and decreasing in a lineal ramp, were performed on each avocado puree sample. Data was plotted and analyzed for different rheological models.

*2.4.2.2* Oscillatory tests.
Viscoelastic properties of avocado purees were analized using small amplitude oscillatory shear; lineal viscoelastic region was determined at 10 Hz using an amplitude sweep for torque from 1 to 50 mNm. Data was collected with Rheowin 4 Job Manager. Five repetitions were used.

## 3. Results and Discussion

### *3.1. Apparent viscosity and flow properties*

Figures 2, 3 and 4 show the data obtained for apparent viscosity and flow properties, respectively. Apparent viscosity changes for avocado pulp stored at 0; 20 and 40 days at 5°C and the time for subsequent aging at 20°C (10; 4 and 5 days) are shown in Figure 1 . For maturation at 5°C it was observed that the apparent viscosity decreases after 40 days from 6332 to 1146 Pas, according to the advance of maturity, indicating that there was a change in the avocado pulp rheological properties resulting in increased flowability of avocado pulp. For maturation at 20°C, generated by sampling (T0, T1 and T2) of the fruit previously stored at 5°C it was observed that the time to maturity of consumption (given by a texture equivalent to 2 lbf) were 10; 4 and 5 days resulting in a decrease in apparent viscosity in relation to the counterpart at 5°C, which reached similar values which varied between 758 and 988 Pas.

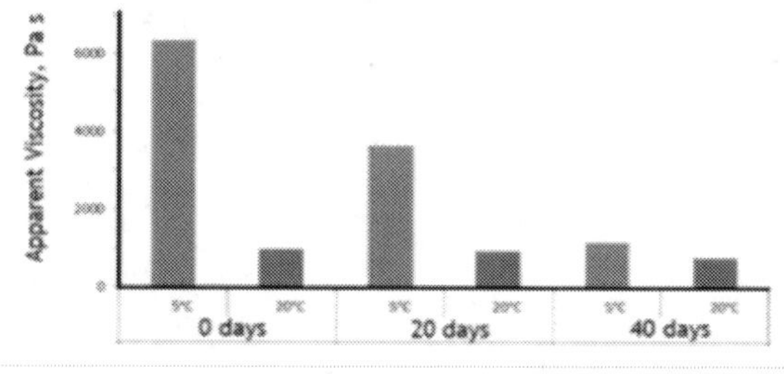


**Figure 2.** Maximum apparent viscosity changes with avocado storage time. Storage temperatures 5C and 20C.

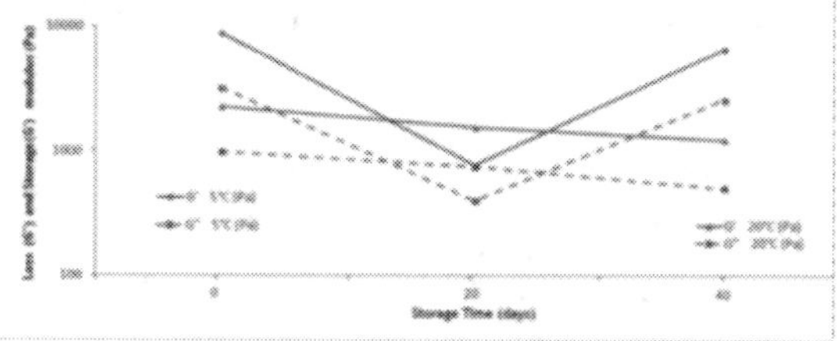

**Figure 3.** Changes in Loss (G") and Storage (G') modulus of avocado pulp for different storage times at 5C and 20C.

From Figure 3 it appears the behavior of the pulp to be Bingham plastic, the model has only been recognized by the shape of the curves obtained, but it could not be confirmed by the equation because more tests are needed to make a conclusive statement.

Figure 4 shows that in the frequency range studied (between 1 to 10 Hz), that samples behave more solid-like than liquid-like, because the elastic module is above the viscous modulus, behavior that is maintained over the progress of avocado ripening.

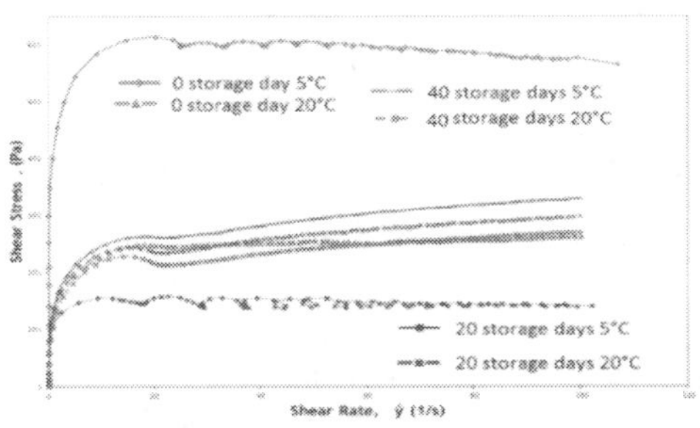


**Figure 4.** Shear stress versus shear rate for avocado pulp at different storage times at 5C and 20C.

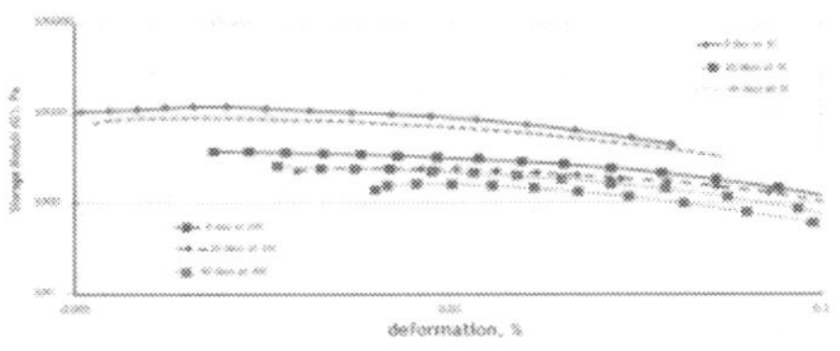


**Figure 5.** Changes in Storage (G') modulus for different storage times of avocado pulp at 5C and 20C at different deformations

## 4. Conclusion

The changes in the viscoelastic behavior of the samples of avocado pulp is dependent on phenomena related to the damage of the cell structure produced by the maturation on both at cold storage at 5°C and at the consumption temperature at 20°C. During maturation changes would generate due to temperature and time, which modify the internal cellular structure of the fruit and its associated composition to physical changes produced on its components.

Preliminary results of the rheological characteristics of avocado puree show a Bingham plastic behavior.

**Acknowledgements**

The authors would like to acknowledge FONDECYT CHILE (National Commission of Scientific and Technological Research) project 1120312 for their financial support.

**References**

[1] Tabilo-Munizaga G 2005 *Journal of Food Processing and Preservation* **29** 196
[2] Logaraj T V, Bhattacharya S, Sankar U, Venkateswaran G 2007 *Food Chemistry* **106** 937
[3] Landahl S, Meyer M, Terry L 2009 *J. Agric. Food Chem* **57** (15):7039
[4] Wang W, Bostic T, Gu L 2010 *Food Chemistry* **122** 1193

# Preliminary observations on the effects of milk fortification with conjugated linoleic acid in yogurt preparation

**RV Salamon[1], I Albert[2], CD András[1], J Csapó[1,3] and C Ibănescu[4]**
[1]Sapientia – Hungarian University of Transylvania, Faculty of Science, Department of Food Science, RO-530104, Miercurea Ciuc, Libertăţii sq 1, Romania
[2]Rofarom, RO-530231, Miercurea Ciuc, Revoluţiei din Decembrie str 34B/B/1, Romania
[3]Kaposvár University, Faculty of Animal Science, Department of Chemistry and Biochemistry, H-7400, Kaposvár, Guba S. str 40, Hungary
[4] Tehnical University "Gheorghe Asachi", Faculty of Chemical Engineering and Environmental Protection, Department of Natural and Synthetic Polymers, RO-700050, Iaşi, D. Mangeron bld 63, Romania

email: salamonrozalia@sapientia.siculorum.ro

**Abstract:** The fortification and enrichment of food with health benefic natural or natural identical substances creating new functional foods became an important issue for food researchers and processors. However, often occurs that the obtained products (despite of their health benefic activity) cannot be marketed due to strange or accustomed taste and/or texture. The aim of the research was to elucidate the effect of conjugated linoleic acid (CLA) enrichment of raw milk on the rheological properties of the obtained yogurt. The results show that the values of the complex viscosity at 50 rad.s$^{-1}$ (correlated with the thickness and sliminess of the food gel structures) of the CLA-enriched yogurt was the lowest among the studied samples, meaning the enriched yogurt is more creamy than the commercial products. These observations gave us the hope that, in this case, the texture of enriched product will not present any drawback related to consumer quality judgment.

## 1. Introduction

Conjugated linoleic acids (CLAs) form a group of positional and geometric isomers of conjugated dienoic derivatives of linoleic acid. The cis-9,trans-11 isomer of CLA, also known as rumenic acid, accounts for 90% of the total CLA intake and is mostly found in beef and dairy products. The cis-9,trans-11 isomer in CLA has been considered to be the most biologically important in terms of anticarcinogenic activity, protection against athero-scolerosis, cachexia, and treatment of non-insulin-dependent diabetes [1]. The level of CLA in milk fat commonly falls between 3 and 6 mg/g [2]. To receive the recommended 3 g/day, approximately 30 servings per day (about 250 mL) of milk with naturally increased levels of CLA would be necessary. Direct fortification of milk with CLA oil is a more feasible and realistic solution to obtain maximum health benefits [3]. The aim of the paper is to determinate the influence of the CLA fortification on the rheological properties of the obtained yogurt products and to correlate the rheological data with the yogurt sensory attributes (thickness, stickiness).

## 2. Materials and methods

### *2.1. Yogurt production and control*

A volume of 400 mL of whole milk (pH = 6.79, fat content 2.7%, total solid content 12.9%, density 1032 kg.m$^{-3}$) was used in each sample. A Hanna HI 99161 instrument was used for pH measurements, fat content was determined with Gerber method and the total solid content with Kern MLB 50-3N thermobalance. For CLA fortification 1 g of CLA supplement (Vitking) from the market was used. After fortification the milk was pasteurized at 72 °C, 14 s, cooled to 43 °C and inoculated with 0.2% mixed-starter culture of *Lactobacillus delbrueckii* ssp. *bulgaricus* and *Streptococcus thermophilus* YC-380 (Chr. Hansen's). The inoculated milk was mixed and incubated at 43 °C. Fermentation was stopped at a pH=4.52 then stored at 4–5 °C for 12 h, before evaluation. Control samples obtained without a starter culture and without CLA addition were prepared in parallel with fortified and starter culture added yogurt samples.

### *2.2. Rheological measurements*

The rheometric investigations were performed on an Anton Paar, Physica MCR 501 modular rheometer provided with a Peltier-system for the temperature control. A solvent trap was used to avoid the solvent evaporation. Parallel plate geometry with serrated plates was used to avoid slippage of the sample. The upper plate from stainless steel was 50 mm in diameter. The working temperature was 5 °C, and the gap between the plates 0.5 mm. The obtained data were processed with the Rheoplus software. Amplitude sweep tests were carried out for all four samples in the range of strain 0.1–100%. Based on the obtained data the linear-viscoelastic range (LVE) was estimated and the critical share stress was calculated (both in terms of deformation and shear stress).

**Table 1.** The properties of the studied yogurt varieties

| Sample | Sample description |
|---|---|
| $P_1$ | Set yogurt fortified with 0.25 g CLA /100 g, inoculated with mixed-starter culture, 4 h incubation |
| $P_2$ | Set yogurt without pasteurization, mixed-starter culture and CLA, 24 h incubation |
| $P_3$ | Set yogurt only with mixed-starter culture, 4 h incubation |
| $P_4$ | Commercial set yogurt |

## 3. Results and discussion

Firstly a stain test (amplitude sweep – AS) (constant frequency, $\omega$ = 10 rad.s$^{-1}$ and variable oscillation amplitude between 0.01–100%) was performed to determine the LVE and to study the mechanical stability of the yogurt structures [4,5,6]. The variation of the dynamic moduli (G' – storage modulus; G" – loss modulus) is presented in figure 1.

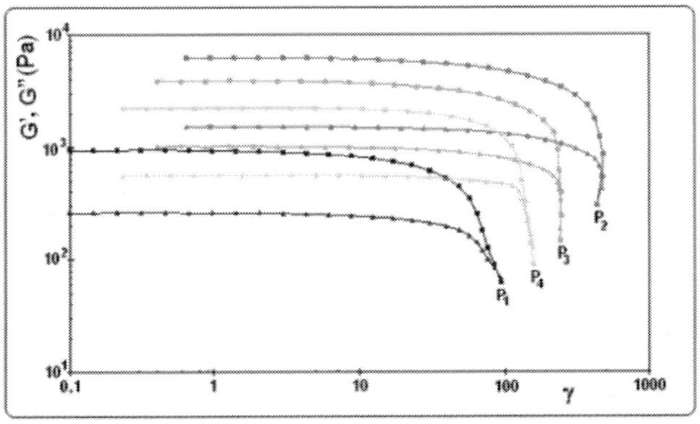


**Figure 1**. The variation of the dynamic moduli with strain for the yogurt samples

From the graphs (figure 1) the LVE limits were determined and the values of the critical shear stress was calculated. The limit of the LVE region indicates the maximum deformation tolerated by the sample before the internal super-structure is destroyed. All samples exhibit a well developed and stable structure even for higher deformations. Starting with the crossover point of the curves G'(γ) and G"(γ), hence for G'= G", the gel character of a system turns into a liquid character and this point is referred to as "flow point". The sample $P_1$ with lower values both for G' and G" has the lowest rigidity and consequently is smoother and creamy with a pleasant sensation for the consumer. The frequency sweep (figure 2) is widely used as a standard test in polymer rheology. In this test a sinusoidal strain with a constant amplitude (depending on the LVE range limits for each sample, $\gamma$ = 10% in our case) was applied and the oscillation frequency was varied (between $10^{-2}$ and $10^{1}$ rad.s$^{-1}$). The evolution of the storage modulus with frequency evidenced the smoothness and creamy structure of the sample $P_1$.

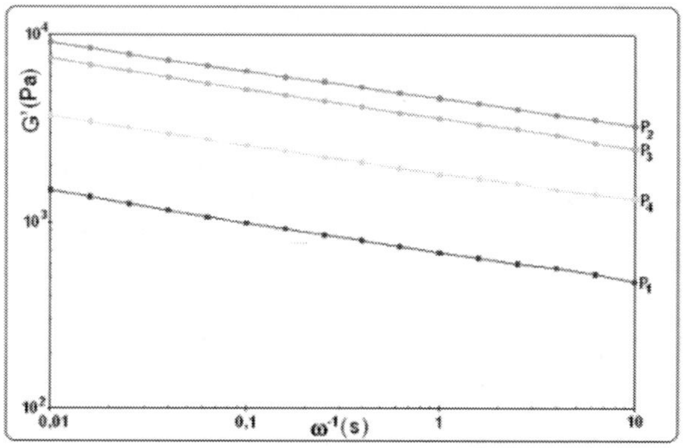


**Figure 2.** Evolution of the storage modulus in the frequency sweep test of the yogurt samples

The complex viscosity ($\eta$*) is related to the consistency of the gel structure [7,8]. The results of the measurements at low amplitude are presented in figure 3. This measurement has a particular importance, as the mimics the oral conditions during mastication of foods and the value of the complex viscosity correlate well with the slimness, creaminess and the mouth feel of the food products. Furthermore the oral release of the taste and aroma compounds from jellified structures highly depends on the gel fragmentation [9]. The values of the complex viscosity at 50 rad.s$^{-1}$ are presented (figure 4). Because the $P_1$ sample possess the lowest value for complex viscosity can be concluded that the $P_1$ sample is the best from the rheological point of view. Complex viscosity at the angular frequency of 50 rad.s$^{-1}$ obtained in small amplitude tests correlates with the thickness, sliminess verbal description of the in-mouth texture.

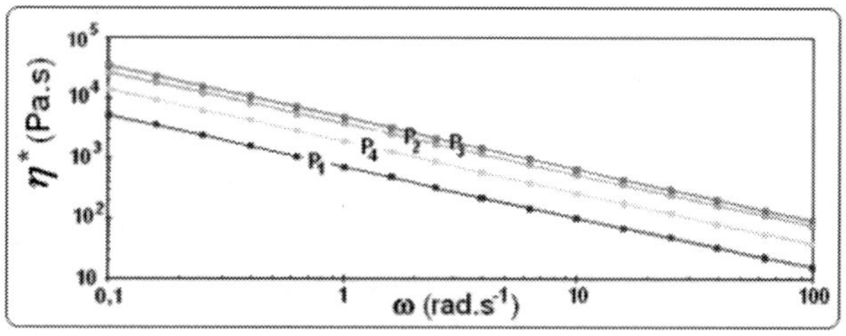


**Figure 3.** Complex viscosity in function of angular frequency

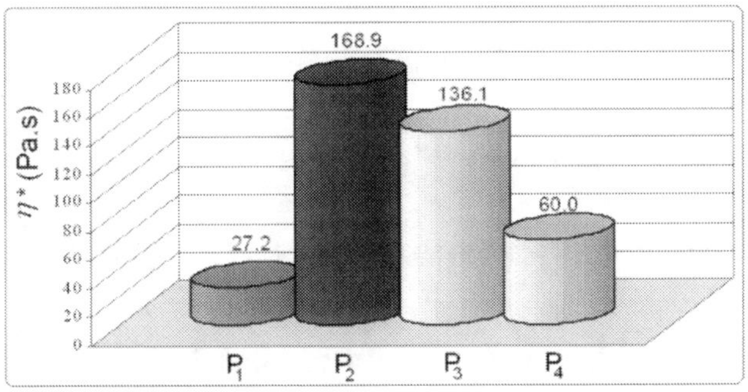


**Figure 4.** The values for complex viscosity at 50 rad.s$^{-1}$ angular frequency

## 4. Conclusions

The rheological studies allow to establish good correspondence between the rheological parameters and the sensory properties of the CLA-fortified yogurts which exhibited better mouth feeling characteristics as compared with the control probes. To elucidate the effects of processing parameters on the quality of the final products more studies are needed.

## Acknowledgements

The project „*Obtaining of functional food with semi-synthetic conjugated linoleate additivation*" was financed by the Institute of Research Programs (Contract no.: 59/9/08.11.2012)

## References

[1] Belury M A 2002 *Annu. Rev. Nutr.* **22** 505
[2] Dianóczki Cs *et al.* 2003 *Oil Soap Cosmet.* **52** 41
[3] Nazare J A *et al* 2007 *Br. J. Nutr.* **97** 273
[4] Gunasekaran S and Ak M M 2000 *Trends Food Sci. Tech.* **11** 115
[5] Bhataccharya S and Suva D 2010 Hydrocolloids *J. Food Sci. Technol.* **47** 587
[6] Baretto Penna L A *et al* 2006 *Food Technol. Biotechnol.* **44** 515
[7] Morris E 1993 *Rheological and organoleptic properties of food hydrocolloids* (Nishinari K and Doi E: Food Hydrocolloids: Structures, Properties, and Functions)
[8] Stokes J R 2012 *Oral rheology* (Chen J and Engelen L: Food Oral Processing)
[9] Salles C *et al* 2010 *Crit. Rev. Food Sci. Nutr.* **51** 67

# Friction in ferroelastic and martensitic materials

**E.K.H. Salje**
Department of Earth Sciences
University of Cambridge, Downing street, Cambridge UK

e-mail: es10002@esc.cam.ac.uk

**Abstract.** Friction during ferroelastic and martensite phase transformations and under external elastic forcing is related to the appearance of jerks as fingerprint for avalanches of moving domain boundaries. It is argued that such avalanches can be observed if the time scale of the experiment is sufficiently long to allow a statistical evaluation and have a time resolution compatible with the characteristic duration of an avalanche. Typical experimental methods include the observation of acoustic emission, resonance elastic spectroscopy, calorimetry, and advanced phonon spectroscopy. The changes of the internal structure of mobile twin boundaries contribute to changes in their mobility and their friction. The local structures of jammed twin boundary patterns are elucidated by computer simulation, their time evolution is similar to that of observed avalanche dynamics.

## 1. Introduction

The dynamic response of materials measured at high frequencies, such as measured in phonon spectroscopy or resonant Acoustic Spectroscopy, is generally found to be a smooth function of the applied stress in the regime of small stresses [1-4]. These effects have been widely studied in physics and earth sciences over the last 3 decades [5-7]. Non-linear superelastic softening occurs when external stress leads to the movement of interfaces such as twin boundaries or the interface between two phases. Local boundary movements are part of the dynamics of complex patterns with energy dissipation through the emission of phonons. The macroscopic fingerprint of boundary movements, such as a twin boundary displacement, is often the elastic response as seen in the change of stress under changing macroscopic shear of the sample. Note that strain-stress curves are very different for avalanches when the strain is the tuning parameter ( ~ hard boundary conditions) or whether the stress is applied ( ~ soft boundary conditions ). The real part of the elastic response in either case relates to the elastic moduli while the imaginary part measures dissipation or friction. Previous studies of the dissipated energy revealed Debye type relaxations related to the movement of interfaces such as twin boundaries [8-17]. Jerky elasticity, where the dissipation displays singularities such as spikes, was envisaged theoretically and has been reported experimentally for 3-point bending experiments [18]. In general, for avalanches to occur it is required that thermal fluctuations play a minor role in an a-thermal scenarios and that the system is driven slowly enough in order to ensure that avalanches do not overlap. On the other hand, a quantitative study of avalanches requires the use of very sensitive techniques that enable detection of very small transformed fractions characterized on a sub-micrometer length scale and very high time resolution. In the case of structural transitions, including ferroics and multiferroics several model cases were the internal structure of a twin boundary influences the mobility of the boundary were recorded [19-22]. In some cases, other more integral methods including calorimetry have been used to reveal the existence of avalanches [21].

Avalanches have been commonly observed in collapsing porous materials when the applied stress exceeds a small threshold value which allows the first cavities to collapse [23,24]. In additions, the change in the internal structure of twin boundaries also modifies their mobility under elastic forcing [20, 25,26]. In particular, the formation of dipols inside the domain boundaries can lead to pinning and hence may reduce the domain boudnary friction [27].

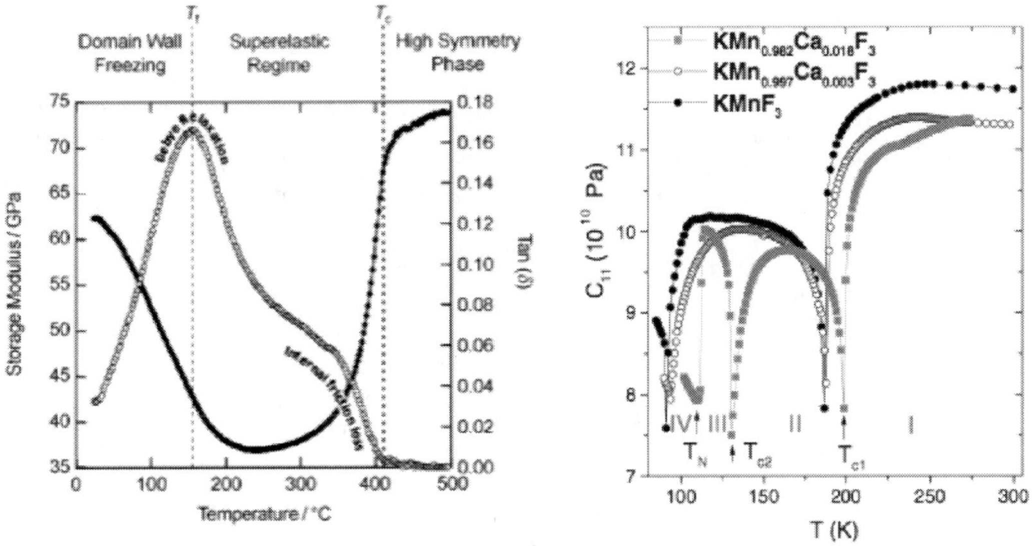


**Figure 1.** Friction and elastic moduli in $LaAlO_3$ and $KMnF_3$-type compounds. The thermal phase transitions are seen as collapse of the elastic moduli and a steep increase of the friction coefficient in the ferroelastic phase [10,11].

## 2. Experimental conditions to observe jerks

We argue in this paper that jerks and avalanches are key ingredients for the understanding of the physical processes underlying friction in ferroelastic materials and martensites. Jerks become observable under experimental conditions when the time scale of the experiment is slow enough to distinguish clearly between individual jerks. This condition is not easily met because the integration time per event is often some $10^{-4}$ seconds while the number of events in a largely a-thermal experiment is related to the total number of particles divided by the number of transforming parts per event. For a sample with $10^{23}$ atoms and on average perhaps $10^{15}$ atoms per avalanche we expect about $10^8$ events for a first order phase transition in ferroelastic materials, such as martensites. If these events were spaced equally in time, we expect the total transformation to happen in $10^4$ seconds. An experiment where a volume transformation is observed as function of the sample temperature requires cooling rates slow enough to detect at least the major events without overlap between them. In this estimate, an appropriate cooling rate for a sample inside the coexistence interval of 10K is 1 milliKelvin per hour, which is an extremely slow cooling rate. Any faster rate would superpose various events and artificially generate the impression of large avalanches, which do not exist in the sample (see Fig. 2).

The avalanche duration is shorter than 1 msec and often reduces to a few phonon times so that the required extremely high time resolution for the measurement of single avalanche profiles is very hard to obtain. Furthermore, the extremely high time resolution has to be maintained over long observation periods to measure a sufficient number of avalanches to determine their statistical distribution. This experimental requirement is hard to match and it becomes clear why so few experimental data can be found in literature.

The situation is somewhat helped if not the full transformation is jerky but where some part involves the continuous propagation of phase fronts through the sample. A typical example is CuZnAl where only 5% of the transformation process involves jerks while the remaining 95% is continuous, Fig 2. Even in this case, the observation of the jerks required heating and cooling rates of some milliKelvin per hour, otherwise signals become corrupted by overlap and cannot be used for meaningful statistical analysis. In summary, good jerk measurements require patience to obtain reliable useful statistical data and a very high time resolution. A typical example for the dissipation by friction in a mechanical shear experiment is shown in Fig. 3.

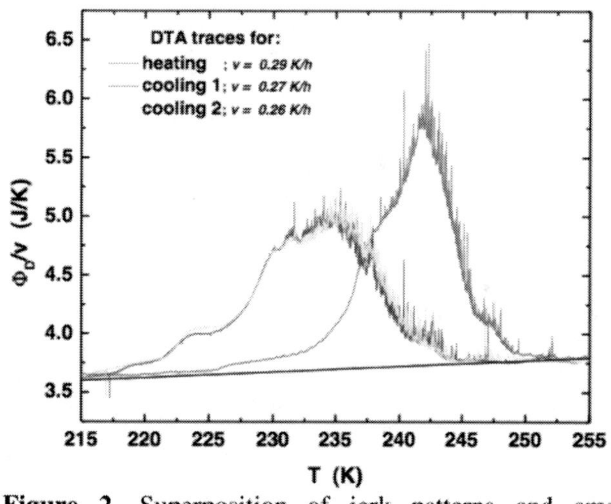


**Figure 2.** Superposition of jerk patterns and smooth background spectra related to the continuous shift of domain structures in CuZnAl [21]

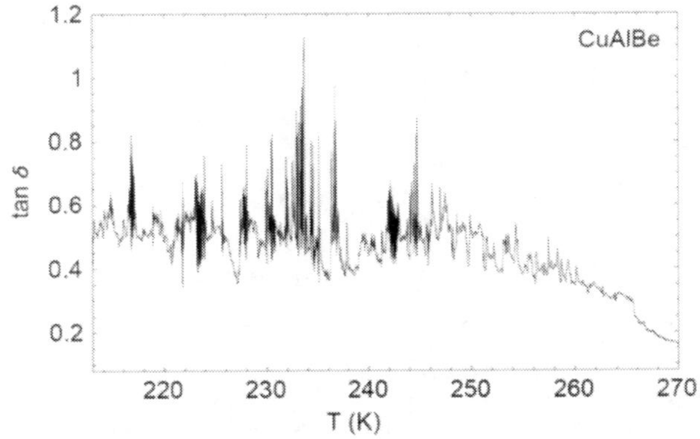


**Figure 3** Friction spectrum of CuAlBe during the austenite-martensite transition. The data were measured by a Dynamical Mechanical Analyzer (DMA) at low frequencies. Tan $\delta$ is a measure for the dissipated elastic energy [18].

## 3. The local domain structure and the size dependence of domain boundary friction

The geometrical movement of twin boundaries in ferroelastic materials and martensites has been investigated systematically by computer simulation by Ding and collaborators [28-30]. They have shown that the pattern formation is strongly size dependent (Fig.4) and that the jerk spectrum follows power law distributions only at low temperatures (a-thermal regime). At higher temperatures these authors found that the noise spectrum is best described by a Vogel-Fulcher dynamics with an stretched exponential regime at the thermal crossover [28,29,30]. These simulations clearly identify two basic domain movements (Fig.5) that characterize the friction behavior of ferroelastics: first we encounter large movement of needle domains which become non-reversible each time a needle collapses at another twin boundary or at the sample surface. Renewed needle movement requires the nucleation of a another needle which often involves large nucleation energies. The second movement is the propagation of kinks inside twin walls which are destroyed when the kink reaches the sample surface. The nucleation energy is rather small in this case and new kinks appear readily under strain.

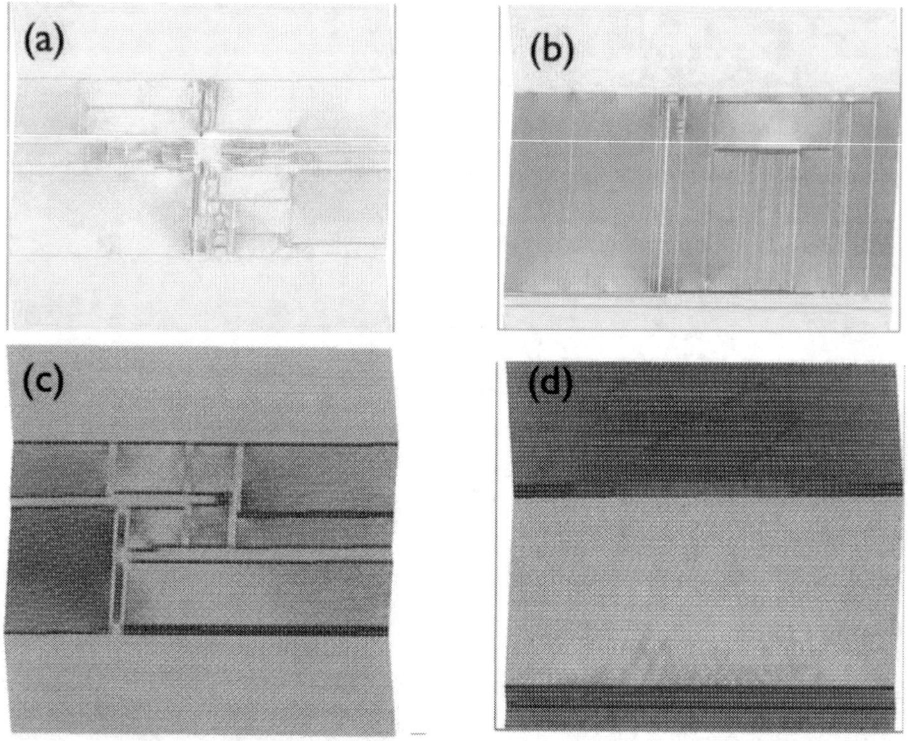


**Figure 4.** Simulated twin pattern under strain forcing. Note that the complexity of the pattern is constraint to large system sizes [28,29].

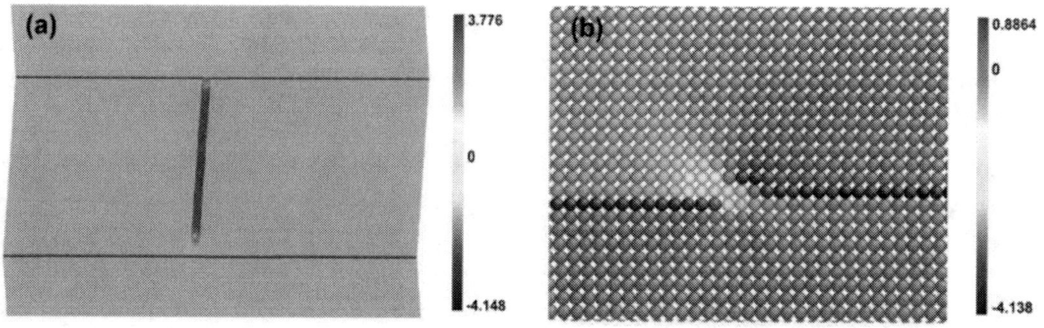


**Fig. 5** Snapshots of computer simulation of a needle domain (a) and a kink in a twin wall (b).

Under applied external strain the needle domain will progress and retract at the lower end of the needle. Friction is related to the emission of phonons during this movement. A large acoustic emission signal is emitted when the retraction of the needles becomes so large that the needle is destroyed and the total self energy of the needle is emitted as friction signal (see Fig.3). The kink in (b) moves inside the twin wall under external strain driving. Dissipation is exclusively related to phonon emission until the kink disappears at the sample surface.

In summary, we find that friction in ferroelastics and martensites is related to the movement of twin boundaries and other microstructures. Smooth movements show little friction and can be observed by elastic spectroscopy while sudden, jerky movements involve large friction and are seen as singularities in a multitude of experiments, including calorimetry, AE, and RUS. The experimental observations are well reproduced by computer simulation. The key physical parameter to characterize domain wall friction is the jerk distribution: we find power law distributions at low temperatures and thermally activated Vogel-Fulcher statistics at high temperatures.

**References**

[1] Salje EKH et al. 1992 J. Phys.:Condensed Matter **4** 571
[2] Salje EKH and Bismayer U 1997 Phase Transitions **63** 1
[3] Zhang M et al. 1996 American Mineralogist **81** 92
[4] Salje EKH et al. Phys. Rev. B **82** 184112
[5] Salje EKH and Wruck B 1983 Phys. Rev. B **28** 6510
[6] Salje EKH and Wruck B 1991 Ferroelectrics **124** 185
[7] Palmer et al. 1989 Phys. Chem. Minerals **16** 714
[8] Salje E 1987 Phys. Chem. Minerals **14** 181
[9]A. V. Kityk AV et al. 2000 Phys. Rev. B **61** 946
[10] Harrison RJ et al. 2004 J. Appl. Phys. **95** 1706
[11] Harrison RJ et al. 2004 Phys. Rev. B **69** 144101
[12] Salje EKH and Zhang HL 2009 J. Phys.: Condens. Matter **21** 035901
[13] Salje EKH 2008 Phys. Chem. Miner. **35** 321
[14] Salje EKH 1988 Phys. Chem. Miner. **16** 298
[15] Goncalves-Ferreira L et al. 2009 Appl. Phys. Lett. **94** 081903
[16] Schranz W et al. 2009 Phys. Rev. B **80** 094110
[17] Goncalves-Ferreira L et al. 2008 Phys. Phys. Lett. **101** 097602
[18] Salje EKH et al. 2009 Appl. Phys. Lett. **95** 231908
[19] Salje EKH 2012 Annual Review of Materials Research **42** 265
[20] Van Aert S et al. 2012 Advanced Materials **24** 523
[21] Gallardo MC et al. 2010 Phys. Rev. B **81** 174102
[22] Salje EKH 2010 Chemphyschem **11** 940
[23] Salje EKH et al. 2013 American Mineralogist **98** 609
[24] Baro J et al. 2013 Phys. Rev. Lett. **110** 088702
[25] Wang Z et al. 2013 Phys. Rev. Lett. **111** 056101
[26] Salje E and Zhang H 2009 Phase Transitions **82** 452
[27] Scott JF et al. 2012 Phys. Rev. Lett. **109** 187601
[28] Zhao Z et al. 2013 Advanced Materials **25** 3244
[29] Ding X et al. 2012 Advanced Materials **24** 5385
[30] Salje EKH et al. Phys. Rev. B **83** 104109

# Structure, phase content and mechanical properties of aluminium with hard particles after shock-wave compaction

**S Kulkov, S Vorozhtsov, I Turuntaev**

Tomsk State University, 36, Lenin Avenue, Tomsk, 634050, Russia
Institute of Strength Physics and Materials Science of Siberian Branch Russian Academy of Sciences 2/4, pr. Akademicheskii, Tomsk, 634021, Russia
Tomsk Polytechnic University, 30, Lenin Avenue, Tomsk, 634050, Russia

E-mail: kulkov@ms.tsc.ru

**Abstract**. The possibilities to combine metal and metal oxide powders in various compositions open a broad range of mechanical and thermal behavior. When using in nanostructured components the resulting materials might exhibit even more interesting properties, like product effectiveness, tensile strength, wear resistance, endurance and corrosion resistance. Intermetallics like TiAl could be obtained as $TiAl_x$ in a quality similar to that obtained from melting where only eutectic mixture can be produced. Similar effects are possible when compacting nanoceramic powders whereas these can be combined with intermetallics. Currently, it is very difficult to produce wires and special shaped parts from high temperature superconducting materials. The compacting by explosives could solve this problem.The present paper uses explosion compacting of Al nanoparticles to create nanocomposite with increased physico-mechanical properties. Russian civil explosive Uglenit was chosen as high energy material (HEM) for shock-wave compaction. The different schemes and conditions were suggested to run the explosion process. Al nanoparticles as produced by electric wire explosion contain 8-10% of aluminum oxide. That aluminum oxide can serve as strengthening material in the final nanocomposite which may be generated in various compositions by explosive compacting. Further modifications of nanocomposites were obtained when including nanodiamonds into the mixture with aluminum nanoparticles with different percentages. The addition of nanodiamonds results in a substantial strengthening effect.

The experiments with compacting aluminum nanoparticles by explosives are described in detail including the process variations and conditions. The physico-mechanical properties of the nanocomposites are determined and discussed by considering the applied conditions. Especially, microstructure and phases of the obtained nanocomposites are analyzed by X-ray diffraction.

## 1. Introduction

It is known [1-5], that material properties are strongly influenced by its internal structure - presence of defects, size of structural elements, presence of reinforcing elements etc. if is not always possible to form desired structure using traditional production schemes. Particularly traditional metallurgical processes are associated with increase of grain size and defect annealing, which require additional mechanical processing of an alloy. At the same time there is a method of dynamic compaction of powder mixtures with desired composition. Such techniques make it possible to obtain high degree of solidity material with density close to theoretical value; along with introduction of desired number of defects and reinforcement elements.

Such technologies are extremely important for production of Al composites by means of dynamic impact on powder aluminum alloys with necessary content of aluminum oxide as a surface film [6], which acts as a source of aluminum oxide reinforcement particles after shock loading. At the same time this method of aluminum powder compaction makes it possible to change the type and quantity of reinforcement material, e.g. aluminum oxide nanoparticles, high-modulus carbon particles in the form of nanodiamonds (further n-diamonds) [7-9] etc. This enables creation of materials with desired properties.

The objective of this work is to study properties of composites produced by explosive compaction of aluminum powder mixed with aluminum oxide nanoparticles and n-diamonds.

## 2. Materials and methods

### *2.1. Selection of explosive for Shock-Wave Compaction of aluminum powders.*

It is known [6], that the pressure required for consolidation of certain material is determined by its hardness. It is pointed out that the pressure required for compaction of Al powder with regular particle size (<160 μm) and hardness Hv=100 is equal to 1 GPa (see Fig.3.13 in [10]). Some industrial explosives such as Ammonit-1, Carbonit [10], Uglenit [11] (see Table 1), do have required characteristics. Calculations performed using the formula [10] indicated that these values must be sufficient to obtain fully compacted material. Moreover, detonation rate calculation for a shock wave according to formula

$$V_d = 2*(1.2*H_v/\rho_e)^{1/2}$$

indicates that required value of detonation rate for an explosive (e.g. Uglenit) comprises 900 m/s, at the same time the value stated in [6] is significantly (more than two times) higher (see table 1). Thus, the simplest explosive "Uglenit" was selected for experiments with Al powders, providing required compaction.

$$P_н \approx ¼*(\rho_{BB}\cdot(D_{BB})^2)\ [10]$$

$P_н$ is functionally related to density and detonation rate of explosive used and according to Prummer's estimation equals to

$$P_н \approx ¼*(\rho_{BB}\cdot(D_{BB})^2)$$

"Uglenit E6" industrial coal mine explosive with bulk density of 1.10-1.25 g/cm$^3$ and detonation rate D=2100-2500 m/s (according to technical specifications), was used as explosive. TNT equivalent was equal to 0.45, fugacity value - 130 cm$^3$. Estimated value of $P_н$=13-15 kBar.

Table 1. Characteristic data of some common explosives.

| **Explosive** | **$\rho_e$, g/cm$^3$** | **$V_d$, m/s** | **P, GPa** | **Ref.** |
|---|---|---|---|---|
| Ammonit-1 | 1.25 | 3500 | 3.8 | [6] |
| Carbonit | 1.05 | 1600 | 0.7 | [6] |
| Uglenit | 1.2 | 2200 | 1.5 | [7] |

### *2.2. Experimental technique*

A 30mm layer of explosive (pillow) compacted to ρ=1,25 g/cm$^3$ was poured into a cardboard glass (Ø=40 mm). A tube was installed onto the pillow using a centering cardboard ring. The gap was also fitted with compacted explosive. A copper tube with powder mixture of designated composition was centered by means of the second ring, and then compacted explosive was put into the glass 50mm above the tube. A standard electric detonator was installed In the middle of this layer at 20 mm depth. Total mass of explosive was comprised 300g. The assembly was put into the blast chamber on a metal plate, detonator was connected to initiation grid, the chamber was dosed and the assembly was blasted. Standard industrial explosive "Uglenit E-6" was used. Design pressure in detonation wave was P=13...15 kBar.

Phase composition and structural parameters of initial powders and obtained materials were studied using diffractometers with Cu*K*α - radiation. Step-by-step collection of X-ray patterns was performed with 0.02-0.1° step in the interval $20°<2\theta<120°$. Phase identification was performed by means of comparison of experimental pattern with ASTM-data. The sizes of coherently diffracting domains (CDD) were determined by using small-angle peaks and micro-distortion of crystal lattice ($\varepsilon$) was calculated using widening of reflection at large diffraction angles [12].

The structure of the composites was examined under a Philips SEM 515 microscope. The hardness of the materials was measured by a Supervikkers hardness meter under a load of 1000g. Compression tests were performed using an Instron 1185 testing machine.

### *2.3. Experimental material*

Compaction of Al powder as well as its mixtures with n-diamond [7-9] and alumina [4, 13] (10 wt% of each) was investigated in the framework of this research. Investigation of such compositions along with practical use for production of particle-reinforced composite materials will make it possible to study influence of high-modulus components on Al compaction process. This issue still remains understudied.

## 3. Results and discussion

According to X-ray diffraction (XRD) analysis average crystallite size for Al powder comprised 90 nm, microdistorsion of crystal lattice – $10^{-3}$. X-ray phase analysis has indicated that carbon powder consists of an X-ray amorphous phase (40%) and diamond phase (45%), the remaining part is represented by crystalline carbon. Average crystallite size for n-diamonds comprises 4 nm, microdistorsion of crystal lattice – $17\times10^{-3}$. According to XRD analysis average crystallite size for alumina powder comprised 30 nm, microdistorsion of crystal lattice – $5\times10^{-3}$.

Metallographic examinations have indicated that shock-wave (SW) treatment [14] of investigated mixtures makes it possible to achieve pore less condition (Fig. 1). The grain size of compacted materials in this case comprises 4-20 μm depending on the type of mixture, which is significantly larger than powder particle size. This appears to be the result of dynamic recrystallization which took place during explosive compaction.

It is obvious from the grain size distributions shown in Fig. 2 that the average grain size in the powders in the initial state is 220 μm and on exposure to shock waves, it reduces down to 10 μm. In this case if grain size for the samples from aluminum nanopowder comprises 20±10 μm, introduction of reinforcement particles significantly reduces it down to 11 μm for n-diamond-containing mixtures, and to 4 μm for mixtures which contain aluminum oxide.

Fig.3 shows X-ray data, obtained for samples after SW-treatment of aluminum powder and aluminum-n-diamonds mixtures as well as mixtures of aluminum with nanocrystalline aluminum oxide.

Obtained data indicate that double-phase state with significantly different structural parameters was formed for samples containing n-diamond and aluminum oxide. This state is indicated with an arrow on Fig. 2 - a small reflection near the main one, which belongs to aluminum. The study of parameters of fine crystalline structure have indicated that sizes of crystalline particles of main aluminum phase in all cases are equal to 80±10 nm, this value is close to the size of the crystallites in the powder of aluminum in the initial state. Sizes of aluminum nanophase for SW-treatment mixtures with cluster diamonds are equal to 13±5 nm, while in case of aluminum oxide powder additions - they are equal to 8±5 nm. Lattice parameter of the nanophase is increased by 0.5% in this case, which indicates its non-equilibrium state. Such an increase may be associated with compressive stresses applied at the radius of compacted cylinder with powder mixture. Estimates have indicated that these stresses are equal to 350 MPa.

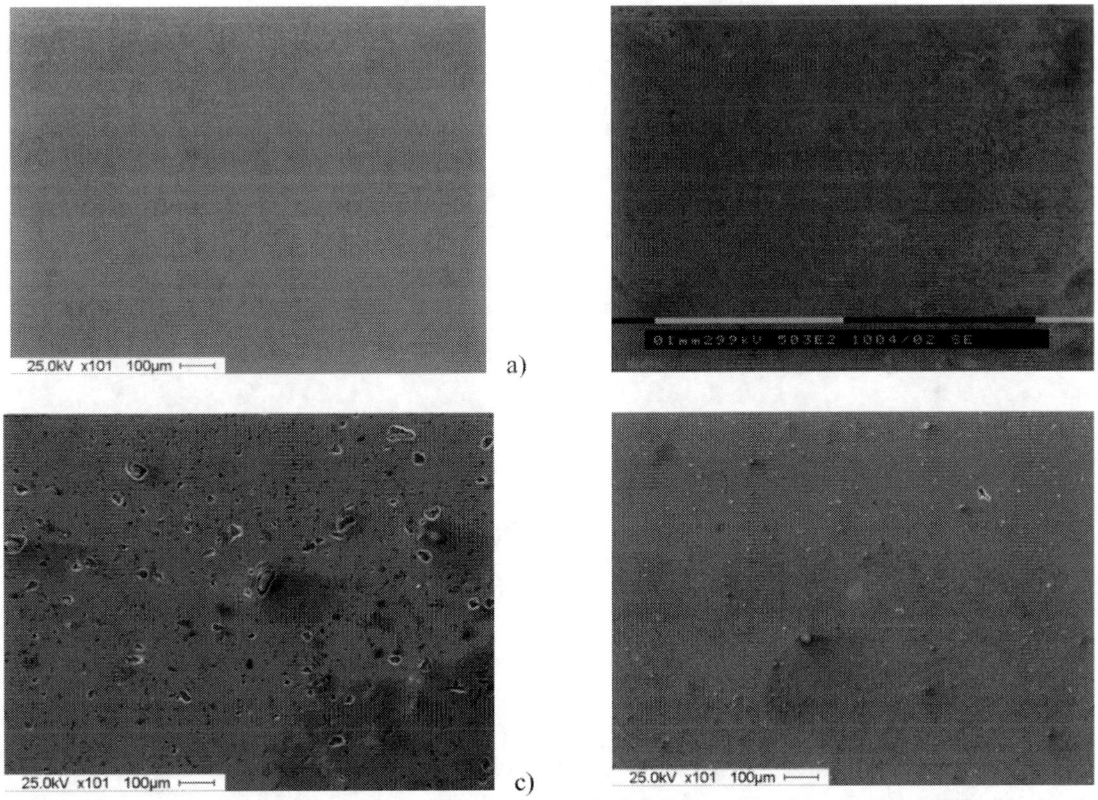


**Figure 1.** Microstructure of the materials: commercial 99.5 % pure aluminum (*a*) and alloysobtained by shock-wave compaction of an aluminum powder (*b*) and Al–C (*c*) and Al–$Al_2O_3$ (*d*).

The fact that there are many compressive stresses influences mechanical properties of SW-treated materials. Indeed, hardness measurements have indicated a 10-time increasing with maximum values for samples containing aluminum oxide. The fact that these internal stresses are determined by shock-wave treatment is confirmed by the data obtained for hardness and strength characteristics after annealing of compacted samples given in brackets in the table. As it can be seen there is a significant stress relaxation, however, it is less significant for samples with n-diamonds.

Microdistorsion of crystal lattice (Table 2) was determined using diffraction patterns for Williamson-Hall relationship. Microdistortion value has increased after adding hard particles. These values match with estimates made for metal powders [8] and make it possible to estimate stored energy [15]. As it can be seen from the table this energy increases when high-modulus particles are added. A shock wave taking place in powder containing ampoule seems to spend its energy on Al activation as well as on activation of particles. Assuming the simplest case of additive contribution to activation of Al particles by hard ones estimates of reserve energy of alumina and n-diamonds can be obtained.

The value for alumina comprised 1.1 J/kg which corresponds to data mentioned in [15], as for n-diamonds, the value was two times lower. This can be explained by complexity of phase composition of carbon mixture consisting of an X-ray amorphous phase, diamond phase and crystalline carbon.

The data regarding the change of parameters of crystal lattice of investigated materials in the process of shock-wave compaction depending on reserve energy are shown on Fig.4.

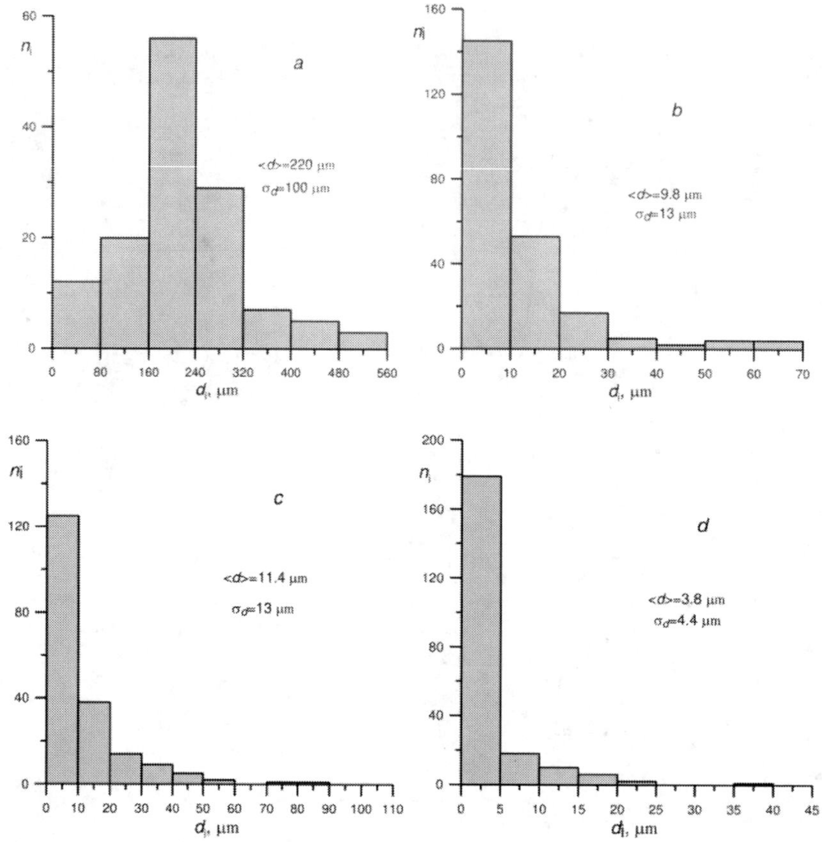


**Figure 2.** Grain size distribution in commercial 99.5 % pure aluminum (*a*) and materials obtained by shock-wave compaction of an aluminum powder (*b*) and Al–C (*c*) and Al–$Al_2O_3$ (*d*).

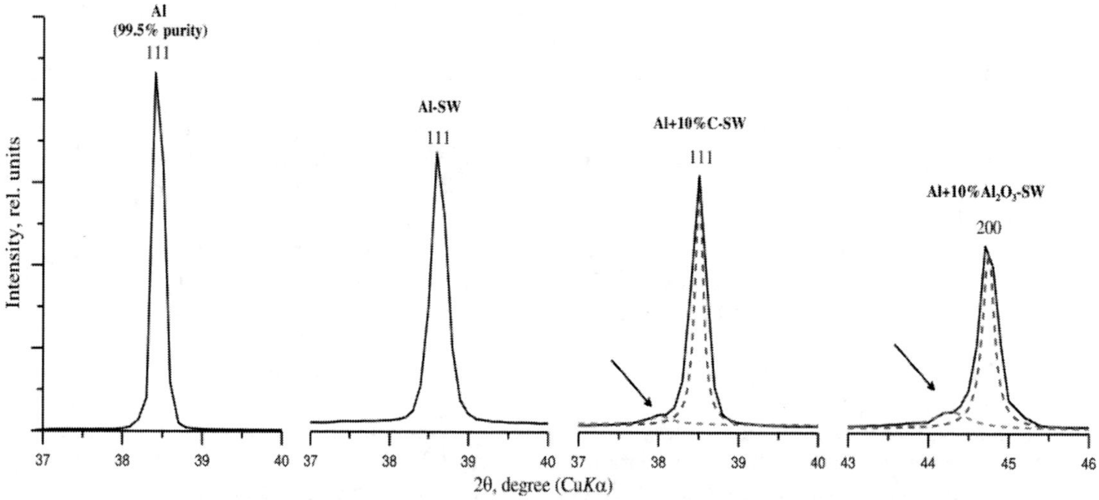


**Figure 3.** X-ray spectra of the samples obtained by shock-wave compaction.

Table 2. The Microstructural Parameters and Mechanical Characteristics of the Specimens

| **Sample** | **Crystallite size of aluminum /nanophase,nm** | **Hardness, HV, MPa** | **Yield strength, MPa** | **Strain on elastic limit, $\varepsilon_y$** | **Microdistortion of Al-lattice, $*10^3$** | **Stored energy for aluminum, J/g** |
|---|---|---|---|---|---|---|
| Al commercial pure | 155 | 190 | 30 | 0.0036 | | |
| Al after SW-compaction | 70 | 870 (35 after annealing) | - | - | 0.75 | 0.042 |
| Al+C after SW-compaction | 90 /13 | 1025 (190 after annealing) | 400 (240 after annealing) | 0.023 (0.014 after annealing) | 1.16 | 0.099 |
| $Al+Al_2O_3$, after SW- compaction | 70 /8 | 1360 (140 after annealing) | 500 (225 after annealing) | 0.0205 (0.0105 after annealing) | 1.43 | 0.151 |

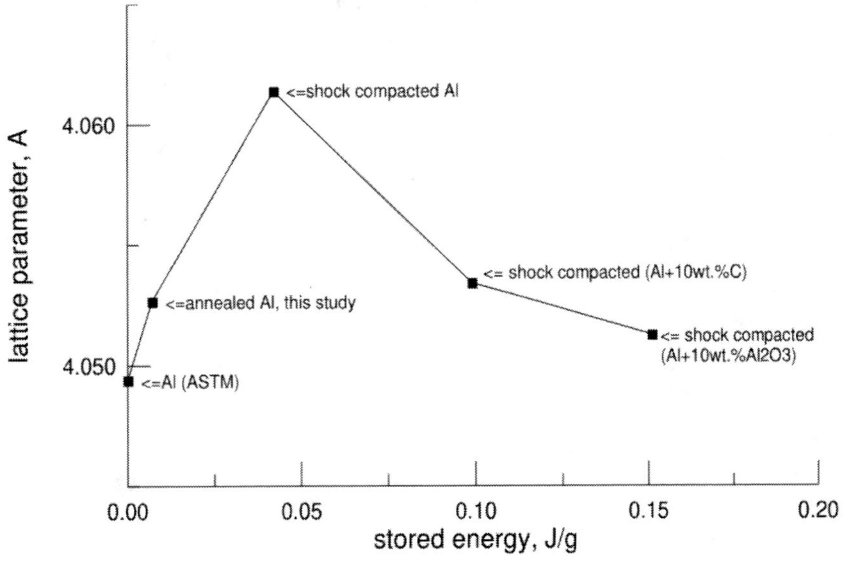


**Figure 4.** The change of parameters of crystal lattice of investigated materials in the process of shock-wave compaction depending on stored energy

As it can be seen this relationship has a maximum peak, which can be explained on one hand by the growth of microstrains (microdeformation of crystal lattice) and on the other hand – by annealing of defects due to more intense heating of Al powder after adding hard particles in comparison with "pure" powder.

## 4. Conclusion

It was shown that shock-wave treatment of aluminum powder mixtures Al+10wt.%C and Al+10 wt.% $Al_2O_3$ in copper capsules makes it possible to obtain samples with almost theoretical density and significantly better mechanical characteristics such as hardness and yield strength. The reason for this is that the part of material changes its state to nanostructured one with the size of structural elements equal approximately 10 nm and formation of compression stresses in a loaded samples up to 350 MPa.

**Acknowledgments**
This work is partial financial supported by Tomsk State University Competitiveness Improvement Program and Ministry of Sciences and Education RF with Agreement #14.578.21.0035-RFMEFI57814X0035.

## References

[1] Kulkov S N, Vorozhtsov S A, Komarov V F, Promakhov V V 2013 *Russian Physics Journal* **56** (**1**) 85
[2] L A Gomze and L N Gomze 2009 *Epitoanyag* **61** 38
[3] J G Miranda-Hernandez, D de la Torre, E Rocna-Rangel 2010 *Epitoanyag* **62** 2
[4] L A Gomze and L N Gomze 2010 *Epitoanyag* **62** 98
[5] L A Gomze 2010 *Mat. Sci Forum* **659** 19
[6] Ivanov Y F, Osmonoliev M N, Sedoi V S, Arkhipov V A, Bondarchuk S S, Vorozhtsov A B at al 2013 *Propellants, Explosive, Pyrotechnics* **28** (**6**) 319
[7] Kulkov S N and Vorozhtsov S A 2011 *Russian Physics Journal* **53** (**11**) 1153
[8] Gromov A A, Vorozhtsov S. A, Komarov V F, Sakovich G V, Pautova Yu I, Offermann M 2013 Materials Letters 91 198
[9] Vorozhtsov S A, Buyakova S P, Kulkov S N 2012 *Russian Journal of Non-Ferrous Metals* **53** (**5**) 420
[10] Prummer R.A. et al. 2006 Explosive compaction of powders and composites 194p. *Science Publishers, Enfield, NH, USA*
[11] Commercial Explosive Uglenit. GOST RU 52036-2003. The All-Russian state standard R52036-2003.
[12] Umansky Ya S, Skakov Yu A, Ivanov A N, Rastorguev L N 1982 Crystallography, X-ray and electron microscopy. *Publishing house: Metallurgy, Moscow*
[13] M V Grigoriev, S N Kulkov 2010 *Epitoanyag* **62** 33
[11] Alba-Baena N G, Salas W and Murr L E 2008 *Materials Characterization* **59** 1152
[15] Faulkner E A 1960 *Phil. Mag.* **5** 519

# FEM simulation of the size- and constraining effect in lead-free solder joints with the theory of strain gradient elasticity

**M Lederer[1,2], J Magnien[1,2], G Khatibi[2] and B Weiss[2]**
[1]Materials Center Leoben, Forschung GmbH, Roseggerstrasse 12, 8700 Leoben, Austria
[2]Faculty of Physics, University of Vienna, Boltzmanngasse 5, A-1090 Wien, Austria

E-mail: martin.lederer@univie.ac.at

**Abstract.** Reliability studies of electronic parts prevalently involve FEM simulations of solder joints under service conditions. However, simulations performed with classical continuum mechanics lead to strain singularities at the surface of material transitions. In consequence, the desired independence of the results from mesh size can usually not be achieved. Therefore, we propose a novel version of strain gradient elasticity which consequently removes strain singularities. Our approach shows similarities to a strain gradient theory which was developed already in the 1960s. But in our version of the theory it is required that the stress tensor of equilibrium states is always symmetric. This approach is implemented in the commercial FEM code ABAQUS through user subroutine UEL. Thus, it is demonstrated that in the new approach mesh convergence is achieved. Furthermore, simulations for solder joints of different sizes predict a mechanical size effect in the sense "smaller is stronger".

## 1. Introduction

The reliability of solder joints is a major issue of microelectronic industry. Lifetime models are usually based on Finite Element simulations revealing the stresses and strains observed in the constituent materials under service conditions. Due to the permanent miniaturization in microelectronics, the material properties of small scaled structures become increasingly important. It is well known that properties of materials in small dimensions cannot simply be deduced from the behaviour of bulk material. However, commercial Finite Element codes do not include a description of this size-effect. Moreover, the desired independence of the results from the mesh size can often not be achieved in the vicinity of material transitions at the sample surface. In order to solve these problems we propose a novel version of strain gradient elasticity which is implemented in the FEM code ABAQUS through user subroutine UEL. Approaching a microscopic length scale, the theory predicts a pronounced size-effect in the sense "smaller is stronger". Furthermore, mesh convergence is achieved by removal of stress singularities.

An earlier version of strain gradient elasticity [1, 2] was based on a publication of E & F Cosserat [3] who developed a theory of a three-dimensional continuous solid where every material point has six degrees of freedom. The additional degrees of freedom were microrotations which were considered supplementary to the displacements. The forces conjugated to microrotations were called couple stresses. This led to an enriched version of continuum mechanics which may for instance be applied to the propagation of acoustic waves in a crystal [4]. Within couple stress theory the stress tensor was treated as asymmetric tensor. In equilibrium the torque arising from asymmetric stresses was compensated by the torque of couple stresses. Mindlin and Tiersten suggested [1] that couple stress theory may be utilized to evaluate size effects. They calculated the stress concentrations around spherical and cylindrical cavities in a field of simple tension. Thereby, it was found that the stress concentrations at small cavities were reduced in comparison to larger ones. Mindlin [2] related the microrotations to a bending of the microstructure. Further, the bending was related to a curvature which resulted from the strain gradient. In this way constitutive equations were obtained where the microrotations were no longer independent from the displacement field. Nevertheless, the stress tensor was still treated as asymmetric tensor. Sternberg and Muki [5] used this theory to calculate stress concentrations at the tip of a sharp crack. In fact, singularities were found for sharp cracks and similar geometries [6]. We here suggest a modification of this theory which leads to a removal of strain singularities.

## 2. A new version of strain gradient elasticity

We propose a theory of strain gradient elasticity which does not emerge from an enriched theory of continuum mechanics. This means that the degrees of freedom considered here are just the displacements. Rotations may only arise according to the polar decomposition of the deformation gradient tensor. In consequence, the stress tensor of our theory is symmetric in static equilibrium. Thus, the elastic energy writes as

$$W = \tfrac{1}{2} \int_V \left(\sigma_{ij}\varepsilon_{ij} + \mu_{ijk}\eta_{ijk}\right) dV \quad \text{with} \quad \mu_{ijk} = B\,\eta_{ijk}, \tag{1}$$

where summation is carried out over repeated indices. The Cauchy stress $\sigma_{ij}$ is related to the linear strain $\varepsilon_{ij}$ by Hooke's law. $\eta_{ijk}$ are the strain gradients $\partial^2 u_i/\partial x_j \partial x_k$, where $u_i$ are the displacements and $x_i$ are the coordinates. $\mu_{ijk}$ are higher order stresses related to strain gradients. The proportionality constant B is called bending modulus and has the dimension of a force. The existence of higher order stresses is explained by the fact that the correlated strain gradients lead to a bending of the crystal lattice. Since any perturbation of the lattice symmetry increases the energy of the system, stiffness against bending occurs. Our constitutive equation (1) represents the simplest approach of two-dimensional isotropic strain gradient elasticity. It should be noticed that this expression for the elastic energy fulfils the requirement of invariance with respect to rotation of the coordinate system.

## 3. Experimental results for the strain distributions observed in solder joints

Every material model developed for Finite Element simulations should be validated by comparison with experiments. However, there are some details where the agreement between theoretical predictions and test results cannot easily be achieved. A typical example is provided by specimens with a transition of materials at the sample surface.

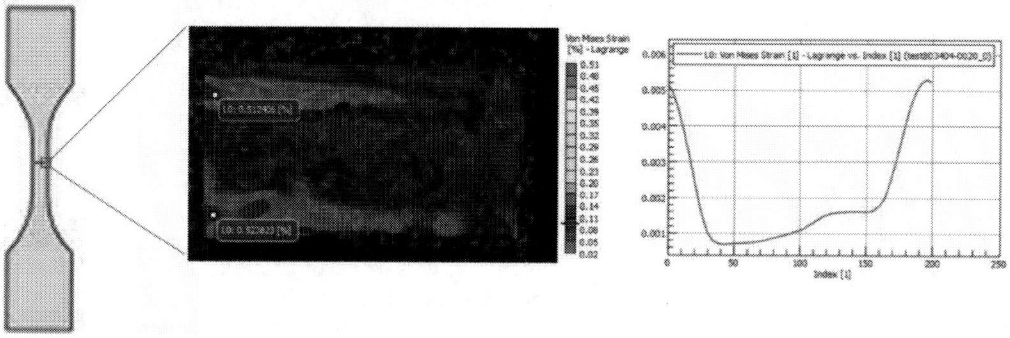


**Figure 1.** Schematic picture of the sample. Plot of the von Mises strain for a 404 μm thick solder gap at a tensile stress of 36.1 MPa.

In this case, the stress singularities predicted by classical continuum theory are not confirmed experimentally. In fact, the discrepancy between experiment and theory is not restricted to an infinitesimal region where the singularity should occur theoretically. Instead, deviations between model predictions and observed strain fields are also found in the neighbourhood of the critical region.

Tensile tests were performed with dumbbell shaped samples as depicted in figure 1, schematically. The specimens consisted of Sn-3.5Ag-0.7Cu solder gaps with cross sections of 3 x 2 $mm^2$ and pieces of 99.9% pure copper base material. The soldering process was carried out in a commercial reflow furnace using a suitable temperature profile. The samples were exposed to uniaxial tensile stress and the deformation of the solder joint was monitored with digital image correlation (DIC). The natural surface of the specimens provided a fine speckle pattern which was appropriate for high resolution DIC measurements. A plot of the von Mises strain showing the strain distribution of a 404 μm thick solder gap at a tensile stress of 36.1 MPa is shown in figure 1. The results of the contour plot along an indicated line between two points with white markings are displayed in the x-y plot on the right.

## 4. Finite Element Analysis

### 4.1 *Simulation with the classical continuum theory*

Three dimensional elasto-plastic simulations were performed for solder joints using cubic element shapes. Thereby, the element size was stepwise reduced. In consequence, the material transition between copper and solder led to the characteristics of a strain singularity within the solder material. At the beginning of mesh refinement the agreement between experiment (see figure 1) and simulation (see figure 2) improved until the best possible agreement was achieved. Thereafter, further mesh refinement deteriorated the results insofar as the strain maximum of the simulation exceeded the experimental values. Figure 2 shows two simulations performed with the material model of reference [7] using material parameters for Sn-3.5Ag-0.7Cu. Due to symmetry conditions, only $^1/_8$ of the sample was simulated. The element sizes used were (a) 28.86 μm and (b) 15.5 μm. In the latter case the mesh size was already too fine to capture the experimentally measured strain distribution at a tensile stress of 36.1 MPa:

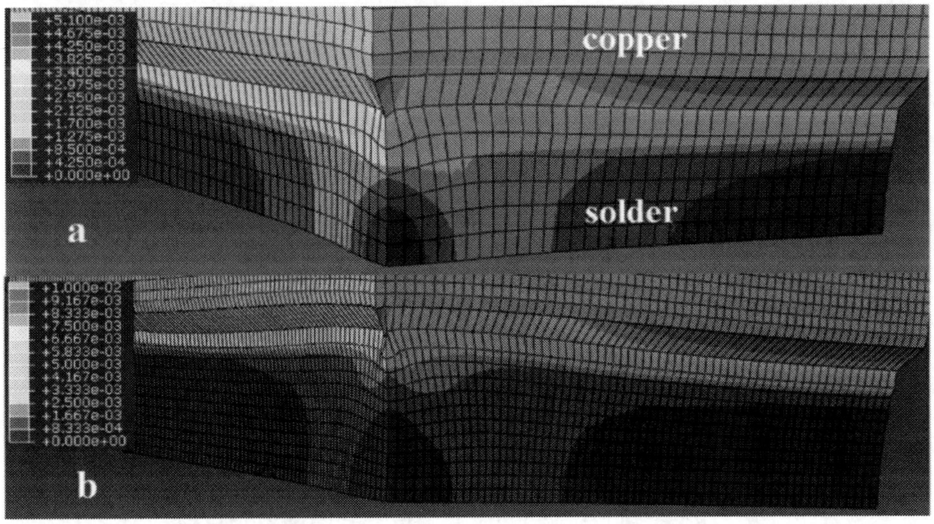


**Figure 2.** Plot of the plastic equivalent strain. Element size (a): 28.86 μm, (b): 15.5 μm

### 4.2 *Implementation of strain gradient elasticity in ABAQUS*

The theory of strain gradient elasticity defined in equation (1) is now implemented in the commercial software ABAQUS through user subroutine UEL. The elements used are 9-node squares consisting of isoparametric 4-node subelements as depicted in figure 3.

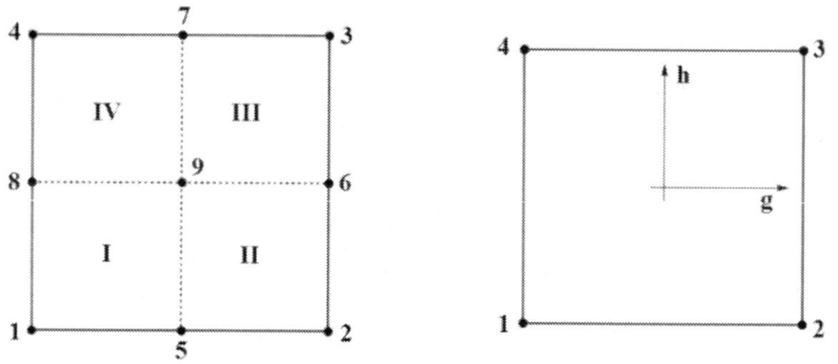


**Figure 3.** (a) left: 9-node elements consist of 4 node subelements which are indicated by italic numbers. (b) right: Subelements are isoparametric 4-node squares.

The interpolation functions for the displacements used on the subelement level are [8]:

$$U = \tfrac{1}{4}(1-g)(1-h)U_1 + \tfrac{1}{4}(1+g)(1-h)U_2 + \tfrac{1}{4}(1+g)(1+h)U_3 + \tfrac{1}{4}(1-g)(1+h)U_4 \quad (2)$$

where g and h are the coordinates of the isoparametric space and $U_1$, $U_2$, $U_3$ and $U_4$ are the displacement vectors of the nodes 1, 2, 3, and 4, respectively. The strains are calculated at integration points which are located in the middle of the subelements. Further, the strain gradients $\eta_{112} = \eta_{121}$ and $\eta_{212} = \eta_{221}$ are obtained by differentiating the interpolation functions for the displacements in the global coordinate system twice at the integration points. However, the strain gradients $\eta_{111}$, $\eta_{122}$, $\eta_{211}$ and $\eta_{222}$ may not adequately be determined within subelements. Instead, they are derived at the level of the entire element by comparison of the strain components of neighbouring subelements.

Next, the residual nodal forces related to stresses and higher order stresses are applied according to the principle of superposition. Higher order stresses are always counteracting the deformation introduced by the corresponding strain gradients. The directions of the nodal forces are illustrated in figure 4. The values of nodal forces are obtained by the condition that the work done by external forces equals the elastic energy stored in the element. In the case of $\eta_{112} = \eta_{121}$ this leads to the equation

$$F_{1,x} = -F_{2,x} = F_{3,x} = -F_{4,x} = -B\,\eta_{112} - B\,\eta_{121} \quad (3)$$

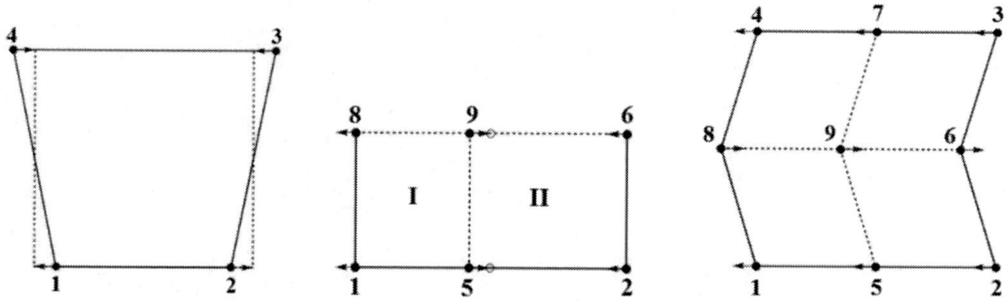


**Figure 4.** Directions of residual nodal forces. (a) left: $\eta_{112} = \eta_{121}$, (b) middle: $\eta_{111}$ and (c) right: $\eta_{122}$.

The case of $\eta_{212} = \eta_{221}$ is analogous to $\eta_{112} = \eta_{121}$. Further, one gets for $\eta_{111}$

$$F_{1,x} = F_{2,x} = -B\,\eta_{111}^{(12)}, \qquad F_{3,x} = F_{4,x} = -B\,\eta_{111}^{(34)},$$
$$F_{5,x} = 2\,B\,\eta_{111}^{(12)}, \qquad F_{6,x} = F_{8,x} = -B\,(\eta_{111}^{(12)} + \eta_{111}^{(34)}),$$
$$F_{7,x} = 2\,B\,\eta_{111}^{(34)} \quad \text{and} \qquad F_{9,x} = 2\,B\,(\eta_{111}^{(12)} + \eta_{111}^{(34)}), \quad (4)$$

where the superscript (ij) indicates that this gradient was obtained by comparison of subelements i and j. The case of $\eta_{222}$ is analogous to $\eta_{111}$. Further, one obtains for $\eta_{122}$:

$$F_{1,x} = F_{4,x} = -B\,\eta_{122}^{(14)}, \qquad F_{2,x} = F_{3,x} = -B\,\eta_{122}^{(23)},$$
$$F_{5,x} = F_{7,x} = -B\,(\eta_{122}^{(14)} + \eta_{122}^{(23)}), \qquad F_{6,x} = 2B\,\eta_{122}^{(23)},$$
$$F_{8,x} = 2B\,\eta_{122}^{(14)} \text{ and } \qquad F_{9,x} = 2B\,(\eta_{122}^{(14)} + \eta_{122}^{(23)}). \qquad (5)$$

The case of $\eta_{211}$ is again analogous.

In strain gradient elasticity, it is necessary to fulfil boundary conditions for the strains at the borders of neighbouring elements. This problem is solved here with an overlapping mesh technique as illustrated in figure 5.

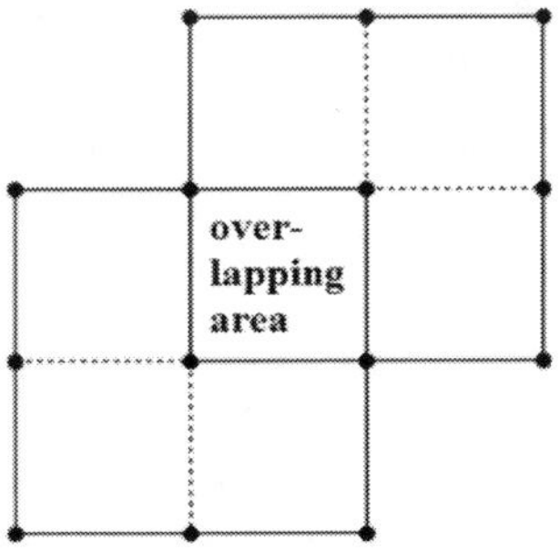


**Figure 5.** Boundary conditions for strains are solved with use of an overlapping mesh technique.

This mesh technique involves the definition of 6-node elements at the sample boundary. In addition, also symmetry elements were defined by assuming virtual nodes which are mirrored at symmetry lines. Since the whole sample is meshed twice, the values for Young's modulus and bending modulus are halved at the element level in order to get the right values for the complete model. The main advantage of the overlapping mesh technique compared to other methods satisfying strain boundary conditions is that we obtain a linear system of equations which may be solved precisely.

### *4.3. Simulation of solder joints with strain gradient elasticity*

In order to prove mesh convergence, we first apply the theory to a simplified model of a solder joint where only the solder material is deformable while the base material is treated as rigid solid. A sketch of the sample and the boundary conditions may be seen in figure 6 (a). The sketch represents ¼ of a sample with 1.5 mm length and 100 μm thickness. The upper line of the model is fully constrained while the bottom line is shifted 20 nm downwards. At the left side of the model and along the bottom line symmetry conditions were applied. Figures 6 (b) and (c) show that the contour plots for the von Mises stress [MPa] are nearly independent of the mesh size. For the simulations, a bending modulus of B = 7.416 N was assumed.

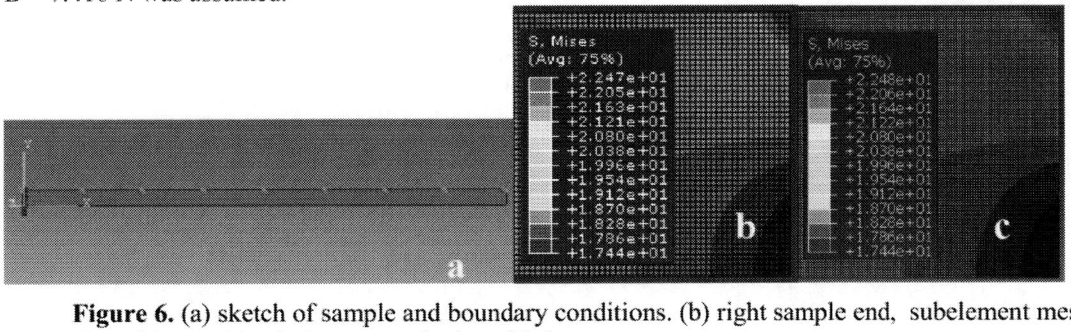


**Figure 6.** (a) sketch of sample and boundary conditions. (b) right sample end, subelement mesh size of 1 μm. (c) subelement mesh size of 0.5 μm.

Figure 7 shows a more realistic model of a solder joint where solder and base material were both simulated as deformable bodies. The comparison of figures 7 (a) and (b) with solder gap dimensions of 1.5 x 0.4 $mm^2$ and 0.75 x 0.2 $mm^2$, respectively, demonstrates the influence of the size effect in the sense "smaller is stronger". In both cases the average strain along the tensile direction was the same, but the maximum of von Mises stress [MPa] at the material transition is reduced for the smaller sample. Therefore, the smaller sample is expected to carry higher tensile stresses when loaded until fracture. In the simulations it was assumed that the bending moduli of copper and solder are proportional to the Young's moduli of these materials.

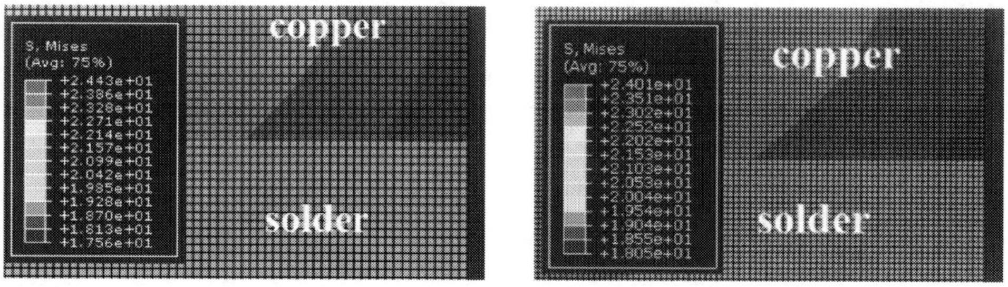


**Figure 7.** Samples with dimensions of 1.5 x 0.4 $mm^2$ (left) and 0.75 x 0.2 $mm^2$ (right), respectively. The material transition at the surface is shown in magnification.

## 5. Summary and conclusions

The modelling of solder joints within classical continuum mechanics leads to a strain singularity at the surface of a material transition. In consequence, the simulation results are not independent from the mesh size. This problem was solved with strain gradient theory. Our approach of strain gradient elasticity is similar to a theory proposed by Mindlin in the 1960s. However, we require that the stress tensor of static equilibrium must be symmetric. In consequence, the strain singularity at the surface of the material transition was removed and independence of the results from the mesh size could be achieved. Moreover, the theory predicts a size effect in the sense "smaller is stronger". Future work on this topic shall include 3-dimensionality, plasticity and time dependent behaviour.

## Acknowledgments

Financial support by the Austrian Federal Government (in particular from Bundesministerium für Verkehr, Innovation und Technologie and Bundesministerium für Wirtschaft, Familie und Jugend) represented by Österreichische Forschungsförderungsgesellschaft mbH and the Styrian and the Tyrolean Provincial Government, represented by Steirische Wirtschaftsförderungsgesellschaft mbH and Standortagentur Tirol, within the framework of the COMET Funding Programme is gratefully acknowledged.

## 6. References

[1] Mindlin R D and Tiersten H 1962 *Arch. Rat. Mech. Anal.* **11** 415
[2] Mindlin R D 1963 *Experimental Mechanics* **3** 1
[3] Cosserat E, Cosserat F 1909 *Théorie des corps dé formables*, Librairie Scientifique A. Hermann et Fils, Paris. (English trans. by Delphenich D, 2007. Theory of deformable bodies.)
[4] Aero E L and Kuvshinskii E V 1961 *Soviet Physics Solid State* **2** 1272
[5] Sternberg E and Muki R 1967 *Int. J. Solids Structures* **3** 69
[6] Muki R and Sternberg E 1965 *Z. angew. Math. Phys.* **16** 611
[7] Lederer M, Khatibi G and Weiss B 2012 *Int. J. Solids Structures* **49** 3453
[8] Bathe K J 2006 *Finite Element Procedures*, Prentice Hall, Pearson Education Inc.

# Drag reduction of nata de coco suspensions in circular pipe flow

**J. Warashina and S. Ogata**

1-1 Minamiosawa, Hachioji, Tokyo, Japan

warashina-jun@ed.tmu.ac.jp

**Abstract**. Reducing pipe friction by adding a drag-reducing agent has attracted interest as a means to reduce energy consumption. In addition to reducing drag, these agents are required to have a low environmental load and conserve natural resources. However, no drag-reducing agent currently satisfies both these conditions. We focused on nata de coco and found that the nata de coco fiber reduced drag by up to 25%. With respect to the mechanism of drag reduction by nata de coco fiber, the relationship between drag-reduction phenomena and the fiber form of nata de coco was investigated by visualization. We also found that the drag-reduction effect appeared to be due to the formation of networks of tangled fibers of nata de coco. However, drag reduction did not occur in the case in which fibers of nata de coco did not form networks.

***Key Words***: Pressure loss, Visualization, Drag reduction, Nata de coco suspension

## 1. Introduction

There has been growing interest in the research on energy consumption [1] with the aim of resolving environmental issues. For instance, one study investigated the effect of friction drag reduction by injecting drag-reducing agents into a pipe flow in a heat transport system. The reduction of pressure loss in a pipeline was found to significantly reduce energy consumption because it reduces the cost of operating pumps in the air conditioning and other systems. Drag-reduction techniques also have the potential to enable smaller and more efficient fluid flow devices. Many techniques can be used to reduce flow resistance. In particular, the addition of a drag-reducing agent (typically a polymer [2]) to the fluid base has been shown to achieve the highest reduction in flow resistance. Drag reduction by adding surfactants [3–5] has been applied to regional heating systems and air-conditioning systems in buildings. However, these techniques require careful disposal to prevent the drained solutions from contaminating rivers and soil because these additives are made of synthetic chemicals. Therefore, these additives are applied to only closed-loop pipelines. In contrast, naturally occurring biopolymer additives have recently been receiving much attention because they are considered to have a low environmental load. For this reason, we focused on nata de coco, which is considered to have a low environmental load, and investigated its drag-reduction effect [6]. Furthermore, the nata de coco has shown drag-reduction behavior similar to that of biopolymers. However, the mechanism of the drag reduction using a nata de coco has not been revealed. The purpose of this study is to investigate the use of nata de coco as a drag-reducing agent in a pipeline-flow loop.

## 2. Experimental setup and procedure

### *2.1. Nata de coco suspensions*

The tested nata de coco suspensions were made from commercial sources. Nata de coco contains over 99% water and less than 1% cellulose. Therefore, we had to remove fluid from the nata de coco and dry it. First, the nata de coco was washed with tap water and then immersed in tap water for 24 hours to remove sugar. Second, the nata de coco was pressed at 10 tons/m$^2$ and then dried in a refrigerator. Fig. 1 shows the dried nata de coco. Finally, nata de coco suspensions were prepared by mixing the dried nata de coco with tap water in a blender. The mixing time of the blender, $T_{blender}$, was 1–10 min. The suspension concentration, $C_w$, was determined based on the mass ratio of dried fiber to tap water, which is defined by

$$C_w[\text{ppm}] = \frac{M_n}{M_n + M_w} \times 10^6, \quad (1)$$

where $M_n$ and $M_w$ denote the masses of fiber and tap water, respectively.

### *2.2. Experimental apparatus*

Fig. 2 shows a schematic of the experimental apparatus used in this study. The experimental apparatus consists of a tank, pump, injection device, pressure transducer, and flow meter. The suspensions were circulated by a pump (25SCD 5.25: Ebara Technologies, Inc.), which was controlled by an inverter (Sysdrive 3G3MV: Omron Electronics, Inc.). The pressure transducer data was recorded by a data logger (NR-1000: Keyence Japan). The pressure drop was determined by averaging 50 data points measured at intervals of 100 ms. The flow rate was measured by an electromagnetic flow meter (COPA-XE: Nippon Flow Cell Co., Ltd). The total length of the pipeline, which was made of stainless steel, was 5100 mm, and the inner diameter of the test section was 15 mm. The volume of water was 4 L. The temperature of the suspension was maintained within ±0.5 °C of the target temperature (20 °C) using a constant-temperature reservoir (LTB-125: As One Corporation). The length of the test section with the pressure drop was 300 mm. Since the entrance length was 1020 mm, the flow was fully developed. The pressure drop was measured by injecting additives into the circulating tap water. Fig. 3 shows the detail of the pipe from the injection port to the measurement part for the pressure drop (test section). Nata de coco suspensions were injected through an injection port (inner diameter $\varphi$ = 3 mm). The flow velocity of injection was 0.7 mm/s, which was about 0.05–0.09% of the flow velocity in the pipe.

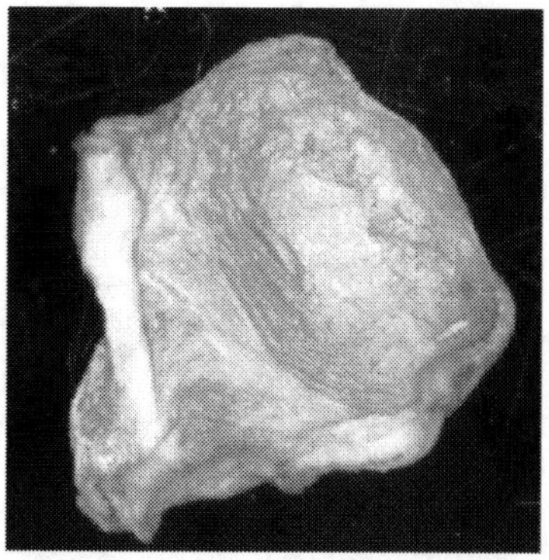

**Figure 1.** Micrograph of dried nata de coco (35×).

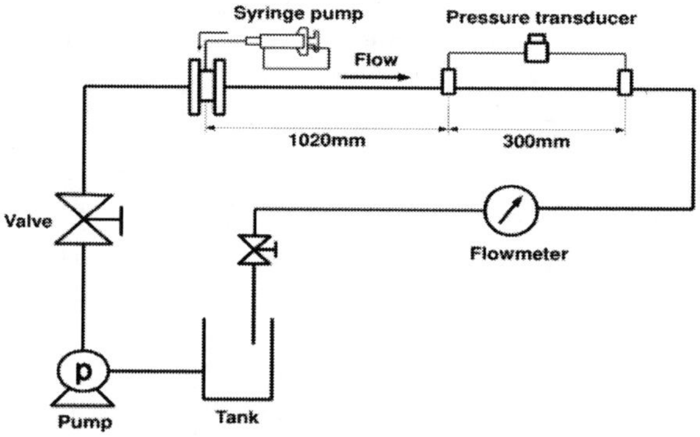


**Figure 2.** Experimental apparatus.

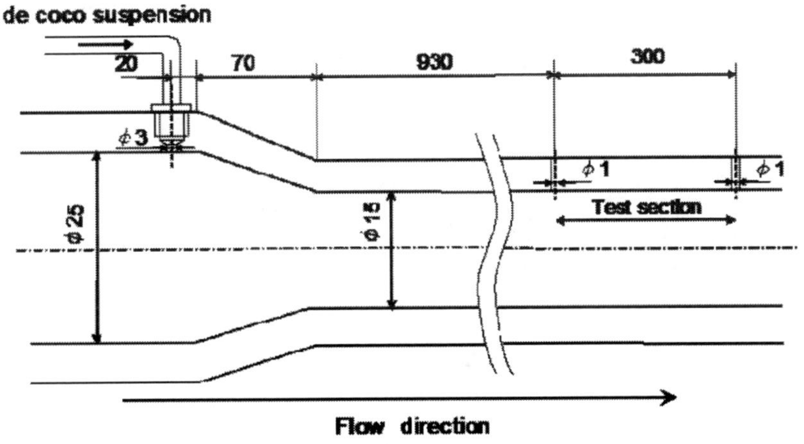


**Figure 3.** Details of test section and injection port.

## 3. Results and discussion

Fig. 4 shows the effect of the injected nata de coco on drag reduction, DR. To quantitatively estimate the amount of drag reduction, the experimental results are expressed in terms of the drag-reduction coefficient *DR*:

$$DR[\%] = \frac{\lambda - \lambda_s}{\lambda_s} \times 100, \tag{2}$$

where $\lambda$ and $\lambda_s$ denote the friction coefficient for tap water and nata de coco suspensions, respectively. The horizontal axis is the injection volume, V, and the vertical axis is the drag reduction. It shows that DR was ~0% when the tap water was injected into the pipe. However, DR increased with increase in the injection volume and concentration of nata de coco suspensions. The concentrations of suspensions in the pipe after injection were 12.5, 25, and 50 ppm. Fig. 5 shows the experimental results for the friction coefficient. The dashed line in the figure was obtained by the Blasius equation for turbulent flow, which is given by

$$\lambda = 0.3164 Re^{-0.25} \tag{3}$$

The measured friction coefficient of tap water was within 2% of that calculated using the Blasius equation. The nata de coco suspensions had lower friction coefficients than tap water for each concentration (~200 ppm). DR increased with the increasing concentration of the suspensions and the slope of this increase was parallel to the line of Blasius equation. Fig. 6 shows the effect of concentration on DR at 0–200 ppm. It shows that DR increased with the increasing concentration and reached up to 25% more than its original value when the concentration was increased to 50 ppm. At a concentration more than 50 ppm, drag reduction was nearly constant.

We researched the relation between the scattering time in the blender and DR because DR was greatly affected by the scattering time during the experiment. Fig. 7 shows the effect of scattering time in the blender, $T_{blender}$, on DR. $T_{blender}$ on the horizontal axis is the mean of the scattering time in the blender when the nata de coco suspensions were prepared. DR was observed for scattering times of ~1–3 min, although a change in DR was not observed in the case of long scattering times (more than 5 min). This shows that DR is greatly affected by scattering time in the blender. In the case where the scattering time was less than 1 min, experiments could not be conducted due to the nata de coco fibers not being sufficiently dispersed.

We observed fiber formation in the case of 1.5 and 10 min of scattering time through the investigation into the mechanisms of DR. Fig. 8 shows a micrograph of nata de coco fibers (a) before and (b) after experiments in the case of 0% DR at 10 min of $T_{blender}$. Significant differences were not observed between (a) before and (b) after the experiments. Fig. 9 shows a micrograph of nata de coco fibers (a) before and (b) after experiments in the case of 20% DR at 1.5 min of $T_{blender}$. The fibers were tangled, and a large network of fibers was formed. This shows that a large network of fibers had formed when drag reduction occurred.

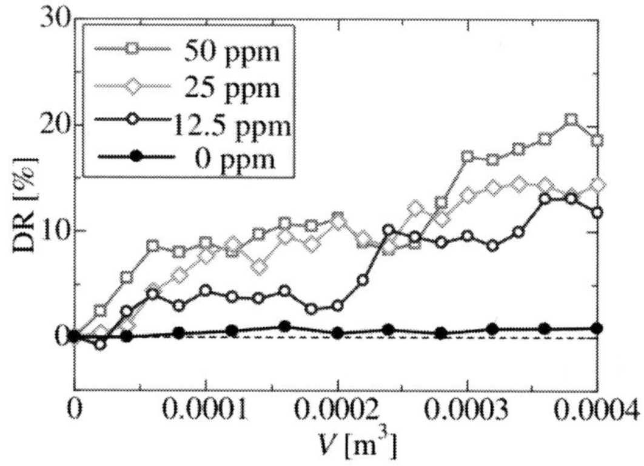


**Figure 4.** Effect of injection on DR ($Re$ = 22,000).

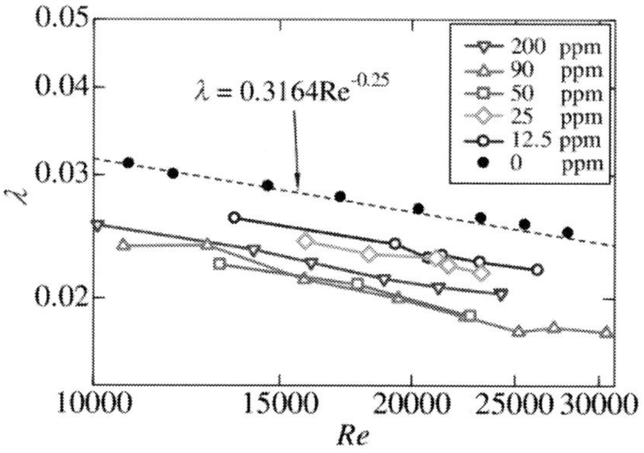


**Figure 5.** Friction coefficient of nata de coco suspensions.

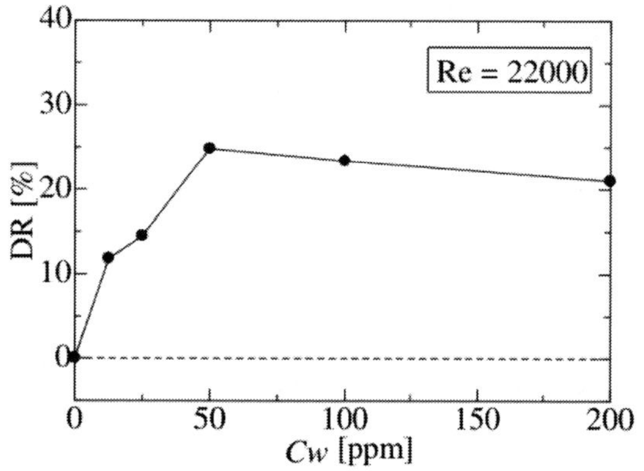


**Figure 6.** Effect of concentration on DR.

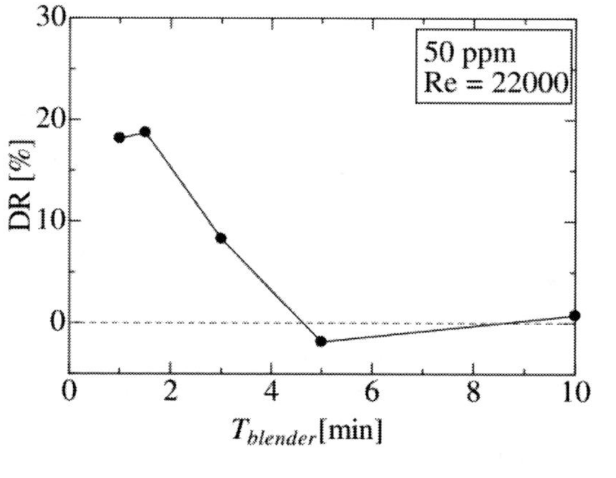


**Figure 7.** Effect of mixing time of blender on DR.

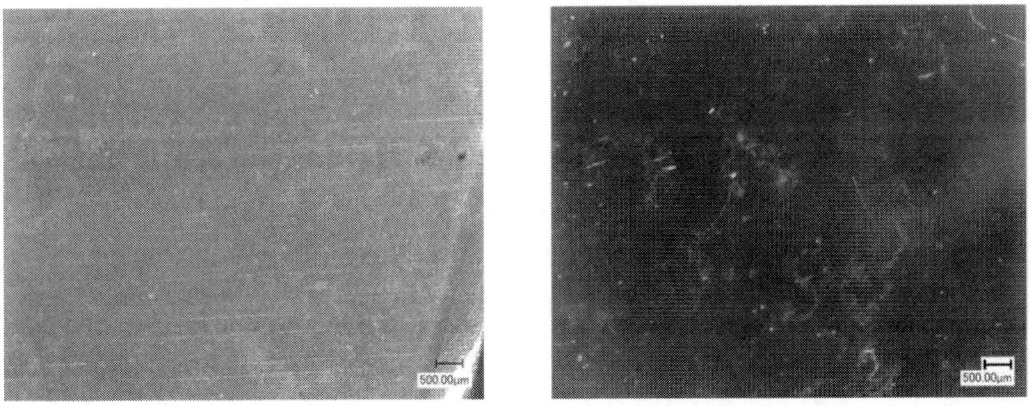

(a) Before experiment (35×). (b) After experiment (35×).

**Figure 8.** Change of fiber form ($T_{blender}$ = 10 min).

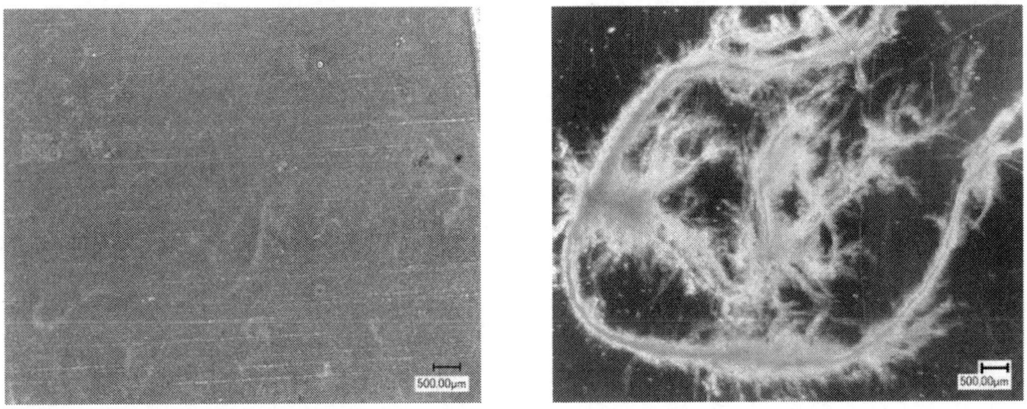

(a) Before experiment (35×). (b) After experiment (35×).

**Figure 9.** Change of fiber form ($T_{blender}$ = 1.5 min).

## 4. Conclusions

In the present work, we investigated the use of nata de coco as a drag-reducing agent in a pipeline flow loop:

(1) Drag reduction increased with increasing concentration and reached up to 25% when the concentration was increased to 50 ppm.
(2) Drag reduction was observed when scattering times were ~1–3 min and a network of fibers was formed.
(3) In the case of scattering times of more than 5 min, drag reduction was not observed and a network of fibers was not formed.

Journal of Physics: Conference Series **602** (2015) 012021 doi:10.1088/1742-6596/602/1/012021

**References**

[1] Gasljevic, K. and Matthys, E.F., Field Test of Drag-Reducing Surfactant Additives in a Hydraulic Cooling System. Proc. ASME Fluid Eng. Div. Summer Meeting, FED-237,(1996),pp.249-260.

[2] Toms, B.A., 1948, "Some Observations on the Flow of Linear Polymer Solutions through Straight Tubes at Large Reynolds Numbers," Proceedings of the First International Congress on Rheology, 2, pp. 135-141.

[3] Wei, J.J.a , Kawaguchi, Y.b , Li, F.C.c , Yu, B.d , Zakin, J.L.e , Hart, D.J.f , Oba, G.f , Zhang, Y.g and Ge, W.e , Reduction and turbulence characteristics in sub-zero temperature range of cationic and zwitterionic surfactants in EG/water solvent. Journal of Turbulence, Volume 10, (2009), Article number N10ferences

[4] Li, F.-C., Kawaguchi, Y., Hishida, K., and Oshima, M.,Investigation of Turbulence Structures in a Drag-Reduced Turbulent Channel Flow with Surfactant Additive by Stereoscopic Particle Image Velocimetry. Experiments in Fluids, Vol. 40 (2006), pp.218-230.

[5] Bewersdorff, H.W., Rheology of Drag Reducing Surfactant Solutions. Proc. ASME Fluid Eng. Div. Summer Meeting, FED-237, No. 2 (1996), pp. 25-29.

[6] Satoshi Ogata, Tetsuya Numakawa, and Takuya Kubo,"Drag reduction of bacterial cellulose suspensions", Advances in Mechanical Engineering Volume 2011, 2011, Article number528373

# A new numerical framework to simulate viscoelastic free-surface flows with the finite-volume method

**R Comminal[1,*], J Spangenberg[1,2] and J H Hattel[1]**

[1] Department of Mechanical Engineering, Technical University of Denmark, 2800 Kgs. Lyngby, Denmark
[2] Civil and Environmental Engineering, Princeton University, Princeton, NJ-08544, USA

[*] E-mail: rcom@mek.dtu.dkdk

**Abstract**. A new method for the simulation of 2D viscoelastic flow is presented. Numerical stability is obtained by the logarithmic-conformation change of variable, and a fully-implicit pure-streamfunction flow formulation, without use of any artificial diffusion. As opposed to other simulation results, our calculations predict a hydrodynamic instability in the 4:1 contraction geometry at a Weissenberg number of order 4. This new result is in qualitative agreement with the prediction of a non-linear subcritical elastic instability in Poiseuille flow. Our viscoelastic flow solver is coupled with a volume-of-fluid solver in order to predict free-surfaces in extrusion.

## 1. Introduction

The numerical simulation of viscoelastic flows with free-surfaces is a relevant problem to the modelling of several shaping processes in the polymer industry, such as extrusion or injection moulding. Polymer melts are typical examples of viscoelastic materials. Such materials can store elastic energy, and have a memory effect. During deformations, the molecular chains get stretched and oriented. As a result, the constitutive behaviour of the material depends on its strain history [1]. In shear flows, the viscoelasticity produces a normal stress difference $\tau_{xx} - \tau_{yy}$. A typical example of a viscoelastic effect happening in the manufacturing industry is the *extrudate swelling*. Another viscoelastic effect is the *elastic turbulence* in curvilinear flows, at very low Reynolds number *Re* (below unity) [2-4]. Recently, Morozov and Saarloos [5] summarized theoretical and experimental evidences of the existence of a *subcritical elastic instability* due to normal stress effects, in planar flows. Linear and non-linear stability analysis [6] showed that the viscoelastic Poiseuille flow is linearly stable at all Weissenberg numbers, but becomes nonlinearly unstable at a Weissenberg number around 4, where the Weissenberg number *Wi* is the dimensionless quantity accounting for the anisotropy created by the normal stress difference:

$$Wi = \left(\tau_{xx} - \tau_{xx}\right)/\tau_{xy}, \qquad (1)$$

The numerical simulation of viscoelastic flows has for a long time been very challenging, because of the so-called *high Weissenberg number problem* [7]. Numerical investigations show that the simulations are prone to *numerical instability* (divergence of the calculation) for *Wi* above the unity. Another numerical difficulty comes from the resolution of the Navier-Stokes equations at low *Re*, where classical fractional-step methods lose their efficiency. In this paper, we present a numerical framework which avoids these two difficulties with two changes of variables: the *logarithmic-conformation representation* [8,9], and the *pure-streamfunction flow formulation* [10,11], as described in the section 3. Both reformulations enhance the robustness of the simulation.

When it comes to the modelling of free-surfaces, Lagrangian methods have the advantage to solve directly the position of the surface (coinciding with the position of mesh), without additional calculations, see for instance [12,13]. However, Lagrangian methods become difficult to use when the free-surfaces experience changes of topology, e.g. when surfaces split or merge. For this reason, Eulerian methods are more suitable in general cases. Thus, free-surface problems are typically solved as bi-phasic flows, where one is of the fluid phase is simply air. In the Eulerian methods, the position of the interface between the two phases is represented through the use of an additional discrete variable. We chose to use the *volume-of-fluid* (VOF) method [14], where the additional variable is the volume fraction of the two phases, which is transported with the flow.

## 2. Governing equations

The governing equations of the viscoelastic flows consist in the continuity equation (conservation of mass)

$$\nabla \cdot \boldsymbol{u} = 0, \tag{2}$$

and the momentum equation (conservation of linear momentum)

$$\rho\left(\frac{\partial \boldsymbol{u}}{\partial t} + \boldsymbol{u}\cdot\nabla\boldsymbol{u}\right) = -\nabla p + \mu_S \nabla^2 \boldsymbol{u} + \nabla\cdot\boldsymbol{\sigma}, \tag{3}$$

where $\boldsymbol{u}$ is the velocity vector, $p$ is the isostatic pressure and $\boldsymbol{\sigma}$ the viscoelastic extra-stress tensor. The left hand side in the equation (3) corresponds to the inertial effects; the three terms on the right hand side of equation (3) are the contributions of the pressure gradient, the Newtonian viscous stress of the solvent, and the viscoelastic stress of the polymers, respectively. Finally, the equations of conservation (2)-(3) are supplemented with a constitutive model which closes the system of equations. We use a generic partial-differential viscoelastic model of general form

$$\frac{\partial \boldsymbol{\sigma}}{\partial t} + \boldsymbol{u}\cdot\nabla\boldsymbol{\sigma} - \left(\boldsymbol{\sigma}\cdot\nabla\boldsymbol{u} + \nabla\boldsymbol{u}^{\mathrm{T}}\cdot\boldsymbol{\sigma}\right) + \frac{f(\boldsymbol{\sigma})}{\lambda}\boldsymbol{\sigma} = \frac{2\mu_P}{\lambda}\dot{\boldsymbol{\varepsilon}}, \tag{4}$$

where $f(\boldsymbol{\sigma})$ is a relaxation function, and $\dot{\boldsymbol{\varepsilon}} = (\nabla\boldsymbol{u} + \nabla\boldsymbol{u}^{\mathrm{T}})/2$ is the strain rate tensor. Depending on the expression of $f(\boldsymbol{\sigma})$, popular viscoelastic models can be recovered [1], see table 1. The material parameters $\rho$, $\mu_S$, $\mu_P$ and $\lambda$ are the density, the solvent viscosity, the polymer viscosity and the relaxation time, respectively.

**Table 1.** Expressions of the relaxation function $f(\boldsymbol{\sigma})$ in the generic constitutive equation (4), for different viscoelastic models.

| Viscoelastic model | Relaxation function $f(\boldsymbol{\sigma})$ |
|---|---|
| Oldroyd-B | $1$ |
| Giesekus | $1+(\alpha\lambda/\mu_1)\boldsymbol{\sigma}$ |
| Linear PTT | $1+(\varepsilon\lambda/\mu_1)\mathrm{tr}(\boldsymbol{\sigma})$ |
| Exponential PTT | $\exp\left[(\varepsilon\lambda/\mu_1)\mathrm{tr}(\boldsymbol{\sigma})\right]$ |
| FENE-CR | $\left[1+(\lambda/\mu_1 L^2)\mathrm{tr}(\boldsymbol{\sigma})\right]^{-1}$ |

## 3. Numerical framework

The continuum problem is discretized with the finite-volume method, where the governing equations translate to balances of the fluxes between discrete control volumes. The fluxes are evaluated by a *quadratic upwind interpolation* scheme with flux limiters (CUBISTA) introduced by Alves et al. [15]. The equations are integrated in time with the *two-level backward differentiation formula* (BDF2). Our implementation of the VOF method uses the second-order accurate algorithm of Pilliod and Puckett [16], where the piecewise linear approximations of the interface are reconstructed with the ELVIRA technique, and the advection is solved explicitly by an operator split algorithm. Because of the non-linearity, the volume fraction, constitutive and conservation equations are solved sequentially, with successive *direct substitution* iterations until a convergence criterion is reached, at each time-step.

### *3.1. The log-conformation representation*

The log-conformation reformulation of the constitutive equation was recently introduced by Fattal and Kupferman [8,9]. It expresses the viscoelastic extra-stress $\boldsymbol{\sigma}$ in the constitutive equation (4) in terms of the logarithmic conformation $\boldsymbol{s} = \log(\boldsymbol{c})$. The conformation tensor, defined as

$$\boldsymbol{c} = (\lambda/\mu_{\mathrm{p}})\boldsymbol{\sigma} + \boldsymbol{I}, \tag{5}$$

is a measure of the extra-stress which has the important property of being *symmetric positive definite*, thus it has a real matrix-logarithm. The matrix-logarithm transformation requires the diagonalization of the conformation tensor, $\boldsymbol{c} = \boldsymbol{QDQ}^{\mathrm{T}}$, where $\boldsymbol{D}$ is the diagonal matrix of the eigenvalues and $\boldsymbol{Q}$ is the orthogonal matrix containing the eigenvector as vector column. The matrix-logarithm is interpreted as

$$\log(\boldsymbol{c}) = \boldsymbol{Q}\log(\boldsymbol{D})\boldsymbol{Q}^{\mathrm{T}}, \tag{6}$$

where the logarithm log($\boldsymbol{D}$) is applied component-wise. The evolution equation of the log-conformation is:

$$\frac{\partial \boldsymbol{s}}{\partial t} + \boldsymbol{u}\cdot\nabla\boldsymbol{s} - (\boldsymbol{\Omega s} - \boldsymbol{s\Omega}) - 2\boldsymbol{B} = \frac{f(e^{s})}{\lambda}e^{-s}, \tag{7}$$

where the $\boldsymbol{\Omega}$ and $\boldsymbol{B}$ are pure rotation and pure extension decompositions of the velocity gradient $\nabla\boldsymbol{u}$. Finally, the divergence of the viscoelastic extra-stress is recovered though the matrix-exponential of $\boldsymbol{s}$. This change of variable ensures by construction the positive definiteness of the conformation tensor. Fattal and Kupferman also showed that the numerical instabilities at high Weissenberg numbers are due to poor resolution of the exponential stress growth in time, and the exponential stress profile near geometrical singularities [17], with quadratic approximations. In the log-conformation representation, such exponential growths/profiles become linear and are accurately approximated by linear and quadratic schemes.

### *3.2. The pure streamfunction formulation*

The main difficulty when solving the conservation equations (2)-(3) with the velocity and pressure as primary unknowns (***u****-p* formulation), comes from the fact that there is no evolution equation for the pressure unknowns. Indeed, the pressure acts as a Lagrange multiplier of the incompressibility constraint which means that errors in the pressure fields directly link to errors in the conservation of mass. Most (if not all) viscoelastic algorithms use velocity-pressure fractional-step decoupling techniques, such as the SIMPLE, PISO or Chorin's projection methods [18-20]. The decoupling is achieved via an approximation of the inverse of the Jacobian matrix in the system of momentum equations. This approximation introduces a decoupling error; e.g. for the standard first order fractional-step method, the dominant error terms are proportional to $\Delta t/Re$ [21]. It means the decoupling technique is accurate at high *Re*, i.e. in flows driven by gravity or inertia, such as in aero-dynamics or hydro-dynamics problems. However, at low *Re* (especially below unity), the decoupling errors becomes very large. These flows are generally driven by pressure gradients, so that the velocity and pressure fields are strongly coupled. As a result, the decoupling techniques may not be the most suitable.

Alternatives to the $\boldsymbol{u}$-$p$ formulation solve the momentum equation in its rotational form, where a transport equation for the vorticity $\boldsymbol{\omega} = \nabla \times \boldsymbol{u}$ is derived [22], e.g. the $\boldsymbol{u}$-$\boldsymbol{\omega}$ formulation and the $\boldsymbol{\psi}$-$\boldsymbol{\omega}$ formulation. As a result, the pressure unknowns are eliminated from the system of equations. Here we use a more robust formulation, the *pure-streamfunction formulation*, recently introduced by Kupferman [10] and by Chang, Giraldo and Perot [11], independently. It also uses the rotational form of the momentum equation, but all the kinematic unknowns are expressed in terms of a streamfunction $\boldsymbol{\psi}$, defined as a vector potential of the velocity field:

$$\boldsymbol{u} = \nabla \times \boldsymbol{\psi}, \tag{8}$$

and linked to the vorticity by a Poisson equation:

$$\boldsymbol{\omega} = \Delta \boldsymbol{\psi}. \tag{9}$$

The advantage of the pure-streamfunction formulation is that the continuity constraint is automatically fulfilled, by construction. In the 2D case, only one component of $\boldsymbol{\psi}$ is non-zero: $\boldsymbol{\psi} = (0,0,\psi)$, therefore the streamfunction vector reduces to a scalar field $\varphi$, where

$$(u,v) = \nabla^{\perp}\psi = \left(-\frac{\partial \psi}{\partial y}, \frac{\partial \psi}{\partial x}\right). \tag{10}$$

Then, the evolution equation for the streamfunction scalar reads:

$$\frac{\partial}{\partial t}\Delta\psi + \left[\left(\nabla^{\perp}\psi\right)\cdot\nabla\right]\Delta\psi = \frac{\mu_S}{\rho}\Delta^2\psi + \nabla\times\left(\nabla\cdot\boldsymbol{\sigma}\right), \tag{11}$$

In general, this formulation is more robust and accurate as it is fully implicit and does not produce errors. For details about the discretization procedure of the 4$^{th}$ order operators, see in reference [11].

## 4. Numerical examples

Our solution method for 2D viscoelastic flow was implemented in Matlab. We first test our algorithm for the simulation of Newtonian and viscoelastic flows at very low *Re* in the 4:1 contraction geometry, without free-surfaces. The geometry is discretised using a uniform orthogonal mesh, with 40 control volumes on the width of the downstream channel. The no-slip boundary condition is applied at the walls, and the fully-developed velocity and stress profiles are imposed at the inlet. We use an Oldroyd-B material with a viscosity ratio $\beta = \mu_S / (\mu_S + \mu_P) = 1/9$. The fluid is initially at rest, and after a short transient response the flow establishes a steady-state solution. In our calculation, the steady-state solution was found stable for $Wi < 4$, while a hydrodynamic instability developed from $Wi = 4$ and above. This instability is not a numerical instability since the simulation does not diverge. The streamlines are perturbed inside the downstream channel. Oscillations of the velocity in the spanwise direction partially separate the stress boundary-layers from the walls, as depicted in the figure 1. The flow perturbation was initiated near to the re-entrant corner and later propagated inside downstream. It is interesting to note that the perturbation started where the streamlines have high curvatures and the material experiences the largest shear deformation. This is an indication that this flow instability could correspond to a (physical) elastic instability [5].

To our knowledge, simulation results of unstable flows in the downstream channel have not been reported in the literature. However, our onset of instability at $Wi = 4$ quantitatively agrees with the non-linear stability analysis in [6], which predicts a subcritical elastic instability in the planar creeping Poiseuille flow of Oldroyd-B fluid occurring at *Wi* around 4. In their stability analysis, a perturbation of a few percentages in the wall shear stress is sufficient to make the flow unstable. The subcritical

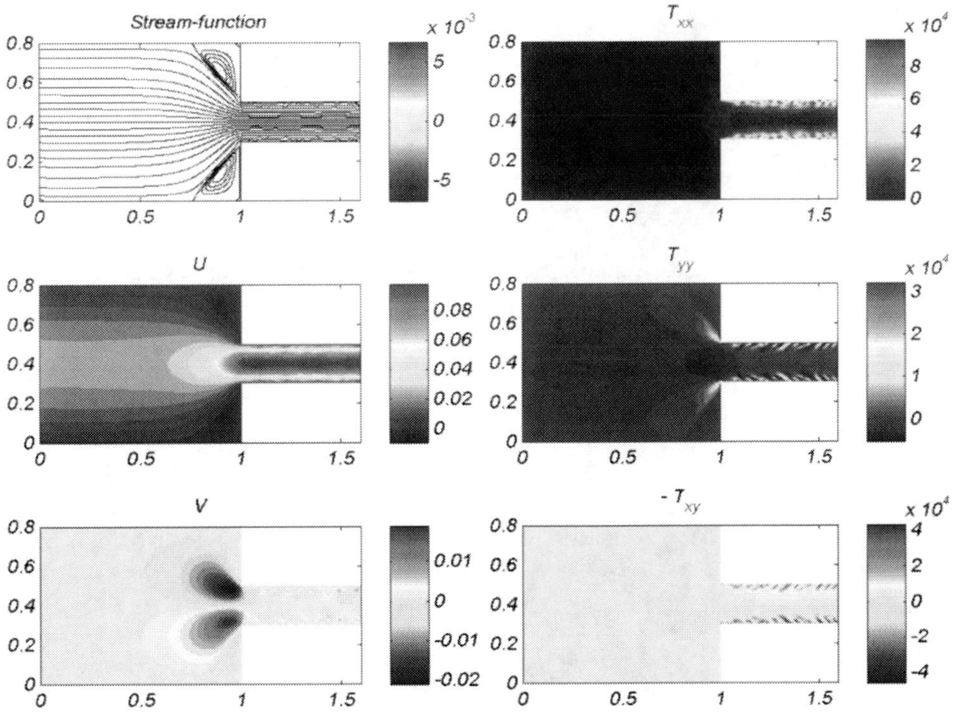


**Figure 1.** Snapshot of the streamlines, velocity and viscoelastic stress components of an unstable flow of Oldroyd-B material in a 4:1 contraction, at $Wi = 4$, $Re = 10^{-3}$ and $\beta = 1/9$.

elastic instability is also identified in [25,26] as an intrinsic mechanism of *melt fracture* defect in polymer extrusion, without stick-slip phenomena. The experimental investigation in [26] estimates the onset of the extrusion defect at $Wi = 4.6\pm0.6$, in cylindrical geometries. This empirical threshold value is also in good agreement with the results of our simulation.

The viscoelastic flow solver has further been coupled with a VOF solver, in order to simulate free-surfaces problems. Preliminary simulations of die-exit flows in planar extrusion were done. Figure 2 shows a snapshot of the free-surface during a simulation of extrusion for a Giesekus material, with the parameters $\alpha = 0.02$, $\beta = 0.5$, $Wi = 2$ and $Re = 1.25\cdot10^{-4}$. The maximum CFL number was set to 0.4. Only half of the domain was simulated, due to symmetry. In this particular flow, the swelling ratio of the planar extrudate is $A = H_{\text{extrudate}}/h_{\text{slit}} = 1.29$.

## 5. Final remarks

Historically, the high Weissenberg number problem has been addressed by developing stabilizing numerical techniques, adding an artificial diffusion to enhance the elliptic operator in the governing equations [19,20,23,24]. However, one may legitimately question whether the use of stabilizing techniques to avoid artificial instabilities does not produce an *artificial stability*. In our method, stabilisation techniques are not used; numerical stability is obtained by the log-conformation change of variable, and the fully-implicit pure-streamfunction formulation. These two reformulations increase the robustness of the numerical method, making it possible to simulate elastic instabilities. The accuracy of the results in the case of a turbulent flow is difficult to assess, because the whole range of spatial and temporal scales of the turbulences might not be resolved with the relatively coarse mesh. However, the prediction of the threshold $Wi$ at which elastic instabilities are initiated is in good agreement with the previous analytical and experimental estimations. Further investigations need to be done on the critical amplitude of perturbation that triggers the subcritical elastic instability.

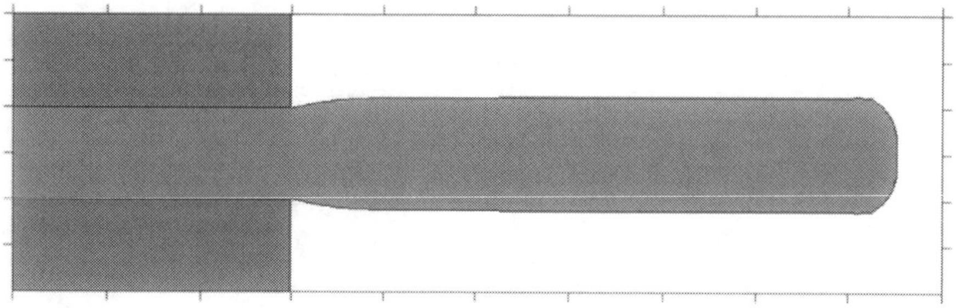

**Figure 2.** Numerical simulation of the planar extrusion die-exit flow of a Giesekus material, with $\alpha = 0.02$ and $\beta = 0.5$, at $Wi = 2$ and $Re = 1.25 \cdot 10^{-4}$.

To our knowledge, numerical simulation of extrusion melt-fracture defects has not been reported in the literature yet, and will be the subject of our future work. Our numerical method is also suitable to simulate interfacial deformations in the co-extrusion of different materials.

**Acknowledgements**
The authors would like to acknowledge the support of the Scientific Research Councils on Technology and Production Sciences (FTP) (Contract No. 09-072888, OPTIMAC), which is part of the Danish Council for Independent Research (DFF).

**References**
[1] Bird R B and Wiest J M 1995 *Annu. Rev. Fluid Mech.* **27** 169-93
[2] Pakdel P and McKinley G H 1996 *Phys. Rev. Lett.* **77** 2459-62
[3] Groisman A and Steinberg V 2000 *Nature* **405** 53-5
[4] Groisman A and Steinberg V 2004 *New J. Phys.* **6** 29-48
[5] Morozov A N and van Saarloos W 2007 *Phys. Rep.* **447** 112-43
[6] Meulenbroek B, Storm C, Morozov A N and van Saarloos W 2004 *J. Non- Newtonian Fluid Mech.* **116** 235-68
[7] Keunings R 1986 *J. Non- Newtonian Fluid Mech.* **20** 209-26
[8] Fattal R and Kupferman R 2004 *J. Non- Newtonian Fluid Mech.* **123** 281-5
[9] Fattal R and Kupferman R 2005 *J. Non- Newtonian Fluid Mech.* **126** 23-37
[10] Kupferman R 2001 *SIAM J. Sci. Comput.* **23** 1-18
[11] Chang W, Giraldo F and Perot J B 2002 *J. Comput. Phys.* **180** 183-99
[12] Keunings R 1986 *J. Comput. Phys.* **62** 199-220
[13] Rasmussen H K and Hassager O 1999 *J. Non- Newtonian Fluid Mech.* **82** 189-202
[14] Hirt C W and Nichols B D 1981 *J. Comput. Phys.* **39** 201-225
[15] Alves M A, Oliveira P J and Pinho F T 2003 *Int. J. Numer. Meth. Fluids* **41** 47-75
[16] Pilliod J E and Puckett E G 2004 *J. Comput. Phys.* **199** 465-502
[17] Rallison J M and Hinch E J 2004 *J. Non- Newtonian Fluid Mech.* **116** 141-62
[18] Oliveira P J, Pinho F T and Pinto G A 1998 *J. Non- Newtonian Fluid Mech.* **79** 1-43
[19] Xue S C, Tanner R I and Phan-Thien N 2004 *J. Non-Newtonian. Fluid Mech.* **123** 33-58
[20] Fiétier N and Deville M O 2003 *J. Comput. Phys.* **186** 93-121
[21] Perot J B 1993 *J. Comput. Phys.* **108** 51-8
[22] Gresho P M 1991 *Annu. Rev. Fluid Mech.* **23** 413-53
[23] Sureshkumar R and Beris A N 1995 *J. Non- Newtonian Fluid Mech.* **60** 53-80
[24] Baaijens F P T 1998 *J. Non- Newtonian Fluid Mech.* **79** 361-85
[25] Meulenbroek B, Storm C, Bertola V, Wagner C, Bonn D and van Saarloos W 2003 *Phy. Rev. Lett.* **90** 024502
[26] Bertola V, Meulenbroek B, Wagner C, Storm C, Morozov A, van Saarloos W and Bonn D 2003 *Phy. Rev. Lett.* **90** 114502

# Qualification Methods of $Al_2O_3$ Injection Molding Raw Materials

**Á Egész[1,2] , L A Gömze[2]**

[1] Department of Ceramics and Silicate Engineering, University of Miskolc, Miskolc, Hungary
[2] GE Hungary Ltd., Budapest, Hungary
E-mail: adam.egesz@ge.com

**Abtract.** For producing ceramic arc tube parts (plugs), there are used two different major components for producing injection molding raw material (feedstock): high purity alumina powder as the main component, and an organic paraffin wax as a binder material. It is expressly important to know the material, physical and chemical properties of these components, since mainly these have effect on the homogenity of feedstock, and therefore on the quality of end product. In this research, both of the main components and the moldable raw material was investigated by visual, physical, and thermal methods. As most important and main statement, the researchers found that the dynamic viscosity of raw material depends more on the applied temperature, than on the deformation speed gradient.
Applied analitycal methods were laser granulometry, sieve analysis, differential thermal analysis and rheology analysis.
**Keywords:** alumina, wax, injection molding, laser granulometry, thermal analysis, rheology

## 1. Introduction

Nowadays, the sophisticated ceramic industry and technologies require higher and higher assumptions against quality, volume and yield of end product [1-6]. In the illuminant industry, for producing arc tube parts for High Intensity Discharge lamps the applied method is the ceramic injection molding. The ceramic arc tube parts are made of high purity alumina powder. By producing ceramic parts, one of the most critical step is to optimizing the injection molding process, determining and controlling the influential machine paramteres, which have effect on the quality of end product [7]. Nevertheless, before we fall to doing the optimization of injection molding process, we need to know the properties of injection molding raw material, because later the molding process will be optimized for this material, to decrease the amount of cracked ceramics [8]. Powder injection molding (PIM) is a technology for manufacturing complex, precision, netshape components from either metal or ceramic powder. The potential of PIM lies in its ability to combine the design flexibility of plastic injection molding and the nearly unlimited choice of material offered by powder metallurgy, making it possible to combine multiple parts into a single one. Furthermore, PIM overcomes the dimensional and productivity limits of isostatic pressing and slip casting, the defects and tolerance limitations of investment casting, the mechanical strength of die-cast parts, and the shape limitation of traditional powder compacts [4, 9].

In the illuminant industry for producing ceramic arc tube parts (plugs) for the High Intensity Discharge lamps, there are used two different major components for producing injection molding raw material (feedstock): high purity alumina powder as the main component, and an organic paraffin wax as a binder material. Producing the feedstock, the wax is heated in a sigma-blade mixer, and the alumina powder is sequentially dosaged to the molten wax material during mixing. The given, properly homogenized mixture is cooled down, and broke in jaw crusher to get granules from it. This granules now is usable for the ceramic injection molding process. (figure 1.)

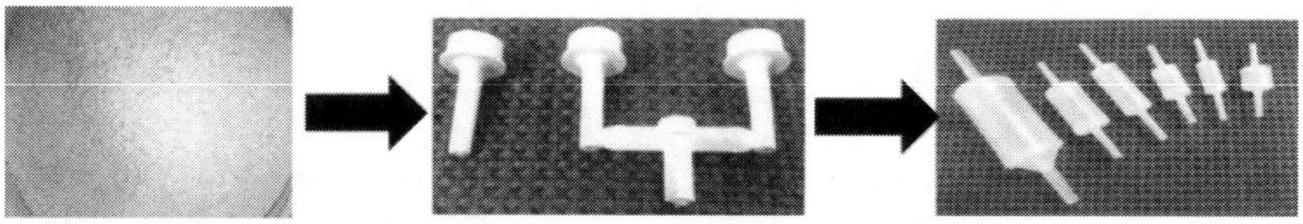

**Figure 1.** Granules raw material, injection molding green- and end product (arc tube)

To know the physical, chemical and thermal properties of the given feedstock granules is necessary, if we would like to set up an efficient injection molding process, which results a quality end product. The most commonly occurring failures in the end product are the cracks, voids and different material discontinuities, and to avoid these failures, the best way is the qualification of raw materials [1, 10].

## 2. Experimental

To qualify the alumina powder and feedstock material, the following measuring methods were tested:

### *2.1. Laser granulometry*

By qualifying the alumina powder, absolutely necessary to use laser granulometry method to determine the grain size distribution of the choosen powder. In this way, we can predestine the quality of end product [11, 12]. In addition, using this method, if we know the grade of polydispersity, and the given grain distribution (figure 2.), we can conclude to the compaction and volume filling properties of the powder [13, 14, 19], and thus the density of the end product [21-26].

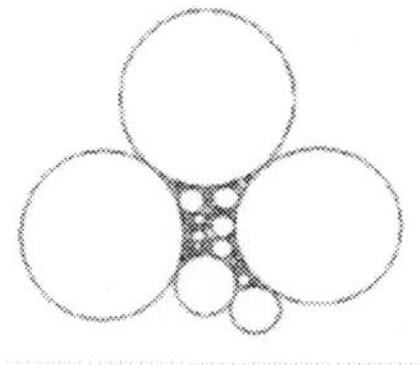

**Figure 2.** Schematic of polydisperse alumina powder
From: Tamás F 1970 *Laboratory test sin silicate Industry* (in Hungarian) Budapest

### *2.2. Sieve analysis*

The sieve analysis is the one of the oldest methode to classify powders and granules based on its particle size distribution. Using woven screen material, the sieving classifies the particles according to its middle dimension, namely the width. The mechanical sieving is the most suitable in the case, when the most of the particles larger than 75 μm. In the case of sieving smaller particles its slight weight does not ensure the proper force to overcome the cohesion and adhesion forces [15-18].

In our case, the prepared feedstock granules can be classified using sieve analysis. This qualification method is very important, because during the injection molding the too small grains can cause feeding difficulties, they can stick on the feeder and so abort the injection molding process. In addition, the too large grains can cause inhomogeneous material melting, and therefore material incontinuities, voids crack can be in the end product [8, 9].

Using the sieve analysis, the securely usable grain size of granules is determinable, and we can identify and eliminate the inadequate grain fractions from the material.

*2.3. Differential thermo analytics*

The effect of heat occuring endotherm and exotherm reactions in solid materials can be measured using the well known differential thermo analytics method. The investigated material heated together with a comperative material (inert material). In the inert material does not occurs any endotherm or exotherm process or transformation in the used temperature range. Applying this measurement, the temperatures of phase transformations are determinable in different materials, like the melting, evaporation or inflammation temperatures [18, 20].

In our case the method seems to be very important, when we would like to identify the injection molding work temperature of the used raw material. Since the process have to be adjusted to the proper temperature, which is above th melting temperature, but it is under the intensive evaporation temperature, to avoid the material loss during the injection molding process.

*2.4. Rheological test*

By injection molding it is absolutely necessary to know the reological properties and behavior of raw materials, especially its viscosity. Mainly, this property specifies the flow behavior in the injection molding tool, thus it can have effect on the quality of end product [27-29].

To define the viscosity of feedstock, dynamical rotation rheometer was used, the principle of operation can be seen on figure 3.

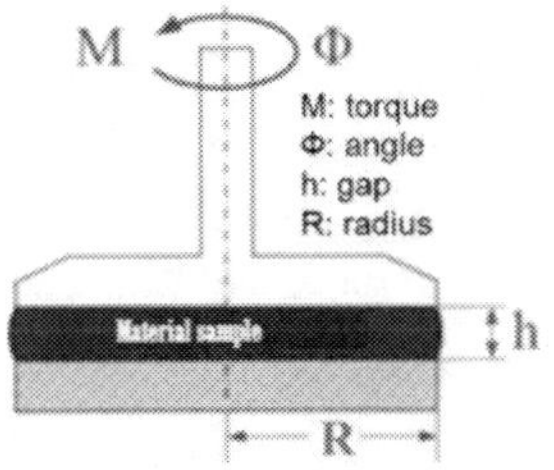


**Figure 3.** Principle of operation of dynamic rotation rheometer

During the test, the rheological parameters are counted from the values of torque and angle using built-in software, with the application of optical encoder [30, 31]. Besides of the raw material, to know the rheological properties of wax binder material can be also important, since mainly the binder material defines the rheological behavior and viscosity of feedstock.

## 3. Results and discussion

*3.1. Laser granulometry*

The first measurement to qualify the properties of alumina powder is the laser granulometry to investigate grain size distribution.

In the following, in my experiment I compared a raw, unmilled and a 10 minutes in vibration ball mill milled alumina powder. On figure 4 we can see the histogram of unmilled powder, where we can observe the two peaks on different grain sizes, which refers to the earlier mentioned polidisperse grain distribution (figure 2.).The polidisperse grain distribution - namely the small grains between the larger grains – results more effective volume filling, than the unified grain distribution.

The shape of histogram of milled powder is clearly different (figure 5.). We can see, that the secound peak is lower, which refers to the less larger particles and less polidisperse properties of powder, but better compaction propertiesand higher density, thanks to the more small particles.

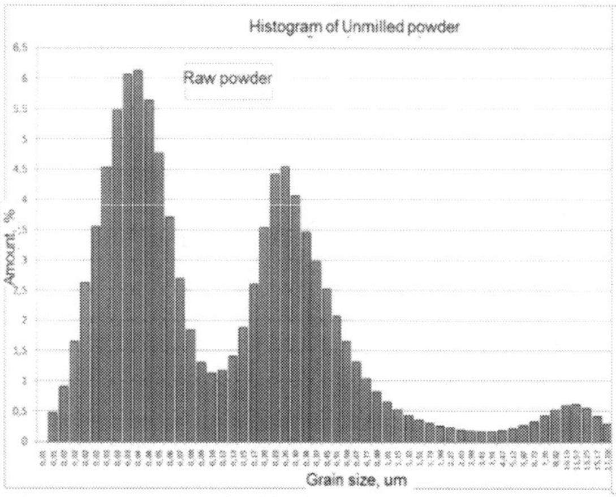


**Figure 4.**Grain distribution of unmilled alumina powder

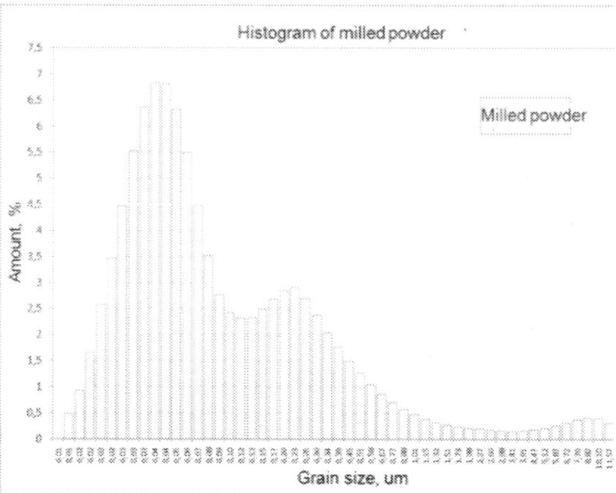


**Figure 5.** Grain distribution of 10 minutes milled alumina powder

*3.2. Sieve analisys*

The injection molding process is very sensitive for the grain size of granules, becouse the not proper grains can couse dosage and compaction difficulties during injection molding. Therefore it is necessary to qualificate the mixed raw material according to its grain size, using shieve analysis. In this way, befor usage we can eliminate the inproper grains from the material.

In the experiment, the feedstock material was classified using this method, where the amount of usable fraction was ~55% between 0,1 and 1,5 mm, the smaller grains had to rejected, the larger grains had to regrinded.

*3.3. Differential thermo analytics*

The next measurement, what can be used for the IM raw material is the well known differential thermo amalythics. This method is usable to investigate the behavior of injection molding raw material inthe function of temperature. For the played out processes we can conclude from the weight changing of material. The feedstock material was experimentally tested, the derivatograph of thermal analysis can be seen on the figure 6.

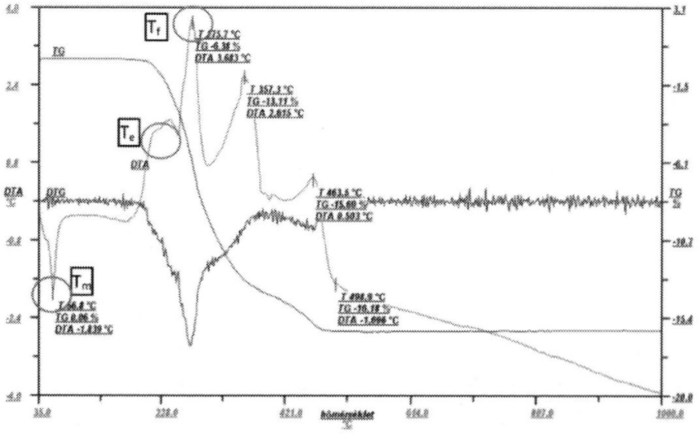


**Figure 6.** Derivatograph of injection molding raw material

We can observe the melting point ($T_m$) on the first negative peak, the evaporation point ($T_e$) on the first positive peak and the flashpoint ($T_f$) on the secound positive peak of the material on the DTA curve.

### *3.4. Rheological test*

The reological properties of raw material we can investigate using dynamical rotation rheometer to define the flow properties of material on different temperatures and shear velocities. This measurement is very important becouse the flow properties of the material has influence onthe injection molding process, and thus onthe quality of end product. The principle of the dynamic rotation rheometer can be seen on the schematic picture (figure 3). The feedstock material was investigated using this method, the registered viscosity curves can be seen on the figure 7.

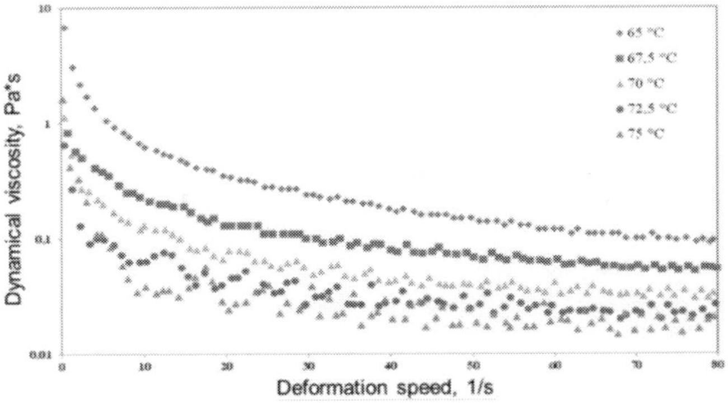


**Figure 7.**The viscosity of injection molding raw material depending on the deformation speed and the applied temperature

On this plot we can see how the viscosity of raw material depends onthe applied deformation speed and temperature. The viscosity exponentially decrease with the increasing deformation speed, and decreases with the increasing temperature. We can observe, that the temperature have a huge effect on the viscosity.

## 4. Conclusions

Using the demonstrated investigations, the quality of the raw materials are determinable, and possible the classification before usage. Testing these methods, very important statements were determined in the optimization of injection molding process point of view.

Measuring the alumina powder using laser granulometry it is possible to determinate the grade of polydispersity of the powder, and in addition, the grade of polydispersity can be reduced in the aluminum-oxide powder using vibration ball mill, to enhance volume filling.

It is necessary to qualify the raw material according to its grain size, to eliminate the inappropriate powder fractions from the granules, which can couse injection molding difficulties.

The injection molding material is well-characterized by DTA, to establish evaporation, melting and inflammation point, and to determinate the temperature work point during injection molding of the usable material.

The dynamic viscosity of the feedstock exponentially decreases with the increasing deformation speed and decreases with the increasing temperature. In addition, it was observed, that the rheological behavior of raw material depends more on the applied temperature, than on the deformation speed gradient.

## References

[1] Mutsuddy B C 1995 *Ceramic Injection Molding* Chapman and Hall,
[2] Gomze A L, Gomze N L 2009 *Epitoanyag* **61** 38
[3] Gomze A L, Gomze N L 2013 *Mat. Sci. Forum* **729** 43
[4] Hausnerova B Marcanikova L 2011 *Rheological Characterization of Powder Injection Molding using Feedstock Based on Alumina* Proceedings of Recent Advances in Fluid Mechanics and Heat & Mass Transfer 245
[5] L A Gomze and L N Gomze 2013 *IOP Conf. Ser.: Mater. Sci. Eng.* **47** 012033
[6] Kalatur S E, Buyakova P S, Kulkov N S, Gotman I, Kocserha I 2014 *Epitoanyag* **66** 31
[7] Krauss V A, Pires E N, Klein A N 2005 *Materials Research* **8** 187
[8] Rak Z S 1999 *Powder Metallurgy and Metal Ceramics* **38** 126
[9] Stanimirovic Z 2010 *Injection Molded Mn-Zn Ferrite Ceramics* Proceedings of 27th International Conference on Microelectronics MIEL
[10] Wei W C J, Wu R Y Ho S J 2000 *J. of the Eur. Cer. Soc.* **20** 1301
[11] Yin H Q, Jia C C, Qu X H 2008 *Science in China Series E: Technological Sci.* **51** 121
[12] Gitzen W H 1970 *Alumina as a ceramic material* The American Ceramic Society Ohio
[13] Messinig G L, Mccauley W, Mazdiyasni K S, Haber R A 1987 *Advances in Ceramics* (The Am. Cer. Soc., Colombus) **21** 495
[14] Kok M 2005 *Mater. Proc. Tech.* **161** 381
[15] Sevik H, Kurnaz S C 2006 *Mater. Des.* **27** 676
[16] Filser F, Guackler L J 2006 *Keramische Werkstoffe Beispiele für strukturkeramische Werkstoffe* ETH-Zürich Departement Materials
[17] Tisza M 2001 *Anyagvizsgálat,* Miskolci Egyetemi Kiadó (in Hungarian)
[18] Somiya S 2003 *Handbook of Advanced Ceramics Vol. 1,* Elsevier
[19] Tamás F 1970 *Szilikátipari laboratóriumi vizsgálatok,* Műszaki Könyvkiadó Budapest (in Hungarian)
[20] Enrique J E, Ochandio E, Gazulla M F 1991 *„Chemical Analysis" in Engeneered Materials Handbook* ASM International Materials Park **4**
[21] Funk J E, Dinger D R *Interceram* **41 (3)** 176
[22] Stenger F Mende S 2005 *Chemical Engineering Science* **60** 4557
[23] Peukert W 2004 *Int. J. Miner. Process.* **74** S3
[24] Barth H 1985 *Modern Methods of Particle Size Analisys* Wiley-Interscience, New York
[25] Allen T 1981 *Particle Size Measurement* Wiley-Interscience, New York
[26] Cotica L F, Paesano A Jr. at al. 2006 *Journal of Alloys and Compounds* **413** 265
[27] Mutsuddy B C 1983 *Fabrication Science* **3** 117
[28] Mamata P, Parag B 2005 *J. of the Am. Cer. Soc.* **88** 833
[29] Edirisinghe M J, Evans J R G 1987 *Ceram. Trans. J.* **18** 86
[30] Tanaka T 1985 Yogyo-Kyokai-shi, **93** 572
[31] Cima M J, Lewis J A 1989 *Binder J. Am Ceram. Soc.***72** **(7)** 1192

# Study of Shape Memory Alloys and the Phase Transition by DMTA and DSC Measurements

**Kis Dávid, Béres Gábor , Dugár Zsolt** and **Hansághy Pál**
Kecskemét College - Faculty of Mechanical Engineering and Automation, Hungary

E-mail: david.istvan.kis@gmail.com

**Abstract**
The aim of this research is the examination of the DMTA technique to measure the phase transition temperatures of shape memory alloys. These parameters are difficult to detect by traditional material testing techniques, meanwhile quick and reliable results can be provided by DMTA measurements.
Keywords: DMTA, phase transition, shape memory alloy, complex modulus, Nitinol.

## 1. Introduction

The shape memory alloys are shape memory intelligent materials. Their main functional property is the shape memory effect. It means, that a permanent macroscopic deformation of the specimen can be ceased after heating it above a certain temperature. The basis of the shape memory effect is the martensitic phase transition. The alloys are usually adjusted such, that they have martensitic phase in room temperature and their austenitic phase transition temperature is below 100 °C. When the deformed martensitic specimen is heated to its austenitic phase transition temperature, the grid transformation occurs, and the specimen recovers its original shape.

The problem of the production of shape memory alloys is the determination of the phase transition temperatures. The most important parameters of manufacturing these materials are the temperature range of shape memory effect and superelasticity.

There have been many examples of measuring transition temperatures of metallic materials by DSC method, which was technically designed for plastic specimens. This method has also become well known in metal technology for the last ten years. However, the DSC measurement can only provide relevant results in case of higher heating speed, which worsens the reliability of the results.

The other possible method to measure the phase transition temperatures of such materials is DMTA.
Our goal was to study if more exact results can be recovered by DMTA measurements.The other part of our research was studying a special kind of shape memory alloy, the Nitinol. It can be used as a material of a new kind of actuator in prosthesis hands. These kinds of actuators are compact and silent, so they make it possible to build a moving prosthesis. To demonstrate what these actuators are capable of, a prosthesis hand was built based on our design. [1, 3, 4, 7, 8]

Journal of Physics: Conference Series **602** (2015) 012024 doi:10.1088/1742-6596/602/1/012024

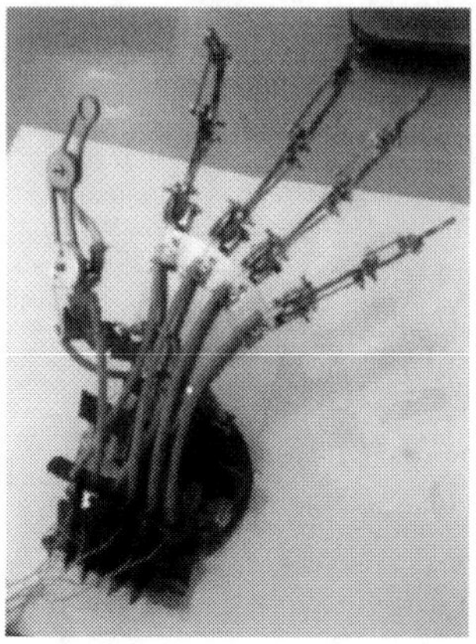

**Figure 1.** The first prototype of the prosthesis hand which is actuated by Nitinol springs

## 2. Method

When a sinusoidal load is applied to a viscoelastic material (e.g.: polimers), a loss angle will exist between the load and the responding deformation. Its value will be between $0^{o}$ and $90^{o}$. The complex modulus (E*) expresses the properties of viscoelastic materials. The real part is the storage (E') and the imaginary part is the loss modulus (E''). The ratio of loss and storage modulus (E"/E') is tand which describes the elasticity and the plasticity of the samples. When the value of tand is high the material acts more viscous, while low tand means a more elastic behaviour. The DMTA monitors the changes of tand as a function of temperature.
The main unit of a DMTA equipment is the furnace, the moving and the fix clamp. During the measurement the temperature changes in the furnace, while stress activation is applied on the sample with the moving clamp. That creates a direct link between the material's chemical structure and mechanical properties. Meanwhile the drawback of DSC is that the change of mechanical properties can not be monitored. [2, 6, 7]

## 3. Measurements results

The specimens were CuAlNi and CuAlNiMnFe rods in martensitic phase. The rods outer diameter were 2 mm and they were cut into 20 mm long pieces. A TA instruments Q800 DMTA and a TA Instruments 2920 Modulated DSC equipments were used in our experiments. Our research focused on the DMTA measurements and the DSC results came from the University of Miskolc, Faculty of Material Science. The test frequency was 5 Hz, the amplitude was 30 μm and the heating rate was 3 °C. The applied temperature range was between 0 and 200 °C. [2.] Figure 2 shows the DMTA measurement results for CuAlNi specimen. There is a local tand peak in the heating cycles at about 80 °C, which we consider to be in the austenitic transition temperature range. Other peaks can be seen in the cooling cycles at about 100 °C, which is considered to indicate the martensitic phase transition. The change in the tendency of tand indicates that these shape memory alloys showed "viscoelastic" behaviour.

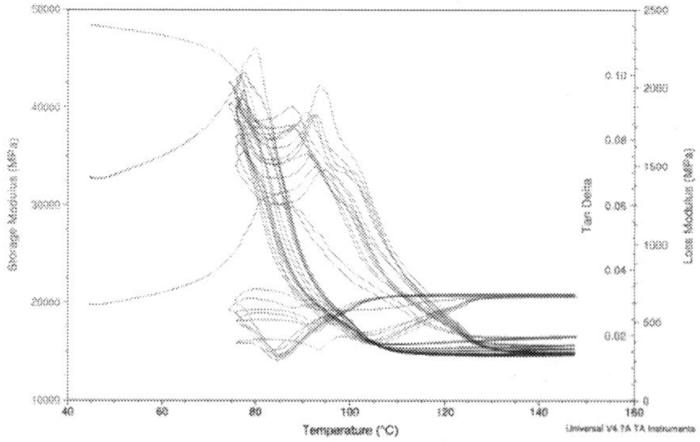


**Figure 2.** Recurrent heating and cooling cycles on CuAlNi specimen indicating the phase transition temperatures of the alloy

Figure 3 shows the DMTA measurement results for CuAlNiMnFe specimen. The aim of this measurement was to compare the results to DSC measurement results which were made in the University of Miskolc, Faculty of Material Science. The tand peak is at about 78 °C in heating cycles and 95 °C in cooling cycles. The DSC curve of the same alloy showed the phase transformation approximately at the same temperatures. These results make DMTA measurement a potential new technique in the investigation of metals.
Discrepancies between the peaks of cooling cycles in case of DMTA and DSC measurements have been found. The reason is, that the specimen used in DMTA is approximately 50 times heavier than those used in DSC, which affects the thermal conductivity. We used liquid nitrogen to cool down the specimen in the DMTA furnace, which might have had an influence on the results, too. [5]

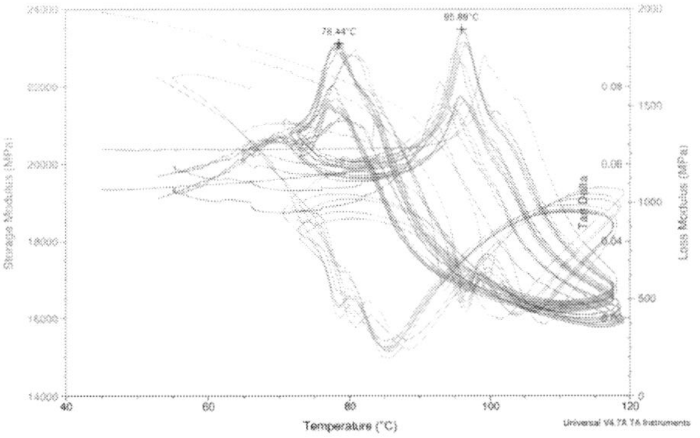


**Figure 3.** DMTA curves of CuAlNiMnFe specimen showing local peak at 95 °C during heating and at 78 °C during cooling

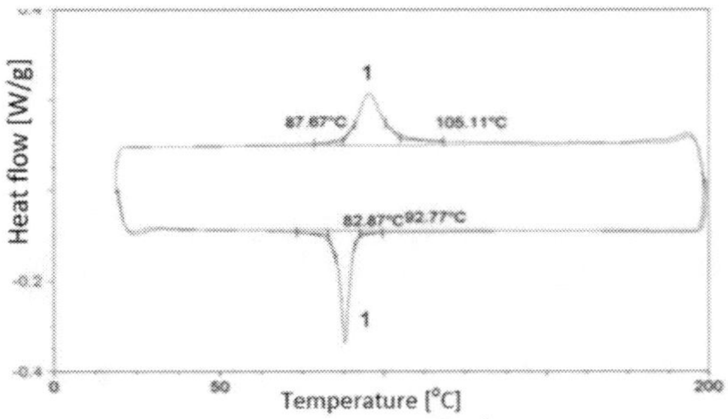


**Figure 4.** DSC curves of CuAlNiMnFe specimen showing local peak between 87 °C and 105 °C during heating and between 82 °C and 92 °C during cooling [2]

## 4. Acknowledgements
The research work presented in this paper/study/etc. was carried out as part of the TÁMOP-4.2.2.A-11/1/KONV-2012-0029 project in the framework of the New Széchenyi Plan. The realization of this project is supported by the European Union, and co-financed by the European Social Fund.

## 5. References
[1.] Ryklina E P, Prokoshkin S D, Khmelevskaya I Y 2008 Proceedings Paper, *Mater. Sci. and Eng. A Structural Materials Properties Microstructure and Processing* **481** 134
[2.] Benke M, Mertinger V, Daroczy L 2009 Proceedings Paper, ESOMAT 2009 - *8th European Symposium on Martensitic Transformations* 06003
[3.] Kiss Attila 1997 *Fiatal Műszakiak Tudományos Ülésszaka* **2** (in Hungarian)

Journal of Physics: Conference Series **602** (2015) 012024 doi:10.1088/1742-6596/602/1/012024

[4.] Huang W 1999 *Journal of Intelligent Material Systems and Structures* **10** (**3**) 221
[5] Tian Q C, Wu J S 2002 *Intermetallics* **10** (**7**) 675
[6] Foreman J 1997 *American Laboratory* **14** (**1**) 21
[7] Varhegyi G, Groma G, Lengyel M 1979 *Thermochimica Acta* **30** (**1-2**) 311
[8] Bundhoo V, Haslam E, Birch B, Park E J 2009 *Robotica* **27** 131

# Modeling strain and dielectric hysteretic type dependences in polycrystalline ferroelectrics by methods of two-level continuum

**A Skaliukh and A Belokon**
Southern Federal University – Mathematic, Mechanic and computer sciences Department– 344090 – Rostov on Don – Russia

a.s.skaliukh@gmail.com

**Abstract.** A one-dimensional model is represented for electrical and mechanical response of polycrystalline ferroelectric material when exposed to it electric field and coaxial him mechanical stress great intensity. The equations describing the residual polarization and deformation are obtained in the form of equations with differentials, which can also be regarded as finite-difference equations. The reversible or induced components par with residual or nonreversible components of polarization and deformation were introduced. The combination of these components allows us to construct the dielectric and deformation hysteresis loops. It is shown that for a specific choice of the model parameters, it is possible to describe the experimental curves not only qualitatively but also quantitatively. Model can be used in finite element analysis to describe the irreversible processes of polarization and depolarization of ferroelectric ceramics.

## 1. Introduction

It is well known that impact of the electric fields and mechanical stresses on ferroelectric polycrystalline sample causes therein the deformation and polarization. If the external forces are of sufficient magnitude, then begins an irreversible process associated with the appearance of an irreversible or residual deformation and polarization, and nonlinear response of the material. The physical basis of the appearance of residual polarization is an integral characteristic of the vectors of spontaneous polarization. In unpolarized state in each crystal the vectors of spontaneous polarization of individual domains can have any direction, differing from each other at fixed angles. For example, in a perovskite type ferroelectric material these angles are 180 and 90 degrees. Each crystallite is oriented also arbitrarily. Therefore, in the unpolarized state in a representative volume the set of all vectors of spontaneous polarization is distributed evenly in all directions. The resulting polarization of the representative volume is equal zero. Application of an electric field leads to the fact that the vectors of spontaneous polarization turn in the direction of the electric field as far as it allows the crystal structure of the ferroelectric. As a result, the total polarization of the representative volume is already different from zero. In this case ordinary say that the process of polarization is begins. Impact of mechanical stress can also lead to the turns of the vectors of spontaneous polarization that in the final result affects on the polarization of a representative volume. Mechanical stress, being a tensor of second rank, can both promote and inhibit of polarization. In the latter case we speak about the depolarization ceramic by mechanical stresses. In this regard, the main objects of the experimental studies are the dependencies between the external loads in the form of electric fields and mechanical stresses on the one hand, and arising internal parameters of the deformation and polarization, on the other hand. In irreversible processes, such relationships are presented in the form of deformation and dielectric hysteresis loops. For circular processes the response of material is described by the big hysteresis loop. For stepwise increasing and decreasing external loads is observed a small hysteresis loop. The process of appearance of the residual polarization by the electric field is called polarization. In homogeneous electric field the ceramics is brought as generally to a state of saturation polarized state, i.e. a state where a further increase of the electric field do not alter the polarization state. Application, for example, a mechanical compressive stresses can depolarize ceramics partially or completely. The appearance of polarization in a representative volume leads to polar direction and change electrical and mechanical properties of the material. As a result, the class of material anisotropy is changed. If the material is not polarized, it is isotropic, both with respect to the electrical and mechanical properties. Polarized ceramics is locally transversely isotropic material with the axis of isotropy coinciding with the direction of the residual polarization. Based on the above we can conclude that the external electric field and mechanical stress can alter polarized and deformed state, i.e. change the class of local anisotropy of the material. Therefore, the modeling of nonlinear behavior irreversible process of polarization and depolarization such materials are of considerable interest for engineering and technological developments. To problems of construction mathematical models of nonlinear irreversible polarization at recently been given more attention. We note some directions of modeling such problems. In [1-4] presented mathematical models to describe the polarization based on the mathematical theory of plasticity. In [5-9] given the averaging methods by which one can construct a hysteresis dependence. Another focus is the method of micro- mechanical model in which a representative volume is considered as a set of unit cells, where the vector of spontaneous polarization can take place their position in a certain direction. The transition from one position to another is given by certain geometric or energetic criteria. These models are presented in [10-17]. Closely adjacent to this the model is described in [18-21]. Here the density distribution of the vectors of spontaneous polarization based on Boltzmann statistics was introduced. Further the energy balance between the energy needed to breaking the locking mechanisms of domain walls and work of external electric fields was formulated. More general list of published works in these directions can be found in [22, 23]. Our model is closest to the last of the models. But it has significant differences concerning the simultaneous action of both electrical and mechanical fields. Besides this we have a completely different view on induced polarization and strain components. This approach allowed us to describe the strain and dielectric small hysteresis loops not only qualitatively but also quantitatively.

## 2. Formulation of the problem and method of solution

### *2.1. Subject of investigation*

We consider the representative volume of polycrystalline ferroelectric material (ceramics), which contains a many crystallites and each of the crystallites contains many domains. The walls of domains are fixed existing inhomogeneities presenting in the material. Without going into the physical nature, we say that there are mechanisms for fixing domain walls. And only the application of intense external loads can destroy these mechanisms, and rotate vectors of spontaneous polarization. It is required to construct a mathematical model that allowed us to determine the distribution of the integral characteristic of vectors of the spontaneous polarization, or in other words it is necessary to find the vector of the polarization and tensor of the strain for representative volume.

### *2.2. Stages of constructing a model*

Assuming that the domain walls are not fixed and domains do not affect each other, we derived the expressions for the residual polarization and deformation from the action of Weiss electric field and mechanical stresses. This case is called an ideal. Let $E_{ef} = E + \alpha P_0$ - electric Weiss field. Quite similarly, we introduce $\sigma_{ef} = \sigma + \beta\varepsilon_0$ - Weiss mechanical field, assuming that the residual deformation of a representative volume affects on mechanical stress on the micro level. The energy of electric dipole in Weiss fields can be represented as $U = -\chi(\varepsilon_s \sigma_{ef} + P_s E_{ef})$. Therefore, according to the findings presented in [22], we write the expression for the limiting polarization and strain in a representative volume in the following form

$$\varepsilon_\infty = \frac{\int e^{-U/kT} \varepsilon_s \, dS}{\int e^{-U/kT} \, dS}, \qquad P_\infty = \frac{\int e^{-U/kT} p_s \, dS}{\int e^{-U/kT} \, dS}, \tag{1}$$

Then we calculate the energy required to rotate the domain, which consists of electrical and mechanical parts

$$\Delta U = \int_\Omega \left( k_1 \frac{dP_0 \, dE_{ef}}{|dE_{ef}|} + k_2 \frac{d\varepsilon_0 \, d\sigma_{ef}}{|d\sigma_{ef}|} \right) d\Omega \tag{2}$$

Next we calculate the work of the electric field and mechanical stresses and for a cyclic process established that part of the work that is dissipated by the hysteresis loss

$$\Delta A = -\int_\Omega \varepsilon_0 \, d\sigma_{ef} \, d\Omega - \int_\Omega P_0 \, dE_{ef} \, d\Omega \tag{3}$$

The image and likeness, we can specify the work needed to turn the domain in the ideal case

$$\Delta A_\infty = -\int_\Omega \varepsilon_\infty \, d\sigma_{ef} \, d\Omega - \int_\Omega P_\infty \, dE_{ef} \, d\Omega \tag{4}$$

Using (1) - (3), we can set the energy balance, according to which the actual loss in the polarization process consist of losses in the ideal case, plus the energy costs required for breaking the locking mechanisms of the domain walls

$$\Delta A = \Delta A_\infty + \Delta U \tag{5}$$

Substituting (2) - (4) in (5) and taking into account that these relations must be satisfied for any representative volume and any increment of the effective field, we obtain two desired equations to determine the residual strain and polarization

$$-\varepsilon_0 = -\varepsilon_\infty + k_2 \frac{d\varepsilon_0}{|d\sigma_{ef}|}, \qquad -P_0 = -P_\infty + k_1 \frac{dP_0}{|dE_{ef}|} \tag{6}$$

It is easy to see that the obtained relations are equations in differentials, which can also be regarded as a finite-difference equation. Included in these relations the ultimate strain and polarization in 1D case according to (1) can be written as

$$\varepsilon_\infty = \varepsilon_S \frac{\int_{-1}^{1} e^{\varphi(t)} (3/2t^2 - 1/2) dt}{\int_{-1}^{1} e^{\varphi(t)} dt}, \quad P_\infty = p_S \frac{\int_{-1}^{1} e^{\varphi(t)} t dt}{\int_{-1}^{1} e^{\varphi(t)} dt}, \quad \varphi(t) = \frac{\varepsilon_S / p_S (\sigma_{ef} (3/2t^2 - 1/2) + E_{ef} t}{a}. \tag{7}$$

*2.3. Method of solution*

Examine, for example, the case when only the mechanical stress affects on a ceramic sample. Then $E_{ef} = 0$. The solution of first equation of (6) is proposed to search for using the method of successive approximations. To do this, we rewrite it in finite differences. Let the increment of the mechanical stresses is $\Delta\sigma$ and organize an iterative process as follows

$$(\Delta\varepsilon_0)_m = \frac{\varepsilon_\infty[(\varepsilon_0)_m] - (\varepsilon_0)_m}{k_2} | \Delta\sigma + \beta(\Delta\varepsilon_0)_{m-1} |, \qquad (\varepsilon_0)_m = (\varepsilon_0)_{m-1} + (\Delta\varepsilon_0)_m \quad . \tag{8}$$

Once $\Delta\varepsilon_0$ will be found, we give a new increment of the mechanical stresses $\Delta\sigma$ and the process repeats as long as the stresses does not reach a predetermined value.

Similarly, consider the case where the electric field is set and mechanical stress is absent. Let the electric field got the increment $\Delta E$. Here also, as in the previous case, it is possible to organize the following iterative process

$$(\Delta P_0)_m = \frac{P_\infty[(P_0)_m] - (P_0)_m}{k_1} | \Delta E + \alpha(\Delta P_0)_{m-1} |, \qquad (P_0)_m = (P_0)_{m-1} + (\Delta P_0)_m \qquad (9)$$

## 3. Results and Discussion

### *3.1. Deformations hysteresis loop*

A constructed mathematical model based on the iterative process (7) allows us to build residual deformations for any kind of uniaxial tensile or compressive stress. However, the experimental detachable deformation hysteresis loops include both residual and elastic deformation. Therefore, in our model we must to introduce the elastic component of deformation. We require that the elastic strain is been proportional to the applied stress i.e. $\Delta\varepsilon_e = c_2\sigma$. Then the full strain will be $\varepsilon = \varepsilon_0 + \varepsilon_e$. As a result, we have 5 parameters: $\varepsilon_S, a_2, \beta, k_2, c_2$, which we can dispose so that the calculated according to our model and experimental hysteresis loops are coincided. The most difficult is the problem of describing small hysteresis loops when mechanical stresses increase and then decrease, and on a new step their value is becoming more and more. As experimental data, we took the results of [ 24 ], where there are small deformations and dielectric hysteresis loops for one type of ceramic. Figure 1 shows the experimental deformations hysteresis loops taken from this work. After that, the nearest auxiliary problem was the task of choosing the model parameters that the calculated curves are coincided with the curves of the experiment. It should be noted that the problem of the choice of parameters is not trivial and represents itself the ill-posed coefficients problem. Fortunately for our operator can say that incorrectness practically does not show itself. This means that in our model manages to select parameters by direct sorting of valid values. I.e. in this case, there is the following trend: small changes of the parameters give little change the hysteresis curves. Following the numerical experiments, we obtained the following values of the model parameters

$\varepsilon_S = 0.01, \beta = 5\cdot10^7 m^2/N, a_2 = 2.5\cdot10^7 N/m^2, k_2 = 2.5\cdot10^8 N/m^2, c_2 = 0.75\cdot10^{-11} m^2/N$.

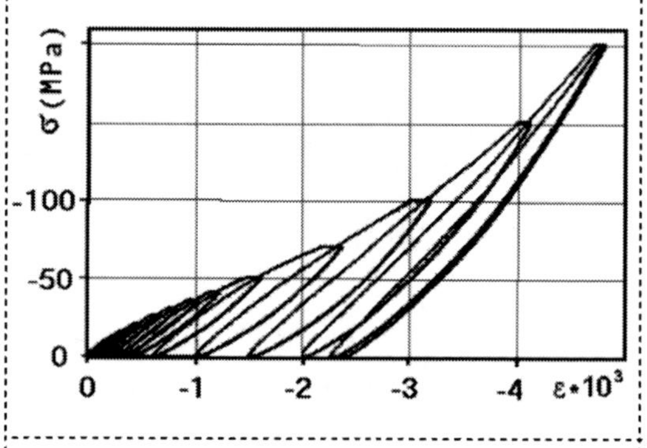


**Figure 1.** Small deformations of the hysteresis loop. Experiment.

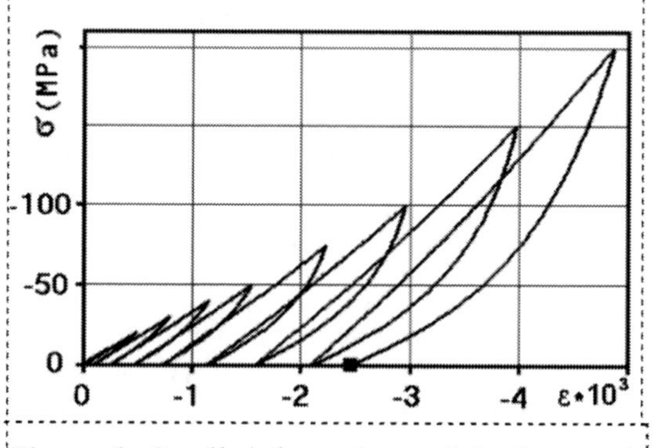


**Figure 2.** Small deformations of the hysteresis loop. Model.

Small deformations hysteresis loop, calculated with help of our model, one can see in Figure 2. Comparing Figures 1 and 2, we can say that these curves coincide not only qualitatively but also quantitatively in some areas.

*3.2. Dielectric hysteresis loop*

Quite similarly we can construct dielectric hysteresis using relations (8). Here also we must introduce a component of the reversible polarization, which, together with the residual, give us a full polarization of the sample. This model also contains five parameters. For compliance of the experimental and calculated curves it is necessary to select these parameters in a certain way. For example, we take as experimental results the results of studies published in [25] which presented in Figure 3. Numerical experiments have shown that if the parameters of the model got values $P_S = 0.43 C/m^2, \alpha = 7.3 \cdot 10^6 Vm/C, a_1 = 9.8 \cdot 10^6 V/m, k_1 = 4.9 \cdot 10^5 V/m, c_1 = 1.4 \cdot 10^{-8} C/Vm$ we can get small dielectric hysteresis loops as shown in Figure 4.

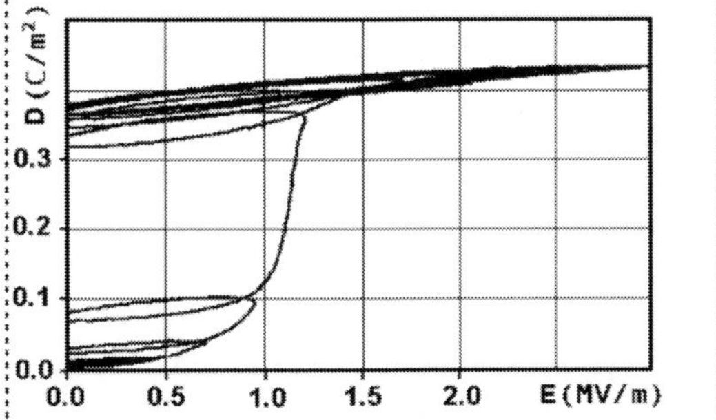


**Figure 1.** Small dielectric of the hysteresis loop. Experiment.

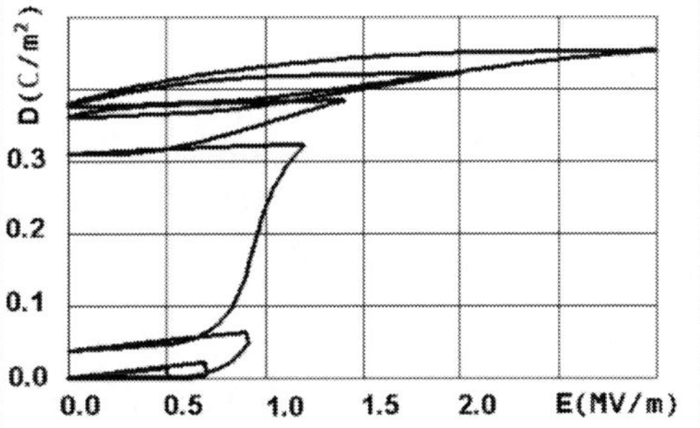


**Figure 2.** Small dielectric of the hysteresis loop. Model.

Comparing Figures 3 and 4, we can say that these curves coincide not only qualitatively but also quantitatively in some areas.

The results show that the model allows us to construct depending strain and dielectric hysteresis only qualitatively. Quantitative coincidence occurs only in certain areas of curves. Most good agreement is obtained for large values of the load. For loads slightly different from zero you can see only qualitative coincidence. In addition, portions of the curve, where the increasing load is replaced by decreasing the experimental curves are rounded but calculated curves have sharp peaks. A next discrepancy is observed in the areas of increasing and decreasing load. For the experimental curves this parts of curves are very close to each other, but for the calculated curves markedly significant discrepancy. Explanation for all of this is may be next fact: when we create a model, we used statistics of Boltzmann distribution, which allow us to construct the density distribution of the domains in the fields of applied loads. In all probability, the exponential distribution is not quite accurately describes the real picture of the behavior of domains in intense fields. This is especially noticeable for loads of small value. In this connection the question of constructing mathematical models to describe the polarization processes remains relevant.

## 4. Conclusion

A one-dimensional model to describe the nonreversible processes of polarization and depolarization of polycrystalline ferroelectric materials was created. Model is based on statistical distribution law in the domain of intense electric and mechanical fields. Using the energy balance between the work of the external loads, actual losses and energy to rotate the domains we built energy relation, from which the equations for the residual strain and polarization in the differentials are obtained. Considering these equations as equations in finite differences we proposed a method of successive approximations to solve them. Introducing in model a reversible strain and polarization components as a quantity proportional to the corresponding loads, we built hysteretic relationship between mechanical stress and strain and between the polarization and electric field. If external loads vary at a circular law of maximum intensity, then we obtain large deformations and dielectric hysteresis loops. If the load increases and decreases incrementally with gradually increasing amplitude, one can obtain the small hysteresis loop. Shown that the model has a set of parameters and we can dispose them of at their discretion. Numerical experiments were performed and the results were compared with experimental data published earlier. It is shown that the optimal choice of parameters can give the coincidence between the small hysteresis curves not only qualitative but also quantitative. The deficiencies of models also were noted and were pointed out on their probable causes. The constructed model can be used in finite element analysis ferroelectric polarization processes.

**Acknowledgements**
The work was supported by the Russian Foundation for Basic Research (grant 12-01-00829-a, grant 13-08-01094-a).

**References**
[1] McMeeking R M. Landis C M 2002 *Int. J. Eng. Sci.* **40** 1553
[2] Huber J E, Fleck N A 2001 *J. Mech. Phys. Solid* **49** 785
[3] Landis C M 2002 *J. Mech. Phys. Solid* **50** 127
[4] Huber J E, Fleck N A, Landis C M, McMeeking R M 1999 *J. Mech. Phys. Solid* **47** 1663
[5] Marutake M J 1956 *Phys. Soc. Japan* **11** 807
[6] Eshelby J D 1957 *Proc. R. Soc. Lond.* A **241** 376
[7] Aleshin V I, Luchaninov A G 2002 *Ferroelectrics* **266** 111
[8] Dunn V L 1995 *J. Appl. Phys.* **78** 1533
[9] Dunn M L, Wienecke H A 1997 *Int. J. Solid. Struct.* **34** 3571
[10] Cheng J, Wang B, Du S 1999 *Acta Mech.* **138** 163
[11] Hwang S C, Lynch C S, McMeeking R M 1995 *Acta Metall. Mater.* **43** 2073
[12] Chen X, Fang D N, Hwang K C 1997 *Acta Mater.* **45** 3181
[13] Huo Y, Jiang Q 1997 *Smart Mater. Struct.* **6** 441
[14] Fotinich Y, Carman G P 2000 J. *Appl. Phys.* **88** 6715
[15] Zhang Z K, Fang D N, Soh A K 2006 *Mech. Mater.* **38** 25
[16] Sun C T, Achuthan A 2001 Domain switching criteria for piezoelectric materials. Proc. SPIE, Smart Structures and Materials
[17] Suo Z, Kuo C M, Barnett D M, Willis J R 1992 *J. Mech. Phys. Solids.* **40** 739
[18] Jiles D C, Atherton D L 1986 *J. Magnet. Magn. Mater.* **61** 48
[19] Smith R C, Hom C L 1999 SPIE Conference on Mathematics and Control in Smart Structures. SPIE: Newport Beach, CA. March 1-4, **3667** 150
[20] Smith R C, Ounaies Z 2000 *J. Intellig. Mater. Syst. Struct.* **11** 62
[21] Hom C L, Shankar N 1998 *IEEE Trans. Ultrason., Ferroelectr. Frequency Control.* **45** (2) 409
[22] Belokon A V, Skaliukh A S 2010 Mathematical modeling of irreversible processes of polarization *FIZMATLIT*, Moscow (Russian addition).
*[23]* Skaliukh A S 2012 in: I A Parinov (Ed.), Piezoelectric materials and devices, *Nova science publishers, Inc. New York*
[24] Salten M, Schneider G A, Knoblauch V, McMeeking R M 2005 *Int. J. Solid. Struct.* **42** 3953

# Investigation of aluminum-steel joint formed by explosion welding

**T Kovacs-Coskun[a,1], B Volgyi[2] and I Sikari-Nagl[3]**
[1,2,3] Obuda University, Banki Donat Faculty of Mechanical and Safety Engineering – Material Science Department – 1081 Budapest – Hungary

kovacs.tunde@bgk.uni-obuda.hu, b.volgyi@freemail.hu, istvan.sikari@gmail.com

**Abstract**. Explosion welding is a solid state welding process that is used for the metallurgical joining of metals. Explosion cladding can be used to join a wide variety of dissimilar or similar metals [1]. This process uses the controlled detonation of explosives to accelerate one or both of the constituent metals into each other in such a manner as to cause the collision to fuse them together [2]. In this study, bonding ability of aluminum and steel with explosion welding was investigated. Experimental studies, microscopy, microhardness, tensile and bend test showed out that, aluminum and steel could be bonded with a good quality of bonding properties with explosion welding.

## 1. Introduction

The explosion welding in our century is a traditional technology even that the mechanism of the joining not well understood. Explosive welding was first recognized as a remote possibility in 1957 in the United States, when it was observed that metal sheets being explosively formed occasionally stuck to metal dies [1]. Bond zone wave formation during explosion cladding is analogous to the formation of vortex streets in fluid flow around an obstacle or in the collision of liquid streams. The fluid flow analogy explains the observed transition from a smooth metal-to-metal bond zone to a wavy bond zone above a critical collision velocity [2].

A metallurgical, high quality bond can be formed between similar metals and between dissimilar metals that are incompatible for fusion or diffusion joining. Examples of these systems are like titanium-steel, tantalum-steel, aluminum-steel, titanium-aluminum and copper-aluminum [3].

The explosion bonded composite like aluminum-steel can be useful in case of the chemical industries, some piece of reactors or vehicles [4].

In this work we wanted test unusual cladded sheets. In case of our experiments the flayer plate was aluminum alloy and the base plate was low carbon contain steel. In case of used adjustment we used empirical formulas to get technology parameters. The joints quality was investigated by microscopy, bend test, special tensile test and microhardness testing.

We observed that the evolved joint profile is different than the conventional cladded joints wave line. It knows that the aluminum alloys makes oxide layer very quickly on the surface. This layer is thin but it's very hard, has a high melting point. During cladding process this layer traverse the conformation of the wave line.

[a] Tunde KOVACS-COSKUN: OBUDA UNIVERSITY, Banki Donat Faculty of Mechanical and Safety Engineering – Material Science Department – 1081 Budapest – Hungary

## 2. Materials and explosive cladding setup

### *2.1. Used metals*

In case of our tests we used steel and aluminum alloy the chemical composition of them is given in Table 1. The base plate material S235JR is a low carbon and alloy contains steel. The flayer plate material AlMgSi1 is an annealed aluminum alloy. The Table 2 shows the basic parameters of the used sheets. The samples size was 50x250 mm.

**Table 1.** Chemical composition of the used materials (at.%)

| DIN; | Si | Fe | Cu | Mn | Mg | Cr | Zn | Ti | |
|---|---|---|---|---|---|---|---|---|---|
| **AlMgSi1** | 0,7-1,3 | ≤0,5 | ≤0,1 | 0,40-1,00 | 0,60-1,20 | ≤0,25 | ≤0,2 | ≤0,1 | |
| | **C** | **Mn** | **Si** | **S** | **P** | **Cu** | **Cr** | **Ni** | **Al** |
| **S235J2** | ≤0,2 | ≤1,4 | ≤0,6 | ≤0,045 | ≤0,045 | ≤0,3 | ≤0,3 | ≤0,4 | ≤0,1 |

**Table 2.** Basic performance of the steel and alumina alloy

| | Sign by DIN | Thickness | Size | HV | Yield stress (MPa) |
|---|---|---|---|---|---|
| Flayer plate | AlMgSi1 (6082) | 3 mm | 50x250 mm | 72 $HV_{0,2}$ | 92 |
| Base plate | S235J2 | 5 mm | 50x270 mm | 120 $HV_{1,2}$ | 235 |

### *2.2. Explosive welding parameters*

The used explosive welding technology was a traditional setup. The base plate and the flayer plate are parallel and the explosive find directly on the surface of the flayer plate without buffer see Figure 1. The explosive force brings the plates surfaces together progressively at collision front [5].

**Table 3.** Setup parameters

| Explosive | PERMON 10T (powder) |
|---|---|
| Volume of the gas | 928 $dm^3/kg$ |
| Detonation rate | 3200 m/s |
| Density | 850 $kg/m^3$ |
| Distance between the plates | 1,5 mm |
| Thickness of the explosive | 30 mm |
| Weight | 319 g |

The velocity of the collision ($v_c$) must be lower than the speed of the sound ($v_s$) (1), that mean it need to use for this technology a low speed explosive. The interfacial pressure at the collision front also must exceed of the materials yield strength to occur a plastic deformation. This is solid state welding under extreme pressure [6,7].

$$\frac{v_c}{v_s} < 1 \tag{1}$$

Before the explosive welding the contacted surfaces were cleaned by mechanical and chemical cleaning to remove the polluted and oxide layer. The results cladded plates show the Figure 2.

In case of the setup parameters optimization it was used some empirical parameters with the density of explosive, base plate and flayer plate (setup see in Figure 1).

The thickness of the explosive powder was optimized on base of practice. It knows that it needs a minimal amount of explosive, that about 0,017 ($g/mm^2$) Permont 10T.

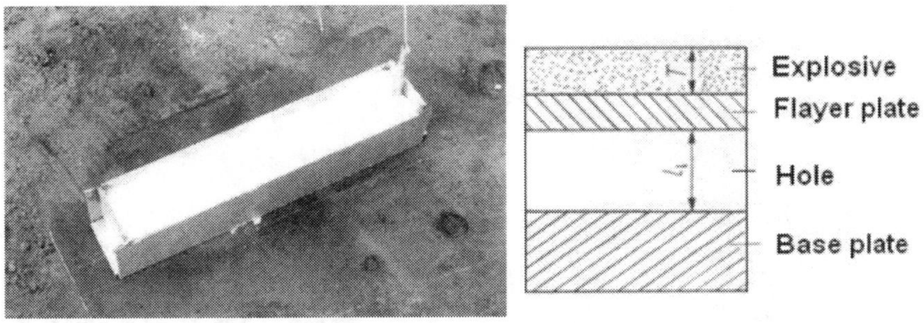


**Figure 1.** Setup of the cladding

The used parameters in case of the setup posed by (2) when $l_b$ is the thickness of the flayer plate and $l_l$ is the distance (hole) between the base and flayer plate. The collision velocity depends on this distance [4].

$$0{,}5 \cdot l_b < l_1 < 1{,}6 \cdot l_b \qquad (2)$$

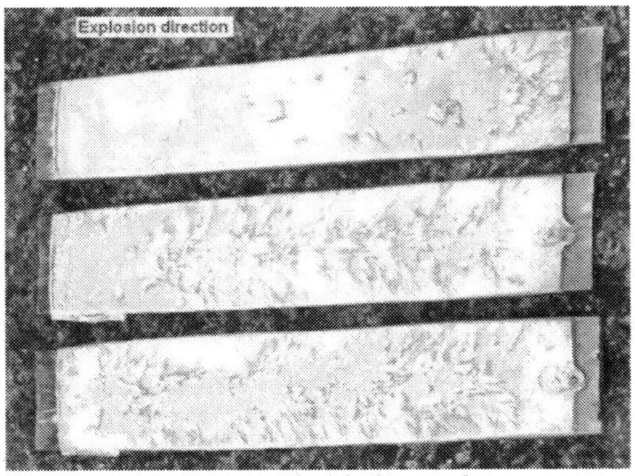


**Figure 2.** Cladded samples

**Figure 3.** Optical micrograph showing the structure of the Al cladded plate cross section

## 3. Experimental procedures and results

### *3.1. Microscopy*

The cross section of the sample shows Figure 3. Usually the materials bond together a wave line form. In case of this procedure the cross section shows also undulating form but it's not really usually wavy like interface (see Figure 3). Visual and light optical inspection of the joint reveals a structure without visible cracks, pores, or separation of the joined components. Though the surface cleaning on the flayer plate surface renewed a thin oxide layer and traversed the wave line formation and supposedly made some intermetallic island.

### *3.2. Microhardness testing*

The used tester was in case of experiments Zwick 3212 microhardness tester by Vickers indenter. Results are shows by Table 4. Cause of the plastic deformation the hardness increased in case of both materials.

**Table 4.** Hardness after explosive welding

| | **Original hardness** | **Hardness after cladding** |
|---|---|---|
| **S235JR** | 120 $HV_{1,2}$ | 178 $HV_{1,2}$ |
| **AlMgSi1** | 72 $HV_{0,2}$ | 85 $HV_{0,2}$ |

*3.3. Bending test*

The bending tests were performed to evaluate the influence of the cracks on the integrity of the joints and the used plates. The tested specimens were deformed 90° bending angle. It was tested both of side of the cladded plate. It can see (Figure 4) that the alumina side of the cladded plate cracked. The hardness of the alumina increased and the plasticity decreased cause of plastic deformation.

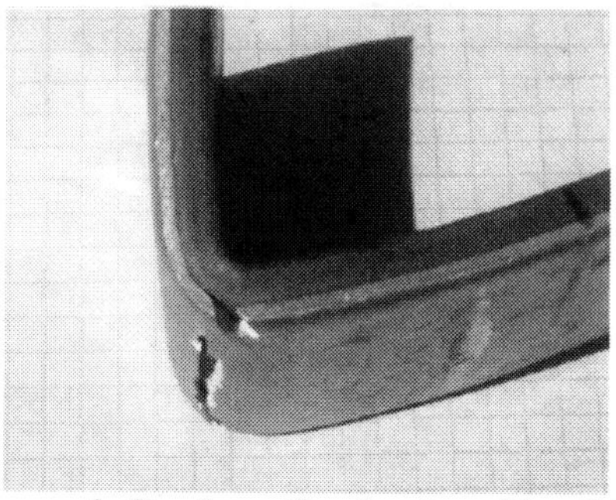

**Figure 4.** Crack on the alumina side of the cladded plate was observed

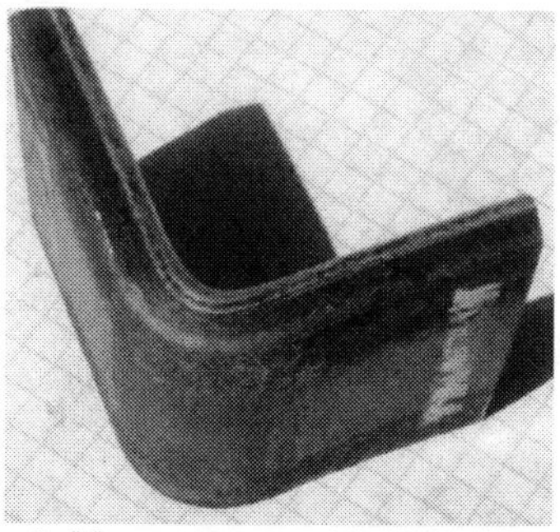

**Figure 5**. No cracking and separation of the components was observed

## 4. Conclusion

Even that it was used a traditional explosion welding technology the results give new information. The low carbon steel platted by alumina is not a very common but very useful composite. In case of explosive cladding the used parameters base some empirical equation what are usually secret. The literature of this process is also poor about the determination of the parameters. To optimize the suitable parameters it was made a many explosive welding tests. The used equations are suitable in case of the technologies parameters calculation.

The results of the tests confirm that the explosion cladding process beside the used parameters is suitable for producing clad plate which will meet or exceed the requirements.

## Acknowledgements

The authors would like to acknowledge all members of this project and especially for Andras Szalay S-Metalltech 98 Kft. and Laszlo Lukacs NKE. The project was realized through the assistance of the European Union, with the co-financing of the European Social Fund (TÁMOP-4.2.1. B -11/2/KMR-2011-0001).

## References

[1] Proceedings of the Select Conference Hove, 18-19 Sept. 1968, *The welding institute 1969* 1
[2] Cowan C R, Bergman O R, and Holtzman A H 1971 *Metallurgical Transactions* **2** 3
[3] Andrew Pocolyko E I 1981 du Pont de Nemours & Compony, Inc. Pompton Lokes, N. J. Metallic Coatings (Explosively Clad) 12
[4] Szalay A, Puskás J 1982 Contact component material made by explosion cladding; *Műszaki Könyvkiadó, Budapest* 172
[5] Kovács-Coskun T, Völgyi B, Sikari-Nágl I 2012 Investigation of explosion cladded plates; Fúrás-robbantástechnika 2012. Balatonkenese, 149-152
[6] Ettaqi S, Langlois L, Bigot R 2008 *Surface & Coatings Technology* **202** 3306
[7] Durgutlu A, Gulenc B, Findik F 2005 Materials and Design **26** 497

# Preparation of functional composite materials based on chemically derived graphene using solution process

**M Kim[1], WJ Hyun[1], SC Mun[1] and OO Park[a,1]**
[1] KAIST, Polymer Nanomaterials Laboratory – Chemical and Biomolecular Engineering Department – Daejeon – Korea

tp1222@kaist.ac.kr, hwj419@kaist.ac.kr, myhighness@kaist.ac.kr, oopark@kaist.ac.kr

**Abstract.** Chemically derived graphenes were assembled into functional composite materials using solution process from stable solvent dispersion. We have developed foldable electronic circuits on paper substrates using vacuum filtration of graphene nanoplates dispersion and a selective transfer process without need for special equipment. The electronic circuits on paper substrates revealed only a small change in conductance under various folding angles and maintained an electronic path after repetitive folding and unfolding. We also prepared flexible, binder-free graphene paper-like materials by addition of graphene oxide as a film stabilizer. This graphene papers showed outstanding electrical conductivity up to 26,000 S/m and high charge capacity as an anode in lithium-ion battery without any post-treatments. For last case, multi-functional thin film structures of graphene nanoplates were fabricated by using layer-by-layer assembly technique, showing optical transparency, electrical conductivity and enhanced gas barrier property.

## 1. Introduction

Graphene, a two-dimensional single atomic sheet of graphite [1,2], is currently attracting much attention for various applications due to its fascinating advantages of high electrical conductivity, optical transparency, gas barrier property, robustness, flexibility and environmental stability [3-6]. The combination of these remarkable electrical and mechanical properties leads us to expect that graphene can be a suitable candidate as a building block for multi-functional composite materials. Recently, even thogh its properties are relatively limited, chemically derived graphene-related materials such as graphene oxides (GOs), reduced graphene oxides (rGOs) and graphene nanoplatelets (GNPs) have drawn massive attention in both academic and industrial areas because of mass production and solution processibility [7,8]. Among them, especially, GNPs are few-layered graphene sheets prepared by chemical exfoliation followed by chemical/thermal reduction of natural graphite, showing excellent electrical and physical properties in itself.

In this paper, three kinds of functional composite materials based on GNPs using solution process are indroduced. Self-assembly from solvent dispersion is an efficient pathway to controlling the organization of nanomaterials. These approaches would suggest potential for development towards the tailored assemblies of graphene-related materials in broad applications.

---

[a] O Ok Park: KAIST, PNML – Chemical and Biomolecular Engineering Department – Daejeon – Korea

## 2. Materials and methods

*2.1. Foldable graphene electronic circuits based on paper substrates*
Commercial GNPs (N002-PDR) prepared by chemical exfoliation and thermal reduction from natural graphite were provided by Angstron Materials. The GNPs were dispersed in ethanol or isopropanol at a concentration of 0.1 mg/ml. Poly(4-styrenesulfonic acid) (PSS, $M_w$ = 75,000, 18 wt% aqueous solution from Sigma-Aldrich) was added into the dispersions with the weight ratio of PSS/graphene, 5/1. Then, the dispersion was treated in an ultrasonic bath for 1h and filtered using a hydrophobic polytetrafluoroethylene (PTFE) membrane filter (pore size: 0.2 μm), resulting in uniform deposition of graphene on the filter. After turning this filter over and placing it on paper substrate, electronic circuits on paper could be obtained by drawing patterns on the back of the filter.

*2.2. Flexible, binder-free graphene paper electrodes for lithium-ion batteries (LIBs)*
GNPs were dispersed in water and stabilized using PSS solution with the weight ratio of PSS/graphene, 5/1. In order to obtain more stable dispersion of GNPs, the dispersion was sonicated for 3hrs followed by addition of GOs in ratio of 3:1 by weight to GNPs in dispersion. Then, the GNP/GO dispersion was filtered using polyvinyldifluoride (PVDF) membrane filter (pore size: 0.45 μm), resulting in free-standing paper-like structure after drying and peeling it off from membrane filter.

*2.3. Multi-functional thin film fabrication by layer-by-layer assembly of graphene*
GNPs and PSS (wt. ratio 1:1) were dispersed in DI water by 30 min sonication. PET films or glass slides were used as substrates for LbL assembly of graphene. After $O_2$ plasma treatment, negatively charged substrates were dipped into poly(diallyldimethylammonium chloride) (PDDA, $M_w$ = 200,000 ~ 350,000) solution, rinsed with DI water and dried with compressed air. Then, PDDA-coated substrates were dipped into poly(vinyl alcohol) (PVA, $M_w$ = 89,000 ~ 98,000) solution and PSS-stabilized GNPs solution, alternatively for n cycles. In each deposition step, rinsing and drying processes were conducted. The assembled film was denoted as $[PVA/PSS\text{-}PDR]_n$. To reduce sheet resistance, as-assembled films were treated by dipping into 95% $H_2SO_4$ for 2 hrs or annealed in a vacuum oven at 160 $^{\circ}C$ for 10 hrs.

## 3. Results and Discussion

*3.1. Foldable graphene electronic circuits based on paper substrates*
Herein, a fabrication process and characterization of foldable graphene circuits based on paper substrates were briefly introduced according to our previous work [9]. Electronic circuits with controlled thickness/width were successfully prepared by vacuum filtration of a dispersion of GNPs and selective transfer printing which is convenient for making the circuit patterns without not only any damage to paper substrate but also pre-patterned masks.

Thickness of the graphene circuits deposited on a paper substrate was easily controlled by varying the volume of the GNPs dispersion for vacuum filtration (Figure 1). As increasing the thickness of the graphene deposition on membrane filter, surface resistance of graphene circuits after transferring to paper substrate was reduced gradually.

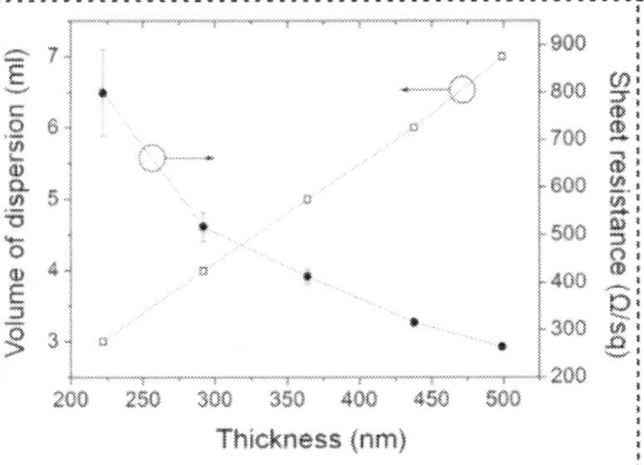


**Figure 1.** Thickness of the graphene circuits controlled by the volume of GNPs dispersion and sheet resistance of graphene circuits with a variance of thickness. Reproduced with permission from ref. 9.

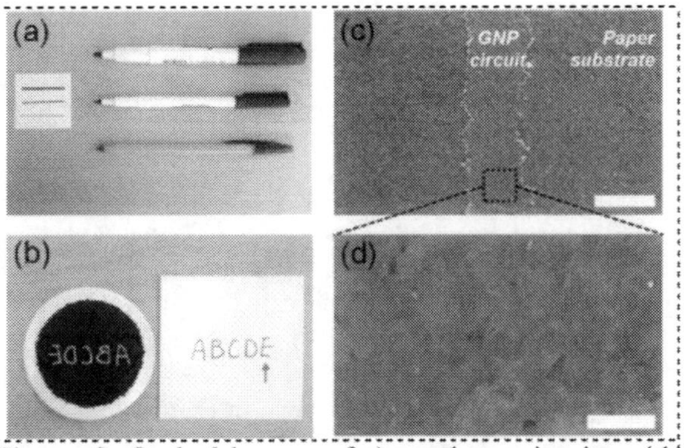


**Figure 2.** Optical images of a) graphene circuits with different widths and b) graphene patterns in the form of alphabet characters. c) Low- and d) high-magnification SEM images of graphene circuits. Scale bars are 500 μm for c and 10 μm for d. Reproduced with permission from ref. 9.

The width of the graphene circuits can be controlled by adjusting the thickness of the tip of the pen. Graphene circuits with different widths of about 2, 1, and 0.5 mm were obtained by employing thick (top) and thin (middle) felt tip pens and a fine (bottom) ball point pen (Figure 2a). Not only straight line patterns but also complicated patterns can be formed on substrates easily. Alphabet characters were written with filtered GNPs on a membrane filter and a pen (Figure 2b). Figure 2c and 2d shows scanning electron microscopy (SEM) images of the graphene line pattern with low- and high-magnification. Owing to good dispersion stability by contribution of amphiphilic surfactant PSS [10], well-organized graphene patterns on paper substrate can be obtained without significant aggregations of GNPs.

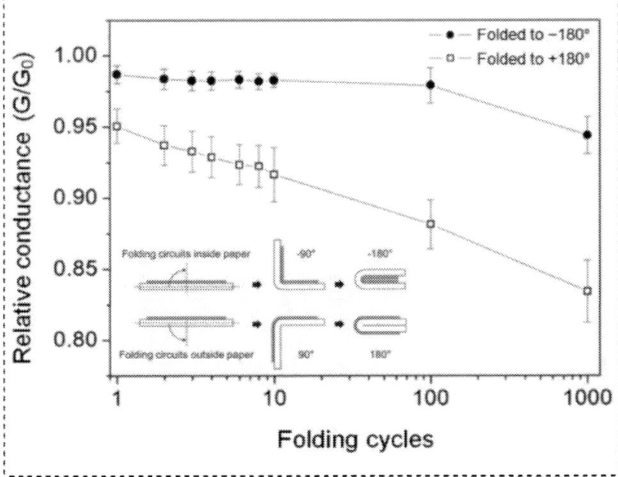


**Figure 3.** Relative conductance of the graphene circuits after repetition of folding and unfolding (inset: schematic diagrams of the folding angles). Reproduced with permission from ref. 9.

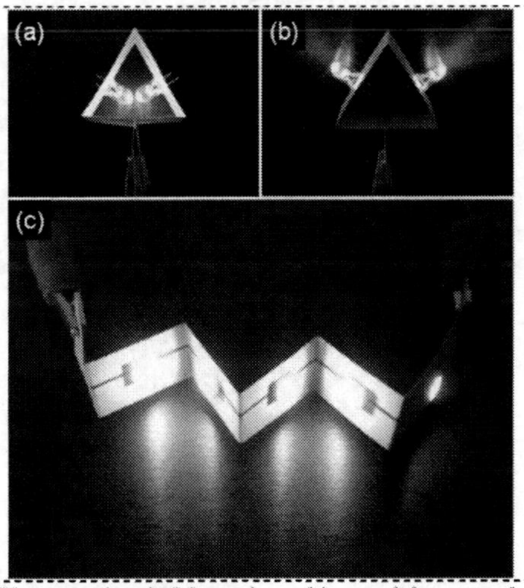


**Figure 4.** a) Negative, b) positive angle folded circuits and c) array of LED chips on a folded paper circuit board. Reproduced with permission from ref. 9.

In order to investigate the folding stability of the graphene circuits on paper substrates, we evaluated the relative conductance of the circuits (the ratio of measured conductance (G) to initial conductance ($G_0$)) with a repetition of folding and unfolding as shown in Figure 3 (negative and positive angles indicate folding of the graphene circuits inside and outside the paper substrate, respectively). The circuits with −180° folding showed higher conductance after folding due to smaller radius of curvature caused by the different deformation behaviours between the graphene circuit and the paper substrate; compression for negative angle folding and stretch for positive angle folding. The graphene circuits maintained about 94% and 83% of conductance compared to the initial conductance after 1000 cycles of −180° and +180° folding, respectively, which is much improved folding stability compared to foldable metallic circuits [11]. The excellent folding stability can be attributed to good adhesion between GNPs and the paper substrate as well as the outstanding mechanical properties of GNPs. This behaviour enables the demonstration of three-dimensional circuit boards including negative and positive angle folding as shown in Figure 4.

### *3.2. Flexible, binder-free graphene paper electrodes for lithium-ion batteries (LIBs)*

In this part, flexible, free-standing graphene paper-like materials were fabricated with GNPs using simple vacuum filtration method. Generally, these graphene-based paper materials can be prepared from GOs or rGOs by chemical reduction as starting materials, resulting in limited electrical properties unless the oxidized domains in each graphene layer are restored by post-treatment such as thermal annealing, additional reduction process [12,13].

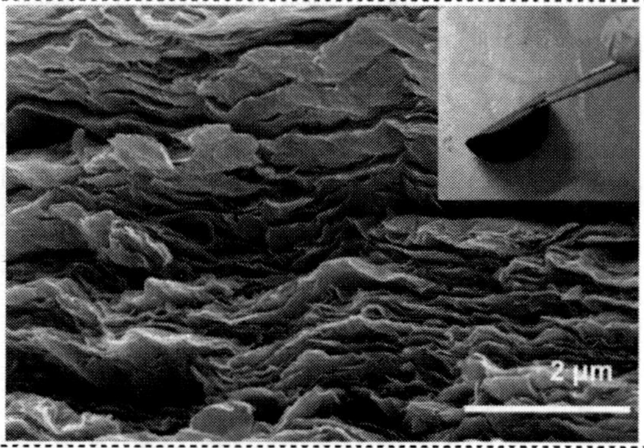


**Figure 5.** SEM image of the prepared graphene paper from film thickness direction (inset: optical image of the as-prepared graphene paper).

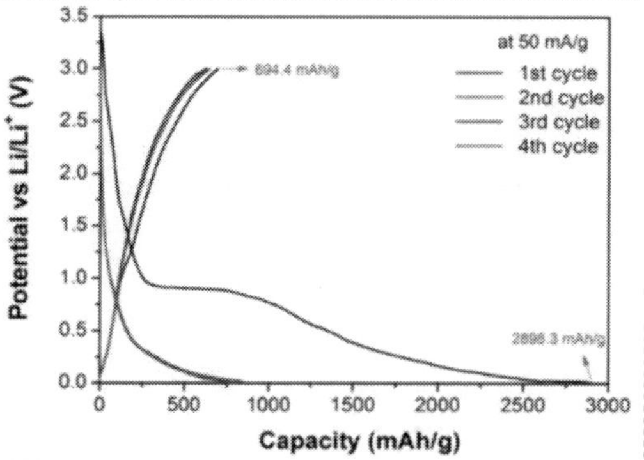


**Figure 6.** Discharge-charge profiles of the prepared graphene paper at a current density of 50 mA/g as a LIB anode.

Since no additional reduction process is required in case of using GNPs as a starting material for graphene paper, the as-prepared graphene papers showed outstanding electrical conductivity up to 26,000 S/m, much higher than that of graphene papers from chemical reduction of GOs [14]. The uneven layered structures (Figure 5) which were formed when the individual GNPs were assembled into layered structure during filtration are favourable to the intercalation/de-intercalation of lithium ions. These features lead to a large reversible capacity as an anode in LIBs about 700 mAh/g as shown in Figure 6, which is much higher than those of reduced graphene papers [15,16].

### *3.3. Multi-functional thin film fabrication by layer-by-layer assembly of graphene*

Layer-by-layer (LbL) assembly is a bottom-up nano-fabrication technique based on alternative adsorption of materials, resulting in fine deposition of molecular multilayers [17,18]. Theoretically, a two-dimensional graphene sheet is a perfect nanoscale barrier for gas molecules, which can be a suitable candidate as a building block for multi-functional composite films. We have successfully fabricated the LbL-assembled graphene thin films with electrical conductivity and enhanced gas barrier property, which is applicable to flexible polymer substrates, using hydrogen-bonding interaction between PVA and PSS-stabilized GNP.

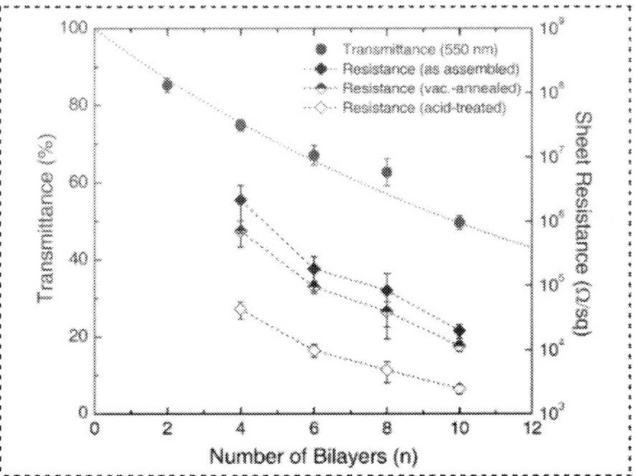


**Figure 7.** Optical transmittance and sheet resistance of LbL-assembled graphene thin films as a function of the number of bilayers.

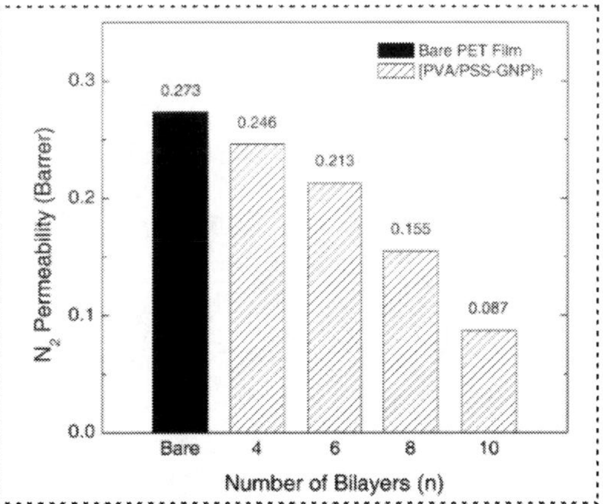


**Figure 8.** Nitrogen gas permeability of LbL-assembled graphene thin films as a function of the number of bilayers.

The optoelectronic properties of LbL-assembled graphene thin films are shown in Figure 7. Both optical transmittance and sheet resistance were reduced by increasing number of bilayers. After post treatments, LbL-assembled graphene thin films present lower sheet resistance compared to as-assembled films. It implies that the microstructure of the films was changed by acid treatment and thermal annealing. Especially, acid treatment could remove insulating polymeric components inside the films, resulting in lower sheet resistance by ~2 orders of magnitude. Figure 8 presents nitrogen permeability of LbL-assembled graphene thin films. As increasing number of bilayers, nitrogen permeability was reduced gradually (~ 68 % reduction for 10 bilayers). It is known that a defect-free monolayer graphene sheet is impermeable to standard gases including helium [4]. Thus, GNPs inside the films act as impermeable walls and make longer diffusion pathway for gas molecules.

## 4. Conclusion

We have successfully demonstrated three kinds of functional composite materials based on GNPs using solution process, which is applicable to flexible electronic devices and multi-functional thin films. The foldable graphene circuits have been simply prepared using vacuum filtration of GNPs dispersion and a selective transfer process without special equipment. The electronic circuits show a small change in conductance under various folding conditions compared to metallic component-based circuits, resulting in realization of foldable circuit boards on paper substrates. Graphene paper-electrodes prepared by simple filtration method using stable dispersion of GNPs with GOs show high electrical conductivity and charge capacity in LIBs as an anode. Graphene thin films based on polymer substrates, assembled with PVA and GNPs, could potentially be both a transparent electrodes and gas barrier for flexible electronics. We believe that these approaches could provide a meaningful method for simple assembling of the dispersion-based graphene sheets to realize highly potential applications in flexible electronics.

**Acknowledgements**
This research was supported by a grant (10037689) from the Fundamental R&D Program for Technology of World Premier Materials (WPM). The authors also would like to acknowledge Prof. C.W. Macosko and Dr. Y.T. Park (Department of Chemical Engineering and Materials Science, University of Minnesota, USA) for their contribution to the work for LbL-assembled graphene films.

**References**
[1] Geim A K and Novoselov K S 2007 *Nat. Mater.* **6** pp 183–191
[2] Kopelevich Y and Esquinazi P 2007 *Adv. Mater.* **19** pp 4559–4563
[3] Bonaccorso F, Sun Z, Hansan T and Ferrari A C 2010 *Nat. Photonics* **4** pp 611–622
[4] Bunch J S et al. 2008 *Nano Lett.* **8** pp 2458–2462
[5] Lee C, Wei X, Kysar J W and Hone J 2008 *Science* **321** pp 385–388
[6] Stankovich S et al. 2006 *Nature* **442** pp 282–286
[7] Lee S H, Lee D H, Lee W J and Kim S O 2011 *Adv. Funct. Mater.* **21** pp 1338–1354
[8] Guo S and Dong S 2011 *Chem. Soc. Rev.* **40** pp 2644–2672
[9] Hyun W J, Park O O and Chin B D 2013 *Adv. Mater.* **25** pp 4729–4734
[10] Stankovich S et al. 2006 *J. Mater. Chem.* **16** pp 155–158
[11] Siegel A C et al. 2010 *Adv. Funct. Mater.* **20** pp 28–35
[12] Chen H et al. 2008 *Adv. Mater.* **20** pp 3557–3561
[13] Huang Z D et al. 2012 *Carbon* **37** pp 4239–4251
[14] Li D et al. 2008 *Nat. Nanotechnol.* **3** pp 101–105
[15] Wang C, Li D, Too C O and Wallace G G 2009 *Chem. Mater.* **21** pp 2604–2606
[16] Abouimrane A, Comton O C, Amine K and Nguyen S T 2010 *J. Phys. Chem. C* **114** pp 12800–12804
[17] Decher G and Schlenoff J B 2012. *Multilayer Thin Films: Sequential Assembly of Nanocomposite Materials.* Wiley-VCH Verlag, Weinheim, Germany
[18] Bertrand P et al. 2000 *Macromol. Rapid Commun.* **21** pp 319–348

# Rheology and porosity effect on the proliferation of pre-osteoblast on zirconia ceramics

**S Kulkov[1,2], L Litvinova[3], E Kalatur[2], A Buyakov[1,2]**
[1]Tomsk State University, 36, Lenin Avenue, Tomsk, 634050, Russia
[2]Institute of Strength Physics and Materials Science of Siberian Branch Russian Academy of Sciences 2/4, pr. Akademicheskii, Tomsk, 634021, Russia
[3]Immanuel Kant Baltic Federal University, Kaliningrad, Russia

e-mail: kulkov@ms.tsc.ru

**Abstract.** It has been studied $ZrO_2(Me_xO_y)$ based porous ceramics, obtained from the powders consisting of hollow spherical particles. It was shown that the structure is represented as a cellular carcass with a bimodal porosity, formed of a large pore close to a spherical shape and the pores that were not filled with the powder particles during the compaction. For such ceramics the increase of pore volume is accompanied by an increase in strain in an elastic area. It was also shown that the porous $ZrO_2$ ceramics had no acute or chronic cytotoxicity. At the same time, ceramics possess the osteoconductive properties: adhesion support, spreading, proliferation and osteogenic differentiation of MSCs.

## 1. Introduction

Porous ceramic materials have been successfully used in various fields, including as heat-insulating building materials, because they are durable, corrosion resistant and they possess stable thermal features [1-6]. Porous ceramics are also a promising material for medical use in the field of traumatology and orthopedics for critical sized bone defect recovery. Thus porous ceramics can act both as osteoplastic material or as 3D scaffold for tissue engineered bone equivalent modeling [7].

Ceramics based on partially stabilized zirconium are the most interesting among the variety of ceramic materials due to their inherent high fracture toughness because of their inherent transformational conversion. It is known that the characteristics are determined by the quality of source ceramic powder (particle shape, particle size distribution), the conditions of compacting and sintering modes and any features that are presented in each phase, and how these phases, including pores, are arranged in relation to each other. The most important factor in the successful application of materials is understanding the features of a structure emerging in them on their behavior under mechanical impact.

The aim of a paper is to examine the pore structure of $ZrO_2(Me_xO_y)$ ceramics and its biocompatibility with multipotent mesenchymal stromal cells (MSCs) *in vitro* assays.

## 2. Materials and methods

The materials for the study were ceramics obtained from powders of $ZrO_2(MgO)$, $ZrO_2(Y_2O_3)$, liquid-phase decomposition of precursors synthesized in high-frequency discharge plasma (the plasma chemistry method). Porous ceramic $ZrO_2(MgO)$, $ZrO_2(Y_2O_3)$ powder was prepared by pressing and subsequent sintering of compacts homologous temperatures ranging from 0.63 to 0.56 during the isothermal holding duration of 1 to 5 hours. The porosity of ceramics $ZrO_2(MgO)$, $ZrO_2(Y_2O_3)$ ranged

Journal of Physics: Conference Series **602** (2015) 012028 doi:10.1088/1742-6596/602/1/012028

from 15 to ≈ 45% and ≈ 30 to 80%, respectively. X-ray studies were carried out on a diffractometer with filtered CuKα radiation. The studies on the ceramic structure were carried out on the scanning electron microscope (SEM) Philips SEM 515.

To assess the biocompatibility of porous ceramics, the adipose-derived MSCs have been used. MSCs were isolated by enzymatic method and cultured in DMEM:F12 supplemented with 2 mM *L*-glutamine and 10% FBS ("Sigma", USA) and incubated with use of integrated continuous live cell imaging and analysis platform Cell IQ® v2 MLF ("CM Technologies"). Third passage MSCs have been used for experiments. Preliminarily MSC culture compliance with minimal criteria was made for phenotype (flow cytometry) and differentiation potential (differentiation assays for adipocytes and osteoblasts) [8]. Prior to seeding over the implants, cell viability in suspension was assessed by Trypan blue staining. [9] To assess the cytotoxicity of the implants and the viability of cultured MSCs over their surfaces, cell combined double staining with fluorescein diacetate (FDA) and propidium iodide (PI) 24h after inoculation and 7 days after culturing has been made [10]. Assessment of cytotoxicity was performed using an inverted fluorescent microscope Axio Observer A1 ("Carl Zeiss", Germany). To further biocompatibility assessment, the MSC osteogenic differentiation assay was performed according to standard protocols [11]. MSCs were cultured in implants or over its surface for 14 days, followed by detection of alkaline phosphatase activity using the BCIP/NBT substrate ("Sigma", USA) [12].

## 3. Results and discussion

### *3.1. Powders*

Figure 1. (a) represents the SEM-picture of $ZrO_2$ powder (3 mol.% $Y_2O_3$), synthesized by the method of plasma chemistry and particle size distribution of the powder size. $ZrO_2$ powders (3 mol.% MgO) and $ZrO_2$(3 mol.% $Y_2O_3$) practically have no difference in morphological structure and they consist of hollow particles of a spherical shape and a large number of units having no regular form. The average particle size of the spherical powders $ZrO_2$(MgO), $ZrO_2$($Y_2O_3$) was 1.8 and 1.5 microns, respectively.

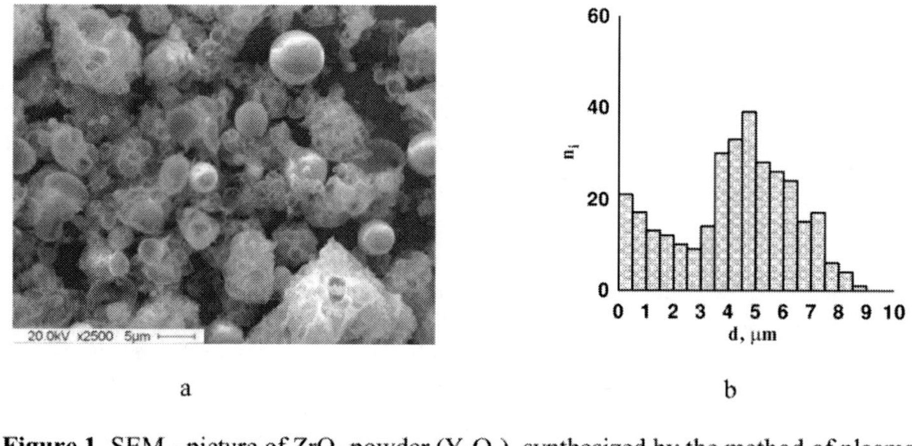


a b

**Figure 1.** SEM - picture of $ZrO_2$ powder ($Y_2O_3$), synthesized by the method of plasma chemistry (a) and particle size distribution of $ZrO_2$ powder ($Y_2O_3$) size (b)

The phase composition of $ZrO_2$ powder ($Y_2O_3$) is presented by tetragonal and monoclinic $ZrO_2$. In the powder $ZrO_2$(MgO) the cubic, tetragonal and monoclinic phases of $ZrO_2$ were present. The rate of tetragonal $ZrO_2$ powder $ZrO_2$($Y_2O_3$) was about 95%, and $ZrO_2$ in the cubic phase $ZrO_2$ powder (MgO)

- 75%. The average size of the coherent scattering regions (SCR) tetragonal $ZrO_2$ in $ZrO_2$ powder ($Y_2O_3$) was 20 nm, and the monoclinic modification - 50 nm. The average size of cubic modification SCR of $ZrO_2$ in $ZrO_2$ powder (MgO) was 20 nm, monoclinic $ZrO_2$ - 30 nm, in the tetragonal phase - 15 nm.

*3.2. Sintered ceramics*

Figure 2 represents the SEM - picture of $ZrO_2$ ceramics structure ($Y_2O_3$) and pore size distribution. $ZrO_2$ ceramics structure (MgO), $ZrO_2(Y_2O_3)$ were represented as a cellular frame. Cells had a nearly spherical shape. The cell size exceeded by many times the thickness of the walls, which was represented as a single $ZrO_2$ layer stacking grain.

Pore size distribution was bimodal. The first maximum pore was formed by interparticle pores that were not filled with powder particles during compaction and the second - with the larger pores close to a spherical shape. From the optical microscopy data we have obtained dependences of interparticles pores and larger spherical pores, it is seen that the increase in the volume of pores in the material from ≈ 30 to 80% was achieved by reducing the sintering temperature of the samples and it was accompanied by an increase in the average size of large pores from 2 to 6 microns. Changing the porosity of the material had practically no influence on the average size of interparticles pores, the average size of which was 0.5 microns. It can be assumed that the presence of large pores close to a spherical shape in the ceramics is due to the presence of hollow spherical particles in source powders, since their average size is commensurate with an average size of presented large pores in the sintered material.

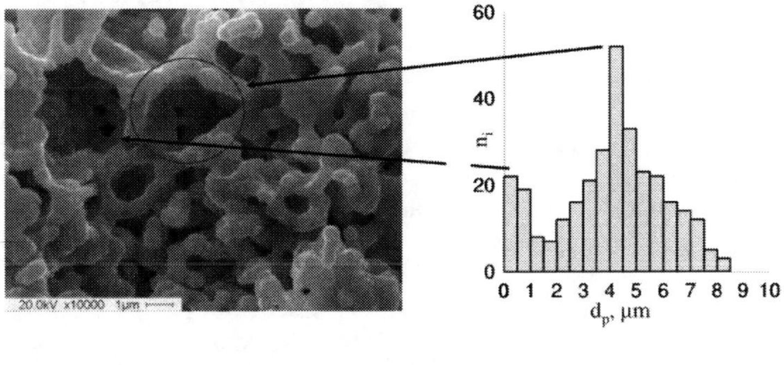


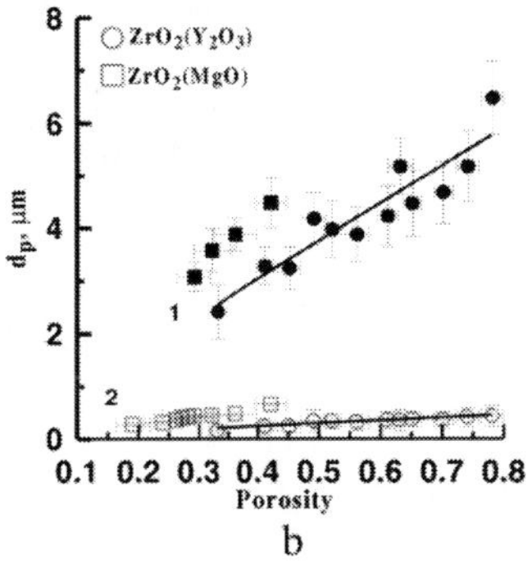


**Figure 2.** SEM-Picture of $ZrO_2(Y_2O_3)$ ceramics structure, the characteristic pore size distribution of $ZrO_2(MgO)$ ceramics with a porosity of ≈ 40% (a) and the dependence of the average pore size vs. porosity of $ZrO_2$ ceramics (b). 1) - the average size of large pores spherical-like shape; 2) - the average size of interparticle pores.

From the data presented in Fig. 2 (b) dependences of interparticles pores and larger spherical pores from porosity in ceramics $ZrO_2(MgO)$ and $ZrO_2(Y_2O_3)$ it is seen that the increase in the volume of pores in the material from ≈ 30 to 80 % was achieved by reducing the sintering temperature of the samples and it was accompanied by an increase in the average size of large pores from 2 to 6 microns.

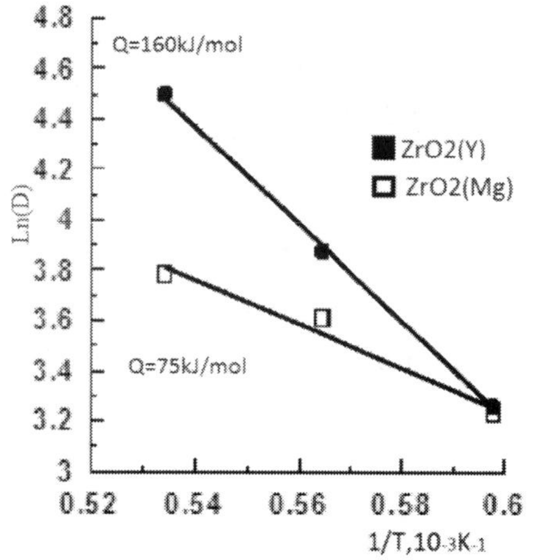


**Figure 3.** The activation energy calculated from dependence of the crystallites size vs sintering temperature of porous ceramics $ZrO_2$

Changing the porosity of the material had practically no influence on the average size of interparticles pores, the average size of which was 0.5 microns. It can be assumed that the presence of large pores close to a spherical shape in the ceramics is due to the presence of hollow spherical particles in source powders, since their average size is commensurate with an average size of presented large pores in the sintered material.

In this study we have determined the activation energy of the crystallites growth for $ZrO_2(Y_2O_3)$ and $ZrO_2(MgO)$ ceramics, Fig.3. It was obtained according to re-plotting of crystallite sizes with increasing sintering temperature, fig.6. Activation energy for growth of crystallites of $ZrO_2(Y_2O_3)$ was 160 kJ/mol, for system $ZrO_2(MgO)$ – 75 kJ/mol, fig. 6, these values are well agree with literature data [13], suggest that the predominant mechanism in the sintering $ZrO_2(MgO)$ is the surface diffusion and for the system $ZrO_2$ $(Y_2O_3)$ is a bulk diffusion.

### *3.3. Biocompatibility: cytotoxicity and osteogenic differentiation of MSCs*

Cultured adipose tissue-derived MSCs used for a preliminary assessment of the biocompatibility of porous ceramic implants had the capacity to differentiate into adipogenic and osteogenic directions and had the following phenotype: CD73+ CD90+ CD105+ and CD34- CD45-. Combined staining of FDA/PI cells cultured on the surface or in the implants showed no cytotoxicity of porous ZrO2 ceramics (Fig. 4). The results of evaluation of the viability of cells in suspension with use of Trypan blue staining before seeding and 24h after culturing on implants by staining with FDA/PI revealed similar values (suspension, Trypan blue – 96.42 ± 1.8% viable cells; implant, FDA/PI – 93.78 ± 2.15%). Microscopic observation showed that MSCs 24h after seeding adhered to the surface of the implant and generated intensive and uniform green FDA stain in the absence of red PI stain, which indicates a high metabolic activity of the cells and the integrity of their membrane. At the same time the cells had different spreading degree due to the rough surface of the implant through its physical structure (the presence of pores and composition of hollow particles of a spherical shape and a large number of units having no regular form). Cell viability after 7 days of culture with porous ZrO2 ceramic implants was 92.56 ± 1.44%, which is comparable to cell viability before seeding and after 24h culturing with implants (difference not statistically significant). Moreover, after 7 days of MSC culturing on the surface of porous $ZrO_2$ ceramics, it should be noted the presence of cell clusters due to their proliferation. Thus, porous $ZrO_2$ ceramic implants do not have the acute and chronic cytotoxicity. Detection of alkaline phosphatase activity with use of BCIP/NBT substrate showed that cultured MSCs on the porous surface of $ZrO_2$ ceramic implant retain their ability for osteogenic differentiation (Fig. 5). Based on the results of the osteogenic differentiation of MSCs we can conclude that porous $ZrO_2$ ceramic implants possess osteoconductive properties.

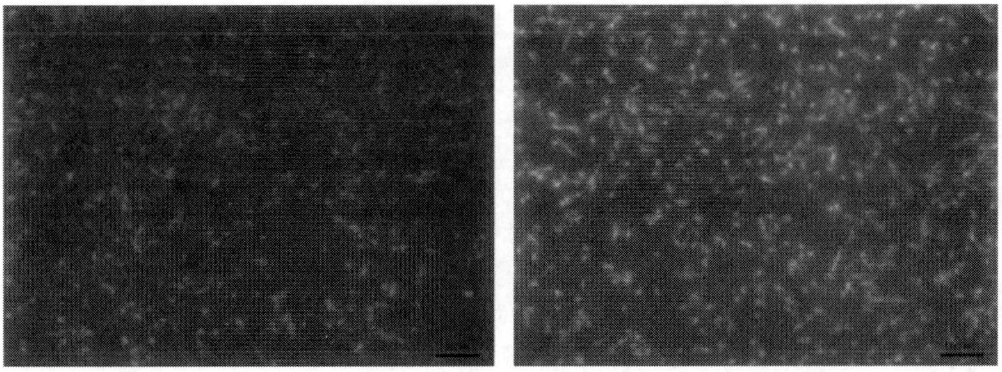

a b

**Figure 4.** Viability assessment of MSCs cultured on the porous surface of the ZrO2 ceramic implants for 24h (a) and for 7 days (b). Combined FDA/PI stain.

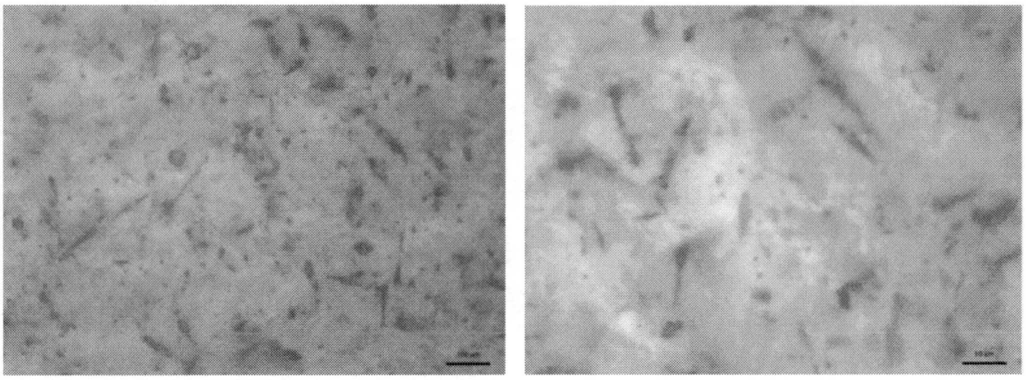

a b

**Figure 5.** Detection of alkaline phosphatase activity after osteogenic differentiation of MSCs cultured on the porous ZrO2 ceramic implants. BCIP/NBT stain.

## 4. Conclusions

It was shown that the structure of $ZrO_2(Me_xO_y)$ ceramics, obtained from the powders consisting of hollow spherical particles with a porosity more 30 % is represented as a cellular carcass with a bimodal porosity, formed of a large pore close to a spherical shape and the pores that were not filled with the powder particles during the compaction.

It was found that in the range of sintering temperatures 0.56-0.63 ceramic $ZrO_2(MgO)$ activation energy of crystallite growth of 75 kJ/mol, which corresponds to the surface diffusion, and for ceramic $ZrO_2(Y_2O_3)$, 160 kJ/mole, which corresponds to the bulk diffusion.

It was also shown that the porous $ZrO_2$ ceramics had no acute or chronic cytotoxicity. At the same time, the porous $ZrO_2$ ceramics possess the osteoconductive properties: adhesion support, spreading, proliferation and osteogenic differentiation of MSCs.

**Acknowledgments**
This work is partial financial supported by Tomsk State University Competitiveness Improvement Program and Grant No 14.607.21.0069-RFMEFI60714X0069 of Ministry of Sciences and Education of RF.

**References**

[1] L A Gomze and L N Gomze 2009 *Epitoanyag* **61** 38
[2] L A Gomze, L N Gomze 2013 *IOP Conf. Ser.: Mater. Sci. Eng.* **47** 012033
[3] S. Kulkov, V. Maslovskii and S. Buyakova 2002 *Technical Physics, The Russian Journal of Applied Physics* **47** (3) 320
[4] S N Kulkov, S P Buyakova, L A Gomze 2014 *Epitoanyag* **66** 2
[5] E S Kalatur, S P Buyakova, S N Kulkov, I Gotman, I Kocserha 2014 *Epitoanyag* **66** 31
[6] E Kalatur, A Kozlova, S Buyakova and S Kulkov 2013 *IOP Conf. Ser.: Mater. Sci. Eng.* **47** 012004
[7] J Will, R Melcher, C Treul, N Travitzky, U Kneser, E Polykandriotis, R Horch and P Greil 2008 J *Mater. Sci. Mater. Med.* **19** 2781
[8] M Dominici, K Le Blanc, I Mueller, I Slaper-Cortenbach, F Marini, D Krause, R Deans, A Keating, D Prockop and E Horwitz 2006 *Cytotherapy* **8** 315
[9] R I Freshney 2010 Culture of animal cells: a manual of basic technique and specialized applications, **6** ed. Wiley-Blackwell, New Jersey
[10] S Johnston, V Nguyen and D Coder 2013 *Curr. Protoc. Cytom.* **64** 9.2.1
[11] D Prockop, D Phinney and B Bunnel 2008 Mesenchymal stem cells: methods and protocols, Humana Press, Totowa, NJ
[12] J D Jang, S J Kim, H M Yoon and D Ch Shin 2011Tissue Eng Regener. Med. **8** 371
[13] P Kofstad 1972 Nonstoichiometry diffusion and electrical conductivity in binary metal oxides / N.Y. Wiley 379 p

Journal of Physics: Conference Series **602** (2015) 012029 doi:10.1088/1742-6596/602/1/012029

# Rheological regional properties of brain tissue studied under cyclic creep/ recovery shear stresses

**F Boudjema[a, 1, 2], M Lounis[1, 2], B Khelidj[2] and N Bessai[1]**
[1] USTO-MB, LAAR Laboratory – Physics Department – 31000 Oran – Algeria
[2] Khemis Miliana University – FIMA Laboratory– 44225 – Khemis Miliana – Algeria

fatiboudj27@yahoo.fr, loumou2000@yahoo.fr, b.khelidj@orange.fr

**Abstract**. The rheological properties of brain tissue were studied by repeated creep-recovery shear tests under static conditions for different regions, Corpus callosum CC, Thalamus Th and Corona radiata CR. Non-linear viscoelastic model was also proposed to characterize the transient/steady states of shear creep results. From the creep-recovery data it was obvious that the brain tissues show high regional anisotropy. However, the both samples exhibit fluid viscoelastic properties in the first shear stress cycle of 100 Pa, while this behaviour evolutes to solid viscoelastic with cyclic effect.

## 1. Introduction

Traumatic brain injury (TBI) is considered as a major worldwide health problem. In developing countries accident rates in general and traumatic brain injury in particular are increasing as traffic increases besides other factors like industrialization, falls and ballistic trauma [1]. The importance of protecting the head from injury during impact and after, is gaining wider recognition, which implies a good knowledge of the mechanical behaviour of the structure to be protected and in particular the cerebral matter. Rheological phenomena may result in shortening of the durability of structural components. Obviously, it is not possible to study in experiments the influence of a violent shock on an *in vivo* human head. The researchers thus had recourse to the technique of modelling by the finite element method (FEM) of the human head to understand and to predict head injuries [2], [3] ... These FE models often contain a detailed geometrical description of the anatomical components but lack accurate descriptions of the mechanical behaviour of the brain tissue. As well, as the bio-fidelity of these objects depends intimately on parameters which characterise the modelled organic material making essential in this way its experimentation. In parallel, *in vitro* studies were realized in small proportion on the man [4-6] and mainly on the animals, mostly porcine brains [6-11], when the availability of mechanical data on *in vivo* tissues remains limited [4], [12-13]. An overview of previous studies on the mechanical properties of brain tissue is giving by Harpko et al. [10]. In these studies brain tissue can be described as viscoelastic material but with different results as in the shear modulus [9]. These considerable differences in results may be caused by variations in material tested, test conditions, testing protocols…

In current study, the inter-regional creep and cycling stresses effects on *in vitro* porcine brain tissue are studied. We will try to determinate a model of a brain shock response based on detailed material

[a] Fatiha BOUDJEMA: USTO-MB, LAAR Laboratory – Physics Department – 31000 Oran – Algeria

data carried out *in vitro* on three different regions testing under shear stress of 100 Pa. In fact, we started the test by short logarithm scale time at the first second to studies the transient state.

## 2. Materials and methods

### *2.1. Samples preparation*
The samples used result from fresh porcine brains; were obtained from a local slaughterhouse. Slices from the corona radiata (CR) and corpus callosum (CC) substances from white matter, thalamus (Th) substance of 2 mm were cut. Porcine brain tissue from six-month-old pig brains was chosen as a substitute for human brains because of availability and the possibility to minimise the *post-mortem* time at testing [6]. To prevent dehydration and to slow down degradation of the tissue, the brains were placed in a physiological solution of Ringer-lactate at 6 °C immediately after acquisition and subsequently were tested within *post mortem* times ranging 5 hours after sacrifice. From the slices, cylindrically shaped samples were obtained from, using a cylindrical sharp tube with a diameter of 20 mm which was applied from the medial to the lateral side.

### *2.2. Creep/recovery test*
Measurements were conducted using a controlled-stress rheometer (Carri-Med CSL100) in parallel plate configuration. The samples are subjected to angular shearing's under constant stress $\sigma_0$ of 100 Pa for 30 s then it was left to recover for 30 s, during which the compliance $J$ (creep/recovery) values versus time are recorded. During the test, samples were covered by a moist chamber to prevent dehydration and tested at (20±0.1) °C. By keeping the same sample to reduce the effect of inter-sample variation, and then tested a second time following the same procedure eight times in order to study the load effects and to characterize the tolerance criteria. Between each cycle, the material was left to recover for up period of 60 s at zero shears. In order to check the sample adhesion, it was glued to the upper and lower plate of the rheometer with cyanoacrylate adhesive to prevent slippage during the shearing of the sample.

## 3. Results and Discussion

### *3.1. Creep/recovery measurements*
In this study, creep-compliance tests were carried out in order to determine the parameters evolution of rheological behaviour model for different brain tissues regions and cyclic effect under static conditions. The viscoelastic properties of the cerebral matter were studied by imposing a constant shear stress in transient and steady state. The porcine brain samples taken from CR, CC and Th regions are tested under constant cyclic shear stresses of 100 Pa. Firstly, the stress was applied for CC sample during *30 s* and the resulting compliance was measured taking logarithmic scale for the first second, then the N same stresses was reapplied each 2 min eight times. A compilation of the shear creep-recovery curves of this sample are presented below in figure 1. Then, from eash first creep phase, inter-regional comparison between three samples: CC and CR sample from white and Th from gray matter is shown in figure 2. The creep evolution of the time deformation for each curve (see the first creep part of CC sample in figure 1 and 2) can be resuming in three quite distinct phases:

- **Instantaneous Elasticity (from O to A)** represented by instantaneous compliance $1/k_0$ with underdamped oscillations. The observed response can be influenced by the application of shear stress at short times [14] or other factors as, it is common for elastic samples to resonate with the applied actuator and transmit transient waves through the sample. These oscillations due to the important shear rate represent typically non-linear viscoelastic behaviour [15] were also observed in the study [16] of living tissue including a stress of 150 Pa; which the use of a nonlinear viscoelastic model showing that the model response did not closely represent the vibration of the experimental result as like Kelvin-Voigt generalized model. So in this work we used this unsteady state to deduce directly the viscoelastic parameters of the brain sample.

- **Primary creep region (from A to B)** this phase presents a time dependent where the shear strain rat decreases with time. This second region can be represented by Kelvin-Voigt body and determinate the characteristic parameters [17].
- **Secondary Creep region (from B to C)** it's a linear phase where the brain sample moves in only viscous flow. Generally the deformation of this phase is irreversible.

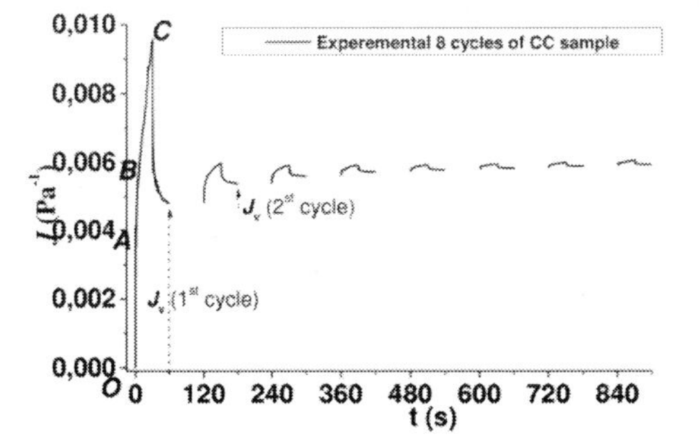


**Figure 1.** Creep-recovery curves of the CC sample response under eight cyclic shear stresses $\sigma_0$ at 100 Pa level in real time.

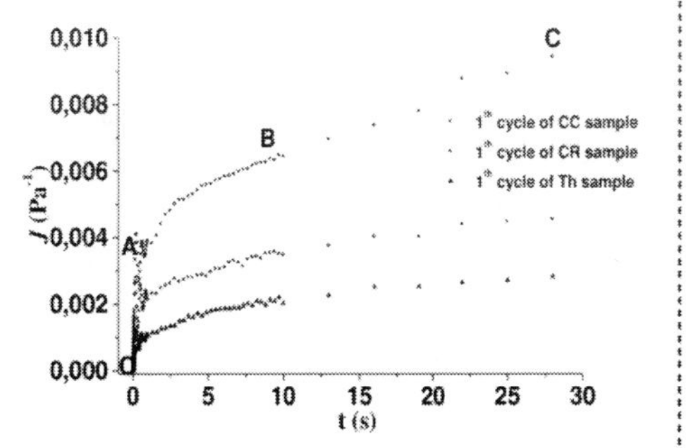


**Figure 2.** Comparative curves of the different region samples: each creep phase consists of instantaneous elastic (OA), retarded elastic (AB) and viscous (BC) deformation.

Once the stress is removed we can see a partial recovery of the deformation indicates the non-linearity of the material behaviour. The sample was destructed and the irrecoverable deformation (represented by $J_v$ in figure 1) can be considered as memory effect. The rest has been dissipated in viscous flow (approximately 50 % of deformation at the first cycle), that confirmed the finding of [17]. The overall pattern of rest white matter curves was similar with total decrease of values: the difference between successive cycles is decayed exponentially, and eventually disappears at the sixth stress for CC sample (as reported in figure 1), due to the straightening of axons fibres. It is distinct from the other central nervous system regions, in that it has longitudinal arrangement of axonal fibers, rather than a more random distribution [18]. It is seen in many others studies that the mechanical properties of the corpus callosum substance is remarkably different from an author to another [7], [10-11], [17]. These differences can be due to many experimental conditions of the tests as the type of the sample cut tested region which presents a high structural anisotropy. The CC substance presented high strong force and become more elastic with cyclic number could be a consequence of the orientation of the axonal fibers. Inter-regional comparison between the creep curves of both samples shearing at 100 Pa, as showing in figure 2 indicated a few difference between CC and CR from white matter and the Th from gray matter properties in shear. These curves showed that the Th region is stiffer than CR region than CC region in shear at this stress level, which coincides with previous comparisons regional studies [11], [19].

*3.2. Creep test mathematical model*

In order to studies the evolution of the rheological behaviour of brain tissue under constant stress; we are interested first to model its response in transient then in steady state. Typically at short time we can observe damped oscillations, which after the deformation take a certain asymptotic variation. The most common mechanical equivalents of rheological behavior are the Maxwell and Kelvin models. Burgers model [20] or $i^{th}$ order generalized Kelvin-Voigt which is a Kelvin and a Maxwell model placed in series can be used to analyze the creep data in steady-state. This model ensures a good description of the polymeric and the biological general observed behaviours [5], [21-22]. But this

Journal of Physics: Conference Series **602** (2015) 012029 doi:10.1088/1742-6596/602/1/012029

standard model presents insufficiencies when we want to characterise the details of the transient which are not correctly fitted. For this reason we associate to the first term of the Kelvin Voigt model an underdamped function. The creep curves of cerebral tissue present a viscoelastic liquid behaviour for a value of the imposed constant shear stress $\sigma_0$ of 100 Pa, which can model by:

$$J(t) = \frac{1}{k_0}(1 - \frac{1}{\sqrt{1-\xi^2}} e^{-\omega_0 \xi . t} \cos(\omega_0 \sqrt{(1-\xi^2)} t - \varphi)) + \frac{1}{k_1}\left(1 - e^{-k_1 t / \eta_1}\right) + \frac{1}{k_2}\left(1 - e^{-k_2 t / \eta_2}\right) + t / \eta_\infty \quad (1)$$

Where $1/k_0$, $1/k_i$ are the instantaneous and the retarded creep compliances. $\eta_i$ , $\eta_\infty$ are the dynamic and the Newtonian viscosities. $\xi$, $\omega_0$, and $\varphi$ are respectevely : the damping factor, the natural frequency and the oscillator phase.

The theoretical model *J(t)* proposed is perfectly adapted to represent the experimental creep curves for all creep stages of each sample, as seen in the figure 3 for Th sample sheared under a constant stress of 100 Pa during 30 s. By adapting for each cycle the parameters values evolution are studied with more detail in other paper submitted to publish.

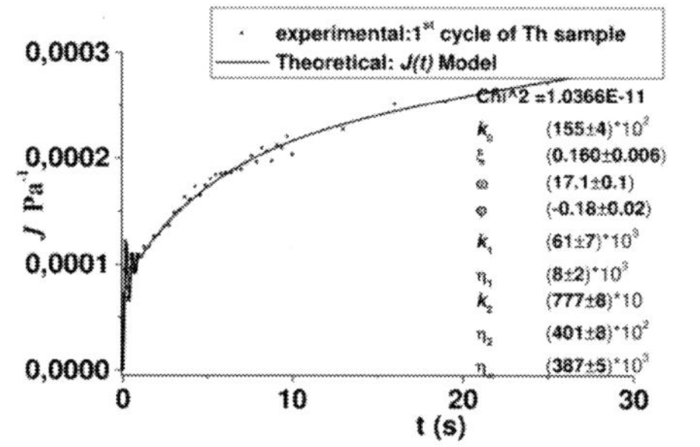


**Figure 3.** Creep model curve of the first cycle experiments results of Th sample sheared under a constant stress of 100 Pa during 30 s

Creep compliance value is mainly associated with softness. The increase of the stress level risk damaging the cerebral matter, over a certain stress value. The degradations result in the loss of elasticity or dissipated flow under 100 Pa stress level as shows it the first incomplete recovery curve (see figure 1). By applying N steps of a same constant shear stress for 30 s on the brain sample, where each step change the internal structure of the brain tissue and transform the mechanical proprieties.

## 4. Conclusion

The present paper outlines around a mechanical characterization of three different regions in the brain from white and gray matter, which are the corpus callosum, corona radiata and thalamus. These regions present a distinct structures and mechanical properties. The results obtained from the creep tests in shearing were performed to determine the evolution of viscoelastic behaviour under static conditions. A constant shear stress of 100 Pa was applied on the CC, CR and Th samples for 30 seconds and the resulting creep compliance *J* was measured. This creep tests proved that the three substances have reversible part due to their elasticity besides an irreversible part due to the viscous properties with differences between their properties in creep. From those results we can showed that the Th region is stiffer than CR region than CC region in shear at this stress level. The stiffness effect of CC samples which present a high structural anisotropy, can explain by crosslinked of the brain fibers with cyclic shearing and the significant liquid loss part, which give a rigid matter. The elasticity parameters in transient, as in steady state can be determined with novel theoretical compliance model. Finally we can easily characters the behaviour of the brain matter under shear stress with a classical creep/recovery tests. The repeated solicitation of the shear stress reveals a memory phenomenon which transforms the behaviour of the brain tissue.

**Acknowledgements**
The authors would like to acknowledge all members of Laboratory of Fluids and Solids Mechanical Institute, Strasbourg University, France for their financial support.

**References**

[1] Langlois J A, Rutland-Brown W and Wald M M 2006 *J. Head. Traum. Rehabil.* **21** pp 375–378
[2] El Sayed T, Mota A, Fraternali F and Ortiz M 2008 *Comput. Methods Appl. Mech. Engrg.* **197** pp 4692–01
[3] Taylor P A and Ford C C 2009 *J. Biomech. Eng.* **131** pp 061007–10
[4] Galford J E and McElhaney J H A 1970 *J. Biomech.* **3** pp 211–221
[5] Shuck L Z and Advani S H J 1972. *J. Basic Eng.* **94** pp 905–911
[6] Nicolle S, Lounis M and Willinger R 2004. *Stapp Car Crash J.*, **48** p 239
[7] Nicolle S et al. 2005 *Biorheology* 42 pp 209–223
[8] Bilston L E, Liu Z Z and Nhan P T 1997 *Biorheology* **34** pp 377–385
[9] Hrapko M et al. 2006 *Biorheology* **43** pp 623–636
[10] Hrapko M et al. 2008 *J. Biomech. Eng–Trans.* ASME **130** pp 031003–10
[11] Prevost TP et al. 2011 *Acta Biomater.* **7** pp 83–95
[12] Gefen A and Margulies S S 2004 *J. Biomech.* 37 pp 1339–52
[13] Prevost T P et al. 2011 *Acta Biomater.* 7 pp 4090–4101
[14] Goodwin J W and Hughes R W 2008. *Rheology for Chemists An Introduction.* 2nd Edition. Published by The Royal Society of Chemistry, Thomas Graham House, Science Park, Milton Road, Cambridge CB4 0WF, UK p 114
[15] Leblanc J L, 1996. *Rheology of elastomer: Engineering of their implementation.* Ed Artel, Namur, Belge.
[16] Kobayashi, Y et al. 2012. *J. Biomech. Sc. Eng.* **7** pp 177–187
[17] Lounis M, Nicolle S and Willinger R 2003 In: *16th French cong. Mech.* (CFM – Nice).
[18] Prange M T and Margulies S S 2002 *J. Biomech. Eng.* **124** pp 244–252
[19] Arbogast K B and Margulies S S 1998 *J. Biomech.* **31** pp 801–807
[20] Burgers J M 1935. Mechanical considerations model systems phenomenological theories of relaxation and viscosity. *In First report on viscosity and plasticity.* NewYork: Nordemann, Publishing Company
[21] Davis S S, 1973 *Bull. Physiopath. Respir.* **9** pp 47–90
[22] Kim G-W et al. 2008 *Food Sc. Tech. Res.* **14** pp 525–532

Journal of Physics: Conference Series **602** (2015) 012030 doi:10.1088/1742-6596/602/1/012030

# Rheological Characterization of Oil Cement Suspensions

**MELLAK Abderrahmane**[1] and **AITOUCHE Moh-Amokrane**[2]

[1]Prof. Dr in Faculty of Hydrocarbons and Chemistry, Boumerdes University, Algeria
Physics Laboratory of Petroleum Engineering. Department Mining and Oil Deposits.
[2]Dr. Earth Physics Laboratory Geophysics Department
Faculty of Hydrocarbons and Chemistry. University of Boumerdes - 35000 – Algeria

Emails: mellakbder@yahoo.fr

**Abstract:** This study is a contribution to the study of the rheological behavior of cement suspensions. An oil well is drilled, cased, cemented and set completion. The well drilling is done in several phases then at various diameters to isolate the following problems like land fragile subsidence and poorly consolidated aquifer formations, loss of the movement in the porous and permeable formations. Therefore, it would go down a casing and cementing to work safely. The materials studied were chosen to satisfy the requirements and the problems encountered in real applications in the oil field (casing cementing wells). So it was used an oil hydraulic binder "G". This systematic study of rheological properties of cement Class "G" standardized API (American Petroleum Institute) deal with a formulation which is compatible with the surrounding environment taking account an optimal efficiency.

## 1. INTRODUCTION.

Cementing an oil well requires the establishment of slurry cement suitable, in the annulus between the borehole and the casing in place to provide a seal between various incompatible layers and between the casing and the geological formations. The cement slurry in water is as solid-liquid two-phases. It constitutes a system having a structure more or less complex depending on its composition parameters. The improved knowledge of these materials, in the physico-chemical and rheological areas is currently the subject of much research works in terms of fundamental subject research and in the field of applications [1, 2]. In order to provide a barrier between aquifer sand salt-bearing in oil wells and sometimes in deep water drills. The choice of grout must be compatible with the conditions inherent in the site. Therefore the knowledge and control of rheological properties are imperative. A physico-chemical study of these materials has been carried out previously [3] with a high use of hydraulic binder "G" which exhibits acquired properties needed for cementing oil. This work focuses on the rheological behavior of different grout in the liquid state.

## 2. MATERIALS USED

### *2.1 The hydraulic binder*

The oil cement class "G", normalized by the API (American Petroleum Institute) was used. Its chemical composition is shown in Table 1. The hydraulic binder "G" is expected to reach great depths under compatibility after adding with various materials. It can be used with accelerators or retarders to cover a wide range of pressures and temperatures. This cement resists to sulfates [4].

Table 1: Chemical composition of cement "G"(in %).

| $SiO_2$ | $Al_2O_3$ | $Fe_2O_3$ | CaO | MgO | $SO_3$ |
|---|---|---|---|---|---|
| 21,30 | 3,60 | 5,40 | 64,20 | 0,90 | 2,20 |

Table 2 shows the average characteristics of the hydraulic binder "G".

Table 2: Characteristics of Cement"G".

| **Specifications** | **Cement class « G »** |
|---|---|
| Particule size (μm) | 1 – 80 |
| Average diameter (μm) | 14 |
| Specific surface ($m^2.g^{-1}$) | 0,3 |
| Real density (μm) | 3,18 |

*2.2 Additions*
The mechanical properties and the rheological behavior of the cement slurry can be modified by adding of various components and other additives. This includes:

*1.2.1. Bentonite*
Recall that the bentonite is a clay rock that has the ability to increase expand its volumes from 10 to 20 times by absorbing about 6 times its own mass of water causing swelling [5, 6]. Clay bentonite is used to reduce cement slurries to avoid fracturing of the low resistance formations.

*1.2.2. Product Weighting*
Baryte (BaSO4) is the primary product used as weighting, with 4.20 as approximate average specific gravity. It is a low cost product. This inert product, also used in drilling mud, shall act through the mass effect to increase the density of cement slurries.

*2.3 Mixing water*
The mixing water used is a water having specific gravity 1.00. In the case of a "hard" water containing a lot of salt, magnesium and calcium, such as the oil fields of Hassi Messaoud (Algerian Sahara), it should be to precipitate the salts of calcium and magnesium by adding 1 to 2 kg NaOH per $m^3$ of water to increase the performance of cement grout . Mixing the cement class "G" with freshwater in the proportions of 44% provides cement slurry with specific gravity of 1.85 [7].

*2.4 Thinning*
The main aim of the thinner is to reduce the shear threshold and the viscosity of the cement paste effect. It works by de flocculating cement grains. This acid organic product is used in the proportion of 1% according to the weight of the dry cement.

*2.5. Anti-foam*
An anti-foam agent is added to prevent excessive foaming (micro bubbles) produced when mixing cement, adversely affecting the operation of the mud pumps. An antifoaming agent is incorporated in the proportion of 0.6% relatively to the weight of the cement and to the cement slurry.

*2.6. Usual used cement*
The cement injected into the formation depends essentially on the density to be compatible with the resistance of the injection formation. According to the API [7], there are currently used cement, slag light cement, (for to not fracture fragile formations) or heavy cement. This gives the following specific gravities summarized in the table 3

Table 3: Type of cement slurries

| Type | Specific Gravity | Addition |
|---|---|---|
| Conventional cement | 1,80 – 1,92 | Cement class « G » |
| Light cement | 1,40 – 1,80 | Cement class « G »+ Bentonite |
| Heavy cement | 1,92 – 2,10 | Cement class « G »+ Baryte |

## 3. MATERIALS AND EXPERIMENTAL PROCEDURE

*3.1 Equipment used*

For the previous rheological studies, rotative rheometer is currently used with parallel plate geometry (Fig. 1) for the characterization of plastic fluids [8, 9]. The measurement geometry is easy to build and to align .It also has the advantage of having an adjustable sensitivity and it requires a very small volume of samples to be studied.

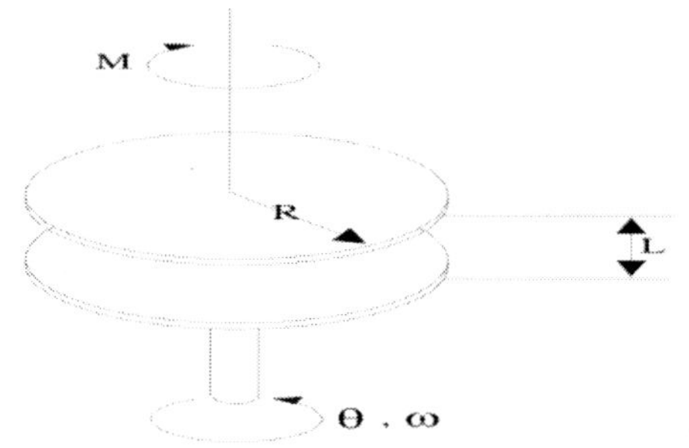


**Figure 1.** Schematic representation of the measurement geometry..

Table 4 shows the measurement geometry used and the main conditions under which these tests were conducted. All samples were studied in a geometry measurement plan/plan protected by an anti-evaporation chamber covering the experimental device to prevent loss of heat in the system.

Table 4: Measurement conditions of rheological parameters.

| Type of device | Carri Med CSL100 |
|---|---|
| Value of the constraint | 100 N/m² |
| Geometry employed | Plan/Plan |
| Gap | 1200 μm |
| Diameter | 4 cm |
| Gradient deformation speed | 0 à 750 $s^{-1}$ |
| Ttemperature | 20° C *(±1°C)* |

*3.2 Procedure:*

The following procedure was applied for the preparation of all the case of cement slurry. Slow introduction of cement in the water, after mixing during two (2) minutes (by the use of a mechanical paddle and simultaneous introduction of other components (plasticizer, anti-foam agent, bentonite clay, barite according the slag to study). All the rheometry tests were performed under the same experimental conditions. For an admissible reliability of the measures, it is important that the fluid keeps its consistency throughout the duration of the test. In addition, the system must remain isothermal (changing the temperature may modify the behavior law), so a closed chamber saturated by water vapor covering the experimental device was used to prevent any evaporation. Temperature (20

°C±1 ° C) is controlled by Peltier effect involved in the system. These materials possess a strongly time-dependent behavior. It is important to obtain a state stationary. This protocol is used to measure the threshold for each grout, defined as the first value of the stress corresponding to a non-zero shear rate.

### *3.3. Rheological measures*

The rheological behavior of the conventional grout was been widely studied then it is relatively well known [10-13]. The problem is fairly not the same with special cement grout. These materials are poorly studied because of their very complex structure and difficulties in their implementation (rapid change of state, rapid transition from slurry to a solid material).

#### *3.3.1 Measures flow*

The rheological characterization according to the density of the slag cement "G" was performed for class three grouts. They are the light cement (L1) with a density equals to 1.50, L2 as conventional slag cement with a density equals to 1.90 and heavy slag cement (L3) for a density equivalent to 2.00. The flow curves obtained are shown in Figures 3, 4 and 5. The rheological characterization of slag cement reduced L1 is characterized by the following flow chart (Fig. 2):

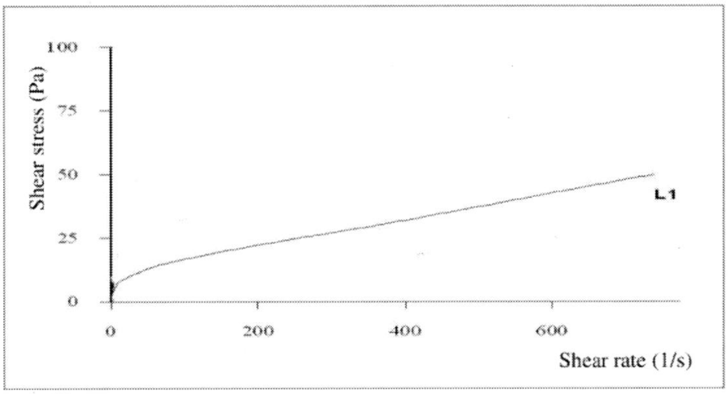


**Figure 2.** Rheogram slag cement case L1 (density1.50).

The rheological behavior of slag cement L1, whose rheological equation is: $\tau = 0.512 + 0.284\ \gamma^{0,72}$ can be described fully by the Herschel-Bulkley model [14] ($\tau = \tau_0 + K\,\gamma^n$), for a shearing speed from 0 to 700 $s^{-1}$.

Where:

τ: is the shear stress (Pa);
τo: the flow threshold (Pa);
γ: the shear rate ($s^{-1}$);
n: behavior index ;
k: the consistency index(Pa.sn).

The rheological characterization of cement slurry currently used l L2, for a density equals to 1.90 is illustrated by the following flow chart (Fig. 3).

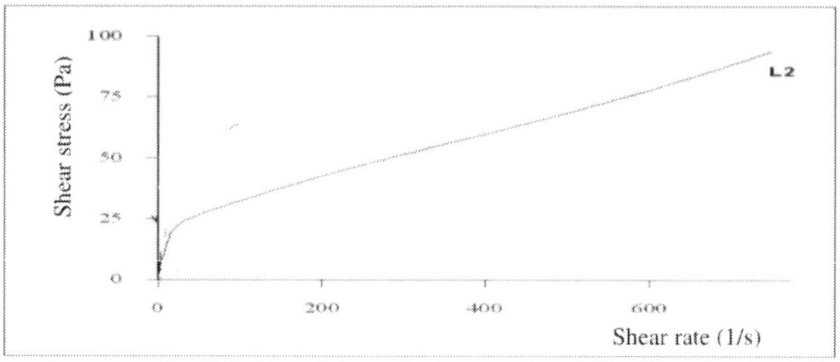


**Figure 3.** Conventional cement slurry rheogram L2 (density =1.90)

Its rheological equation is $\tau = 0.265 + 2.227\ \gamma^{0.47}$ and responds to the Herschel-Bulkley's model in the shear range 0-240 $s^{-1}$ and greater than 240 $s^{-1}$ with the rheological equation $\tau = 12.27 + 0.082\gamma$. It follows the Bingham's model ($\tau = \tau_0 + \mu_{pl}\gamma$) [15], here $\mu_{pl}$ defines the plastic viscosity (Pas).

The rheological characterization of the heavy slag cement L3 is characterized by the following flow chart (Fig. 4).

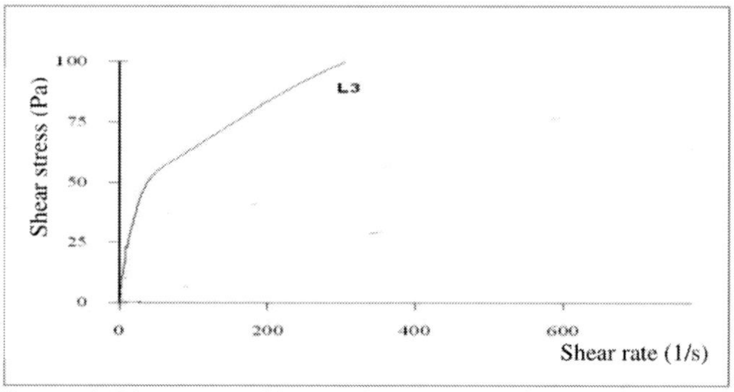


**Figure 4.** Rheogram of heavy slag cement L3 (density =2.00).

In the range of shearing speed greater than 35 $s^{-1}$, the rheological equation $\tau = 24.32 + 0.14\gamma$ verifies the Bingham's model ($\tau = \tau_0 + \mu_{pl}\gamma$).

*3.3.2. Discussion*

The analysis of these curves shows that the rheological behavior of the studied materials can be represented adequately by either the Bingham's model ($\tau = \tau_0 + \mu_{pl}\gamma$), or by the Herschel-Bulkley's model ($\tau = \tau_0 + K^n$).There is an increase in viscosity with the increasing density. Cements L2 and L3 have a rheological behavior of Bingham type from a certain shear rate. In the range of low shear speed, the grout L1 and L2 satisfy the Herschel-Bulkley's model while L3 have a grout rheological behavior undefined by the conventional models. In all cases, we can see that the behavior of the grout as a yield stress (first stress value associated with the existence of a non-zero shear rate) and is shear-thinning ($n < 1$). Then we can argue that the viscosity decreases with the shear rate. This behavior could be explained by the structure of such complex materials. The obtained structure will be resistant to flow at low shear rates, justifying therefore the presence of a threshold stress. When the applied stress exceeds the threshold of stress, the initial structure begins to take another shape and induces a flow. After the destruction of solid aggregates, part of the liquid phase trapped initially t within these clusters of particles is released implying an increase in the liquid phase and thus a reduction of the inter-particle forces This explains the shear thinning nature of these materials. At high shear rates, the flow fully established explains the linear behavior (Bingham-type) of the fluid and thus the constant nature of the plastic viscosity [16].

## 4. CONCLUSION

The experimental results show that the used materials highlight that the viscosity and the threshold stress are greater if the concentration is important and inversely. The comparison with conventional rheological models showed that, in general, that Herschel-Bulkley's models and Bingham's present a better correlation compared to the experimental results. It would be interesting to focus the future works on a structural study to discover the microscopic material factors which are the cause of this behavior, and to provide, by anticipation, the rheological behavior of these materials. In-situ tests situ), taking into account the actual conditions of the deep of the injection medium (mainly temperature and pressure) are needed to develop appropriate methodologies to achieve better control of the use of these materials required in cementing casings of oil drilling.

## REFERENCES

[1] Roussel N. 2005 *Cement and Concrete Research* **35** 1656
[2] Nguyen V. H. 2006 *J. Non Newtonian Fluid Mechanics* **139** 128
[3] Mellak A, Baudeau Ph 1994 "Etudes physico-chimiques sur des coulis de ciment saumurés et microsilicés appropriés aux formations salifères", *Annales de l'ITBTP* **526** Paris
[4] Bensted J 1989 "Oilwel cement. Word cement", *Sunbury Research Center* Middlesex, UK
[5] Hentz A 1970 «Les ciments, leurs caractéristiques et le contrôle de celles ci au moyen d'additifs", in Le Forage Aujourd'hui, opérations speciales. *Publications* de l'IFP, éditions Technip.
[6] A Mellak, M Khodja (née Saber), M Khodja 2007 « Caractérisation rhéologique et physico-mécaniques des ciments allégés par la bentonite et saumurés », Casablanca.
[7] Formulaire du Foreur, 1989 Institut Français du Pétrole, Ed. Technip, Paris.
[8] Krieger I M 1989 "L'écoulement plastique et le rhéomètre rotatif à plateaux parallèles", Cahiers de Rhéologie, VIII, **61**- France.
[9] Mellak A 2007 *Lebanese Science Journal (CNRS)*, **8** (2) 115
[10] Coussot P, Nguyen Q D et al. 2002 *Journal of Rheology* **46** 573
[11] Mellak A 2004 « Caractérisation d'un ciment destiné aux zones à pertes et modélisation de son caractère thixotrope » in 39ème Colloque du Groupe Français de Rhéologie (GFR), **A15** Mulhouse, France.
[12] Gomze A L 1983 *Epitoanyag* **35** 173
[13] Struble L J 1991 "The rheology of fresh cement Paste", International Symposium on Advance in Cementitious Materials, American Ceramic Society, Gaitherburg.
[14] Herschel0 W H abs 1926 *Kolloid Z.* **39** 290
[15] Bingham E C 1922 *Fluidity and plasticity,* McGraw
[16] Quemada D 1977 *Rheol. Acta* **16**, 82

Journal of Physics: Conference Series **602** (2015) 012031 doi:10.1088/1742-6596/602/1/012031

# Rheological Properties of Liquid and Particle Stabilized Foam

**A. Ö. Özarmut and H. Steeb**

Institute of Mechanics, Ruhr-University Bochum, Universitätsstraße 150, D-44801 Bochum, Germany

E-mail: aydin@lkm.rub.de and holger.steeb@ruhr-uni-bochum.de

**Abstract.** In Earth-Pressure-Balance (EPB) tunnelling the excavated ground is used as face support medium to prevent surface settlements. In general, the excavated ground (e.g. for cohesionless soils) does not exhibit suitable conditions to support the tunnelling face. This technical challenge can be solved by adding conditioning agents that are mainly foams. In order to physically understand the rheological properties of the (added) liquid foam and the foam-soil (foam-particle) mixture and to comprehend its influence on the soil, advanced rheological investigations are necessary. Therefore, rheological experiments such as flow curve tests have been performed to determine the effective yield stress. Since the morphology, i.e. the microstructure of the foam accounts for effective rheological properties, size, shape and distribution of the cells of the foam and particle-laden foam were characterized in detail applying imaging techniques. In order to perform the above mentioned experiments, polymer-stabilised shaving foam seems to be a good replacement of tunnelling foam and suitable for laboratory tests due to its time stability, characteristic length scales of the microstructure and accessibility. Glass beads (of different diameter and volume fractions, i.e. specific surface areas) are used to investigate the effective material behaviour of foam-particle mixtures. The experimental results are compared with yield stress models of modified Herschel-Bulkley-Papanastasiou type.

## 1. Introduction

EPB shield machines are heavily used in mechanized tunnelling. In mechanized tunnelling, the excavated ground is used as supporting medium. For an efficient and reliable support of the tunnel face, the face-supporting excavated material has to meet certain criteria. Besides the requirements to support the tunnel face the excavated material should have, sufficient flow properties. In detail, the effective viscosity should be low enough to facilitate flow in the excavation chamber and the transport in the screw conveyor. In addition, a low viscosity of the material decreases the torque of the cutting wheel and screw conveyor; thus, the wear of tools and steel structures in the excavation chamber is reduced. To meet such requirements, the excavated soil is mixed with additives which are often liquid foams. The foam-enhanced soil material shows a similar effective rheological behaviour like soft plastic pastes. In order to evaluate the flow characteristic of soil-foam mixtures, their rheological properties are investigated. So far, many researchers investigated the rheological properties of pure liquid foam. For instance, Herzhaft [1] used a Haake Rotovisco RT 20 rheometer with plate-plate geometry having a diameter of 35 mm. The experimental results (shear flow curves) of this study were in good agreement with the classical Herschel-Bulkley model [9]. Khan et al. [2] obtained in rheological experiments foam viscosity as a function of shear rate by using a plate mode of

Rheometrics Mechanical Spectrometer. Marze et al. [5] used a Anton Paar MCR 300 Rheometer with homemade cone-plate geometries for rheological experiments (flow curve tests) in Plexiglas. They also compare the Herschel-Bulkley rheology with their results and obtained the model-instrincic parameters from their data. The physical characteristic of foam depends on several factors, such as bubble form, polydispersity, i.e. size distribution of bubbles, etc. For a better understanding of the physical behaviour of the foam different imaging characterisation methods are applied to get a better insight into the microstructure of liquid foams. Khan et al. [2] obtained the average bubble diameter for a liquid foam applying light microscopy.

However, the rheological behaviour of particle-foam mixtures has not been investigated intensively. Effective complex moduli (storage and loss modulus) for foam-particle mixtures have been obtained for different particle concentrations but flow curves have not been discussed. The aim of the present contribution is to discuss the foam-particle microstructure and present rheological experiments of liquid foam and glass particle mixtures. Accounting for data of the presented rheological experiments, we apply the modified Herschel-Bulkley-Papanastasiou model which can be fitted in good agreement to our experimental data. Furthermore, the effect of the model parameters was investigated on flow curve tests.

## 2. Materials and Methods

Rheological properties of liquid foam and natural cohesionless soils, like excavated ground in tunneling, cannot be determined in homogeneous laboratory experiments. The reason is the large characteristic lengths of natural sand-gravel mixtures. Typically grain diameters up to several centimeters occur, much to large for (modified) rheometers. In the present investigation rheological experiments have been performed on the basis of a down-scaled substitute material. Then, we use a standard rotational rheometer (Anton Paar, MCR 301) with plate-plate geometry. The diameter of the plate is 50 mm. In order to eliminate wall slip effects, fine grained sand paper (P320) was glued on the upper and lower plates. Instead of the tunneling ground, polymer-stabilised shaving foam (Gillette shaving cream) and glassbeads (Silibeads Glassbeads Type S were used because of their easy accessibility and time stability. The microstructure of the liquid foam and the solid glass particles mixtures have been obtained by imaging techniques applying light microscopy. The mean bubble diameter of the liquid foam is equal to 51.80 μm and has a Gaussian-like distribution [3]. Isarin et al.[6] also determined the bubble size and the distribution of liquid foam by image analysis. They found the mean bubble radius in a range from 30 to 50 μm. Engelsen et al. [7] revealed that the mean bubble diameter could be from 10μm to 1mm.

The effective rheology of the solid particle-foam mixture depends on the volume fraction of dry solid particles $n_0^s = dv_0^s/dv_0$, where $dv_0^s$ is the particles and $dv_0$ is the total volume of the particle-laden foam. The partial volume of the liquid foam is symbolized by $dv_0^f$ and its volume fraction is $n_0^f = dv_0^f/dv_0$. The total volume of the solid particle-foam mixture is depicted in Table 1 and was kept constant in the preparation step at 40 $cm^3$ for each case, while the volume of the solid particles have been increased. The blending time of the mixture is 1 min. The system was driven by rotating the upper plate at a controlled shear rate. The shear stress was measured in the range of shear rate between ($\dot{\gamma}$) from 0.001 and 10 [1/s]. The experimental results for shear stress shown in Figure 2 are the averages of three samples, and the flow measurement time was 8 min. The rheological experiments were conducted under room temperature at 23.27 ± 1°C.

**Table 1.** The volume fraction of dry particle-foam mixture

| Case | 1 | 2 | 3 | 4 | 5 | 6 | 7 |
|---|---|---|---|---|---|---|---|
| $dv_0^s[cm^3]$ | 0 | 2 | 4 | 6 | 8 | 10 | 12 |
| $dv_0^f[cm^3]$ | 40 | 38 | 36 | 34 | 32 | 30 | 28 |
| $n_0^s[1]$ | 0 | 0.05 | 0.1 | 0.15 | 0.20 | 0.25 | 0.30 |
| $n_0^f[1]$ | 1 | 0.95 | 0.9 | 0.85 | 0.80 | 0.75 | 0.70 |

## 3.Results and Discussion

Shaving foam stability is higher than the other liquid foam due to its higher amount of polymers. Non-stabilized foam is not able to preserve its microstructure in time. Coarsening, gravitational drainage and film rupture lead to a change in foam bubbles and therefore in morphology [8]. The bubbles which have different sizes at the initial state emerge together and become one larger single bubble (coalescence) due to pressure difference. As mentioned above, solid glass particles, having a volume fraction of $\boldsymbol{n_0^s = 0.3}$, were mixed with shaving foam. The microstructure of the resulting mixture can be seen in Figure 1. The difference between the following two figures is that the left one was captured after 2 min., whereas the right one after 57 min. Although foam was mixed with solid glass particles, the coarsening (in yellow circle) and coalescence (in red circle) of the foam can be clearly seen. The bubble size increases in time and the small bubbles merge into a larger one.

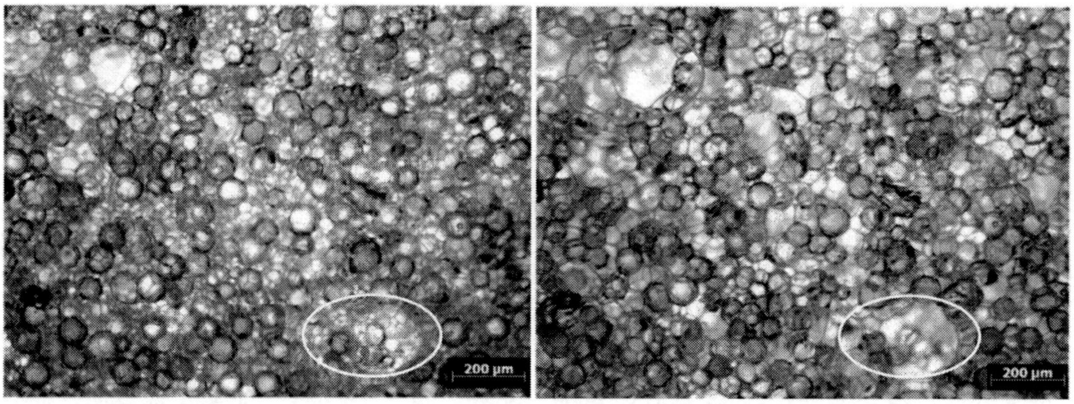


**Figure 1.**Microstructure of dry solid glass beads-liquid foam mixtures captured after 2 min (left) and 57 min (right) (Coalascence and coarsening can be seen in red and yellow circles.)

Flow curve tests were performed to obtain the type of flow behavior for different volume fraction of solid glass particles. As it can be observed in Table 1, volume fraction of the solid particles ($n_0^s$) ranges between 0 and 0.3. At large volume fractions ($n_0^s \geq 0.3$) the mechanical response of the mixture changes completely. We observe a solid-like response which we explain with an adsorption process of the surfactants at the surface of the glass beads destroying the liquid foam morphology and therefore the foam-particle mixture. Thus, we do not further investigate mixture with higher solid volume fractions. To overcome the adsorption phenomenon additional water can be added for larger solid volume fractions. However, in this study, we focus on the mixture of the liquid foam and dry-solid glass particles.

In Figure 2, the flow curve tests for different volume fractions of glass beads are shown. It is worth noting that the approved Herschel-Bulkley model did not fit very well the experimental data at smaller shear rates ( $\dot{\gamma} \leq 10^{-1}[1/s]$). Therefore, i.e. regularized the modified Herschel-Bulkley-Papanastasiou model [4] represented by solid lines in Figure 2, has been applied.

$$\tau = a(\dot{\gamma}^{\frac{1}{2}}) + \tau^{*}(1 - \exp(-m\dot{\gamma})) \tag{1}$$

whereby $a$ is the fluid consistency, $m$ is exponent, $\dot{\gamma}$ and $\tau^{*}$ describes the shear rate and the yield stress, respectively. The solid curves, which are in excellent agreement with experimental data, describe the typical non-Newtonian behavior. According to the modified Herschel-Bulkley model, we do not observe a yield stress anymore, Furthermore, the slope of the shear stress versus shear rate curves is not constant. It is interesting to note, that the mixture shows first a shear thinning effect and for higher shear rates a shear thickening phenomenon. Thus, especially for lower shear rates, we could physically interpret the parameters of the modified Herschel-Bulkley-Papanastasiou model. Comparing our experimental results with the model, we observe that the value of $m$ is constant and the both fitting parameters $a$ and $\tau^{*}$ change for different volume fraction of solid particles (Table 1). Thus, the effect of the parameter $m$ is independent of the composition of the mixture. The values of yield stress ($\tau^{*}$) and consistency ($a$) ascend with descending volume of the solid particles. To investigate the effect of both yield stress and consistency, the first case (pure liquid foam) was taken as a reference. The values of the fluid consistency were changed, while the values of the other parameters were kept constant. The flow curves for three different values of the fluid consistency provide different results for shear rates from $10^{-3}$ to $10^{-2}$ [1/s]. Over $10^{-2}$ [1/s], the shear stress decreases with decreasing the value of the fluid consistency (Figure 3, left). Additionally, it affects the slope of the curves, which represents the viscosity of the mixture. Moreover, the influence of yield stress $\tau^{*}$ parameter was also shown on the flow curve test (Figure 3, right). The flow curve decreases with decreasing the value of yield stress. Hence, yield stress can be defined as a function of the volume fraction of solid particles.

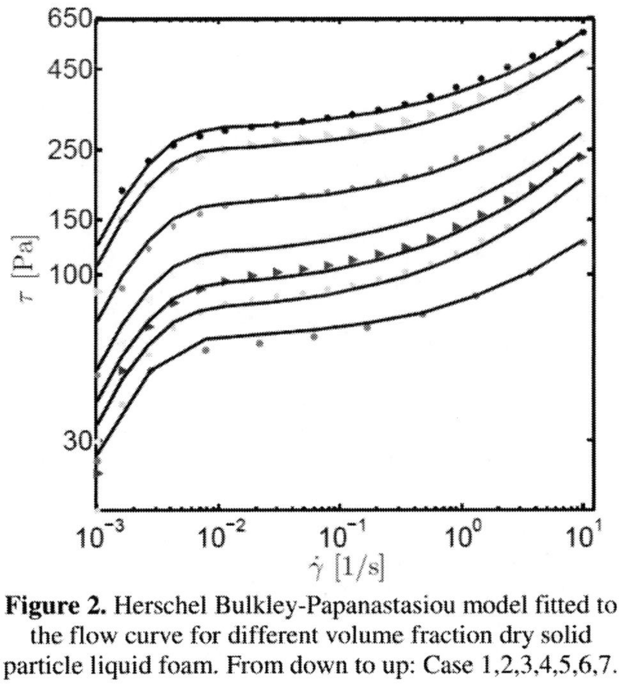


**Figure 2.** Herschel Bulkley-Papanastasiou model fitted to the flow curve for different volume fraction dry solid particle liquid foam. From down to up: Case 1,2,3,4,5,6,7.

**Table 2.** The volume fraction of dry particle-foam mixture

| Case | 1 | 2 | 3 | 4 | 5 | 6 | 7 |
|---|---|---|---|---|---|---|---|
| $a[Pas^{1/2}]$ | 21 | 40 | 49 | 54 | 66 | 86 | 98 |
| $\tau^*[Pa]$ | 61.81 | 75.55 | 89.47 | 113.3 | 161.9 | 242.8 | 283.3 |
| $m[s]$ | 560 | 560 | 560 | 560 | 560 | 560 | 560 |

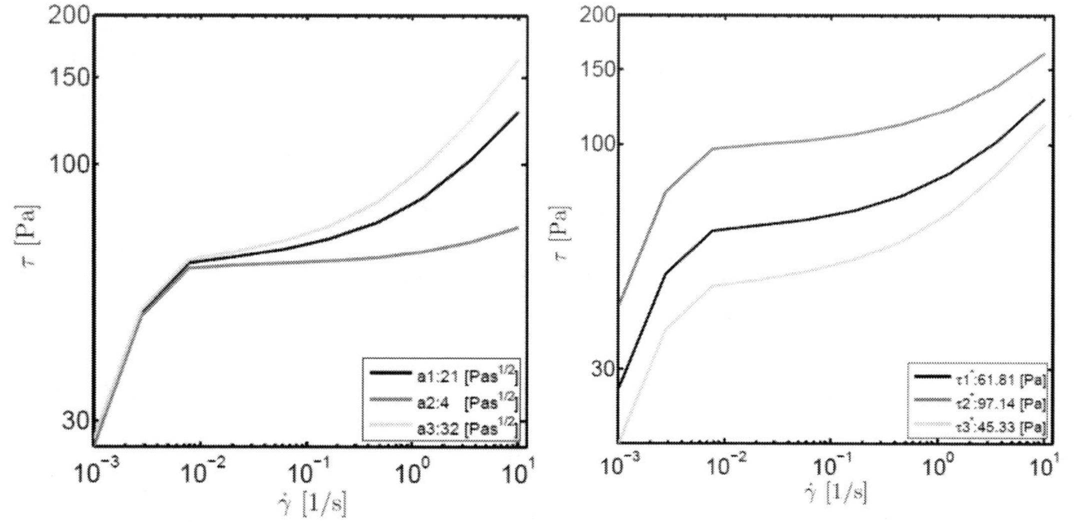


**Figure 3.**The effect of fluid consistency and yield stress on the flow curve test

## 4. Conclusion

The experimental data for flow curves of particle-foam mixtures was obtained by modified plate-plate rheometry. Sand paper (P320) was used to minimize the slip effect on the flow curve test. Comparing the data to the Herschel-Bulkley-Papanastasiou model yield stress, fluid consistency and exponent were determined for different volume fractions of solid particles. It has to be noted that the exponent turned out to be constant for all cases. The effect of fluid consistency and yield stress was also investigated in the experiments. The tested materials exhibit shear thinning and thickening behavior depending on shear rate range. The rheological experiment cannot be performed for high solid volume fractions due to adsorption effects and related solid-like rigidity of the mixture. The microstructure of the mixture analysed by imaging techniques. Foam coalescence and coarsening in time were shown with figures captured at different times of the experiments.

References

[1] Herzhaft B 2002 *Journal of Colloid and Interface Science,* **247** 412

[2] Khan S A , Schnepper C A and Armstrong R C 1988 *Journal of Rheology*, **32**, 1, 69

[3] Özarmut A Ö, Galli M, Steeb H and Thewes M 2013 Two scale investigations on the rheological properties of foam and particle-laden foams.3rd International Conference on computational methods in Tunneling and Subsurface Engineering, Germany

[4] Papanastasiou T C. 1987 *Journal of Rheology* **31** 385

[5] Marze S and Langevin D 2008 *Journal of Rheology* **52** (5), 1091

[6] Isarin J C, Kaasjager A D J and Holweg R B M 1995 *Textile Res. J.*, **64** 61

[7] den Engelsen C W, Isarin J C, Gooijer H, Warmoeskerken M M C G and Wassink J G 2002 *Autex Research Journal*, **2** (1) 14
[8] Weaire D and Hutzler S 1999 The physics of Foams, New York, *Oxford University Press*
[9] Mullineux G 2006 Applied Mathematical Modelling **32** (12) 2538
[10] Cohen-Addad S, Krzan M, Höhler R and Herzhaft B 2007 *Physical Review Letters*, **99**, 168001

Journal of Physics: Conference Series **602** (2015) 012032 doi:10.1088/1742-6596/602/1/012032

# Flow of High Internal Phase Ratio Emulsions through Pipes

**K Kostak, R Özsaygı, I Gündüz, E Yorgancıoğlu, E Tekden, O Güzel, D Sadıklar, S Peker[1] and Ş Ş Helvacı**

Ege University, Faculty of Engineering, Department of Chemical Engineering, Bornova 35100 İzmir, Turkey

E-mail: sumer.peker@ege.edu.tr

**Abstract.** The flow behavior of W/O type of HIPRE stabilized by hydrogen bonds with a sugar (sorbitol) in the aqueous phase, was studied. Two groups of experiments were done in this work: The effect of wall shear stresses were investigated in flow through pipes of different diameters. For this end, HIPREs prestirred at constant rate for the same duration were used to obtain similar drop size distributions. Existence and extent of elongational viscosity were used as a probe to elucidate the effect of drop size distribution on the flow behavior: HIPREs prestirred for the same duration at different rates were subjected to flow through converging pipes. The experimental flow curves for flow through small cylindrical pipes indicated four different stages: 1) initial increase in the flow rate at low pressure difference, 2) subsequent decrease in the flow rate due to capillary flow, 3) pressure increase after reaching the minimum flow rate and 4) slip flow after a critical pressure difference. HIPREs with sufficient external liquid phase in the plateau borders can elongate during passage through converging pipes. In the absence of liquid stored in the plateau borders, the drops rupture during extension and slip flow takes place without elongation.

## 1. Introduction

High internal phase ratio emulsions (HIPRE) have a structured microstructure that is employed in a wide spectrum of applications. The volumetric concentration of the internal phase can be increased to values greater than 74% with the use of appropriate surfactant-solvent systems. The external phase and the surfactants are then present as a thin film surrounding the polyhedral internal phase and in the Plateau borders at the corners. Thus many polyhedral cells of micrometric dimensions are formed separated by thin films of solution making up the continuous phase. HIPRE's were originally designed for the synthesis of polymeric solid foams of high porosity and low density, that were later used as supports for catalytic reactions[1], in the synthesis of porous inorganic metal oxides[2], and enzymatic biotransformations [3,4]. Resistance to diffusion caused by these thin films [5] make these emulsions promising in controlled release applications in drug delivery [6]. The large surface area of the polyhedral cells increases the rate and yield of interfacial reactions between organic and aqueous phases. The compartmentalized volumes of controllable dimensions of the dispersed phase make these emulsions a suitable medium in the synthesis of organic chemicals [4,7].

[1] Sumer Peker: Ege University, Faculty of Engineering, Department of Chemical Engineering, Bornova 35100 İzmir, Turkey

Rheology and flow properties of these emulsions bear significance in these applications. Rheological properties of HIPRE depend on the interfacial properties such as interfacial tension and contact angle[8]; molecular structure of the surfactants used and the microstructure of the emulsion; physical properties such as density and viscosity of the dispersed and continuous phases. Flow behavior of HIPRE through pipes depends in turn, on the rheological properties. This complex interrelationship between physical properties of the aqueous and oil phases used in the composition of the emulsion and the shear history dependence of the rheological properties of these emulsions have precluded a generalized treatment of the flow properties of these emulsions. Consequently theoretical analyses of the rheology of these systems are few [8-11] and the experimental reports on the rheology and flow behavior pertain to the special cases under consideration: Capdevilla et al.[12] report that the shear stress applied in the preparation of the emulsion is responsible for the formation of small drops in the emulsion through shear rate. In cases where the emulsions are formed in a mixer, the angular velocity of the impeller is used as the scale-up criterion [13-16]. Shear and elongational stresses in flow through pipes and converging-diverging nozzles also affect the drop size distribution [17]. Princen [8] predicted an increase in the yield stress that is inversely proportional with the average drop diameter of the emulsion that is also confirmed by Pal [18].

The reports on the flow of HIPRE in pipes are also limited to a few cases. Becu et al. [19] have observed that HIPRE are in a jammed state at wall stresses below the yield stress, slipping undeformed over a lubricating layer of continuous phase along the walls. Above the yield stress, the emulsion flows homogeneously and exhibits slip flow. Masalova, et al. [20] working with W/O emulsions with 94% (v) dispersed phase of liquid explosives of high electrolyte concentration have reported a power-law model of the r heology that later converts into a Herschel-Bulkley model. In a later work Herschel Bulkley model was found to be satisfactory in explaining the pipe-flow behavior [21]. Slip flow was neglected in flow through pipes of 64 and 89 mm internal diameter, but proposed to be significant in flow through small diameter pipes.

The aim of this research was to investigate the flow behavior of W/O type of HIPRE stabilized by hydrogen bonds with a sugar (sorbitol) in the aqueous phase, instead of electrostatic forces with the addition of electrolytes of high ionic strength, as generally used in the stabilization of HIPRE in the literature. Two groups of experiments were done in this work: The effect of wall shear stresses were investigated in flow through pipes of different diameters. For this end, HIPREs prestirred at constant rate and for the same duration were used to obtain similar drop size distributions. In the second group of experiments, existence and extent of elongational viscosity were used as a probe to elucidate the effect of drop size distribution on the flow behavior: HIPREs with different drop size distributions, obtained by prestirring for the same duration at different rates, were subjected to flow through converging pipes.

## 2. Experimental

### *2.1 Preparation of emulsions:*

High internal phase ratio emulsions with 94% volumetric concentration of internal phase were prepared as a function of prestirring rate. The internal phase consisted of 1% (w) sorbitol solution, and the external phase, a mixture of mineral oil (Paraffin oil of average hydrocarbon chain length = 15C) and polyoxyethylene (2) oleyl ether (Brij 92) at a 1/1 volumetric ratio. The emulsions were formed at 60 °C with stirring at a rate of 400 rpm, the minimum rate for the formation of stable HIPRE.

### *2.2 Measurement of emulsion viscosities:*

The emulsions were stirred at the rates of 400, 600, 800 or 1000 rpm for 20 minutes before the viscosity measurements. The viscosities of the emulsions were measured with Haake Mars rheometer (Thermo Electron Co.), using the vane rotors.

*2.3 Flow Experiments:* The flow set-up consisted of a storage tank held under nitrogen atmosphere, with horizontal pipe attachment and pressure gages that could be replaced by a converging pipe. The flow rate was adjusted with the pressure of the nitrogen gas in the storage tank. The glass pipes with a length of 1.5 m in 3.5, 5.6 and 14.7 mm inner diameters were used in the horizontal pipe flow experiments. Two different converging pipes were used for elongational flow of HIPRE. The outlet to inlet diameter ratio of the stainless steel converging pipes was, $D_o/D_i = 9 / 21$, in both cases. Other dimensions of the nozzles are summarized in table 1. The emulsions were prestirred at 400 rpm for 20 minutes before flow through the horizontal cylindrical pipes and at 400, 600, 800 or 1000 rpm rates for flow through converging pipes.

*2.4 Determination of drop diameters:* 1 mL HIPRE samples were diluted with oil-surfactant solution of the same composition as the external continuous phase and optical observations were carried out using an Olympus BX 50 polarizing microscope by spreading the solutions as a thin film on glass slides. Structure evolution was observed under cross-polarized light and was recorded by photography. The drops became spherical when suspended in sufficient amount of the external phase. The drops are polyhedral in shape, separated by films of continuous phase.

**Table 1.** Dimensions of converging pipes used in the experiments.

| Dimension | Short (20 mm) | Long(30 mm) |
|---|---|---|
| Pipe length, mm | 18.4 | 28.6 |
| Wall length, mm | 20.0 | 30.0 |
| Angle of convergence | 23.2° | 18.4° |

## 3. Results and Discussions

*3.1 Microstructure of HIPRE*

High internal phase ratio emulsions (HIPRE) have microstructures that cause these emulsions to exhibit different rheological behavior under different flow conditions. The continuous phase is usually a polymer or a surfactant in solution. The drops are spherical when the volumetric concentration of dispersed phase in the emulsion is less than 74%, as given in figure 1(a). The internal phase concentration of an emulsion can be increased above 74% when appropriate surfactant-solvent systems are used at a high ratio of surfactant/solvent. The external phase and the surfactants are then present as a thin film surrounding the polyhedral internal phase and in the Plateau borders at the corners. At dispersed phase volumetric concentrations greater than 74%, drops can no longer remain spherical and are flattened at contact points with plateau borders at the junction of three drops in 2-D, as shown by the dashed-lined circle in figure 1(b). Hexagonal drops in figure 1 (b) and (c) are in the form of polyhedra in the case of actual 94% internal phase emulsions in 3-D, with the appearance of miniaturized foams.

Surfactants with a linear hydrophobic chain and a small hydrophilic group with approximately the same cross sectional area as the hydrocarbon chain form lyotropic liquid crystals. Molecular structure of polyoxyethylene (2) oleyl ether (Brij 92) has these properties, suitable for the formation of liquid crystals. The compact orientation of the surfactant molecules at the interface are further stabilized by forming hydrogen bonds with sorbitol in the subphase.

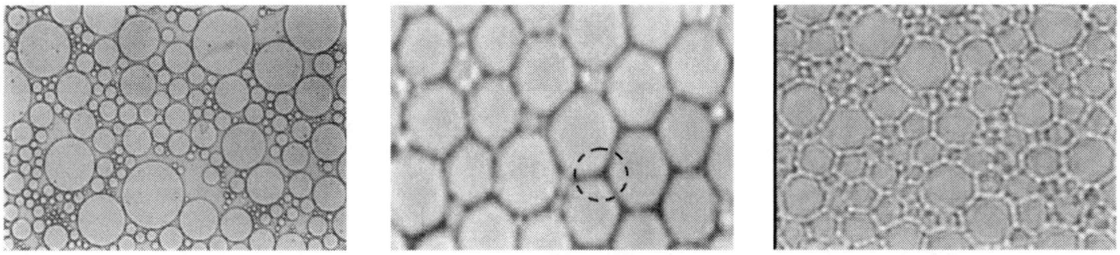

(a) (b) (c)

**Figure 1.** Shape of the dispersed phase drops. a) Free drops with volumetric concentration of dispersed phase less than 74%, b) monodispersed polyhedral drops in HIPRE, c) polydispersed polyhedral drops in HIPRE

*3.2 Rheological behavior of HIPRE*

When the volume of the continuous phase around the drops is sufficient for free motion of the drops, as given in figure 1(a), the drops line up under the action of shear and the emulsion behaves as a power law fluid:

$$\tau = K\dot{\gamma}^n \tag{1}$$

The close-packed polyhedral structures, as given in figure 1(b) and (c), develop a yield stress as any force applied to the structure is transferred throughout the network of bilayers forming the walls of the drops. The emulsion then conforms to Herschel-Bulkley model.

$$\tau = \tau_o + K\dot{\gamma}^n \tag{2}$$

The yield stress is proportional to the compactness of the network evidenced by the yield stress increase with a decrease in drop size [8, 18]. Mechanical energy input, as in the case of stirring of the emulsion is converted into surface energy, reducing the size of the drops, increasing the interfacial area and reducing the amount of bulk liquid in the Plateau borders. On further concentration, the continuous phase film around the drops is reduced to a bilayer. The yield stress of the emulsion increases with the extent of the interfacial contact area among the drops. The rheological constants of the 94% emulsions used in this work are given in table 2 as a function of the prestirring rate.

Due to errors inevitably introduced in handling, the trends rather than absolute values of the rheological constants in table 2 are significant in work dealing with shear history fluids such as high internal phase ratio emulsions: Initially undeformable gel behavior was observed in emulsions prestirred at 1000 and 800 rpm. Aging is very rapid in this case, due to decreased drop diameter with increased total interfacial area among the drops. Aging can be interpreted in terms of the microstructure of the emulsions as follows: As the total amount of the surfactant is constant independent of the prestirring rate, interfacial concentration of the surfactant stabilizing the interfacial film is decreased, facilitating the coalescence of small drops with time. The yield stress decreases when the compactness of the network structure decreases due to coalescence of drops. The enlargement of the drop sizes causes liquid to accumulate in the plateau borders. This causes the rheological response of the emulsion toward shear stress to change, with a reduction in the yield stress and consistency. The low values of yield stress in spite of the high prestirring rates reflect the rapid aging behavior in the time span between the preparation of the emulsion and the initiation of the viscosity measurement. Emulsions prestirred at 400 and 600 rpm conform to the Herschel-Bulkley model throughout the aging period. In general, due to the network structure of HIPRE, the effect of yield stress is observed to be predominant in the rheological behavior, as indicated by the relative values of the yield stress, $\tau_o$, and the consistency index, K. Viscosity index, n, in power law terms reflect the deformability of the emulsions; all emulsions approach Newtonian behavior (n→1) with aging.

**Table 2.** A summary of aging behavior of HIPRE used in this work.

| N (*rpm*) | t(*h*) | $\tau_o$ (*Pa*) | K(*Pa s$^n$*) | n (-) |
|---|---|---|---|---|
| 400 | initial | 107 | 1.4 | 0.77 |
| | 24 | 40 | 1.0 | 0.97 |
| 600 | initial | 101 | 1.1 | 0.41 |
| | 24 | 43 | 0.9 | 0.58 |
| 800 | initial | 97 | - | - |
| | 24 | 41 | 1.4 | 0.37 |
| 1000 | initial | 87 | - | - |
| | 24 | 12 | 1.1 | 0.69 |

*3.3 Flow of HIPRE through pipes*

At the start of the experiments, the paraboloid shape of the moving front of the emulsion could be observed visually in 3.5 and 5.6 mm pipes. This shape could reflect the velocity profile within the pipe; it could also be due to the convex shaped meniscus of the W/O HIPRE that does not wet the glass walls of the pipe. After an entrance length that depended on the flow rate (applied pressure) the walls were smeared by the emulsion. Velocities were determined from the mass flow rates. The experimental flow curves for flow through 3.5 and 5.6 mm pipes indicated four different stages: 1) initial increase in the flow rate at low pressure difference, ΔP, 2) subsequent decrease in the flow rate, 3) pressure increase after the minimum flow rate and 4) slip flow after a critical pressure difference ($\Delta P > \Delta P_{cr}$) as given in figure 2(a) and (b): The flow rate initially increased with increasing pressure difference, then decreased within a small pressure differential range. After the minimum flow rate was reached, the pressure difference increased within a very small range of flow rates. Finally flow rate increased considerably with the applied pressure after a critical pressure difference, wall shear stress as a function of pipe diameter. The first three stages were not observed in flow through 14.7 mm pipe as illustrated in figure 2(c).

In the first step of the experimental curve, the flow may be due to capillary forces only as the wall shear is much less than the yield stress ($\tau_w << \tau_o$) and the emulsion moves as a block. Capillary motion is enhanced by the non-wetting hydrophobic oil flowing within the hydrophilic glass capillary [22]. Evidence of slip due to non-wetting is the convex (paraboloid) meniscus of the moving front of the emulsion at the start. Hydrostatic pressure of the emulsion in the storage tank also contributes to the bulk flow of the sliding emulsion. The pressure difference is the sum of capillary and viscous pressure drops under the hydrostatic head in the storage tank [23],

$$\Delta P = -4\frac{\gamma Cos\theta}{D} + \frac{32VL\mu}{D^2} \tag{3}$$

where, $\gamma$ is the interfacial tension, $\theta$ the contact angle, $\mu$ viscosity of the emulsion, $V$ the velocity of the emulsion, $L$ the length and $D$ the diameter of the pipe. Although capillary forces are dominant in the first stage of the flow, viscous forces additionally start to be effective in the second stage, with subsequent decrease in the flow rate. Wall shear stresses cause deformation and subsequent division of the drops to smaller sizes resulting in an increase in yield stress. Hence the flow rate decreases due to the shift of drop size distribution to smaller sizes. Pressure increases after the minimum flow rate is reached in the experiments because the high-yield-stress fluid cannot deform in spite of the high wall shear stresses along the pipe. Increasing the pressure cannot bring about further increase in the flow rate. When the pressure reaches to a critical pressure difference, $\Delta P_{cr}$ corresponding to the prevalent yield stress with respect to aging time and shear-history of the flowing emulsion, deformation (flow) begins. $\Delta P_{cr}$ values for flow in each pipe were calculated using the yield stress value ($\tau_o = 64$Pa) found by interpolation of the aging trend of the emulsion in the Hershel-Bulkley constitutive equation

$$\tau = 64 + 1.03\dot{\gamma}^{0.85} \tag{4}$$

and equating $\tau_o$ to the wall shear stress $\tau_w$,

$$\tau_w = \tau_0 = \frac{\Delta P_{cr} D}{4L} \tag{5}$$

$\Delta P_{cr}$ value in flow through 3.5mm pipe was found to be 109 714 Pa; 68 571 Pa and 27 430 Pa for the flows through 5.6 mm and 14.7 mm, respectively. These confirm $\Delta P_{cr}$ values observed in the experimental data plotted in figure 2(a) and (b). At greater pressure differences, increase in the number and decrease in the size of the drops produced within a thin film along the wall causes a decrease in the surfactant concentration in the bilayers forming the drop boundaries in this region. Irregularity of the drop shapes resist alignment in the direction of the shear stress, and the drops rupture. As a consequence slip flow takes place at pressure differences higher than $\Delta P_{cr}$. Apparent viscosity was calculated to verify the existence of the slip flow: It was found to be 18. 3 cP for flow through 3.5 mm pipe; 1.1 cP, for 5.6 mm pipe and 1.9 cP, for 14.7 mm pipe that are in the order of magnitude of water-oil mixture. The emulsion viscosity was calculated as 6 Pas from rheological data at 1 $s^{-1}$ shear rate, for comparison, indicating the existence of a high shear film along the walls, conforming observations made in the literature [17, 19].

The experimental flow curves can be interpreted in terms of the microstructure of the emulsions as: In the absence of stored liquid in the corners, stretching of the walls increases the distance between the liquid crystal surfactant molecules. The van der Waals forces among the hydrocarbon chains and the hydrogen bonds among the hydrophilic groups of the closed packed surfactant molecules at the interface resist stretching under the action of shear stresses. When the shear stresses exceed a critical value greater than the attractive forces between the liquid crystal surfactant molecules, the drops rupture. The aqueous phase entrained within the drops forms a film along the hydrophilic pipe walls. The emulsion then slides over the aqueous film without deformation or a change in the microstructure (drop size).

Using the yield stress given in equation (4), a pressure difference $\Delta P$ was assumed, then shear stress at the pipe wall $\tau_w$, theoretical, $Q_{theo}$ and slip, $Q_{slip}$ flow rates were calculated using equations (6), (7) and (8) for all ranges of the experimental pressure differences to compare the experimental flow curves with the theoretical.

$$\tau_w = \frac{\Delta P D}{4L} \tag{6}$$

$$Q_{the} = \frac{\pi D^3}{8\tau_w^3}\left(\frac{1}{K}\right)^n \left\{ \frac{n(\tau_w - \tau_o)^{(1+3n)/n}}{1+3n} + \frac{2n\tau_o(\tau_w - \tau_o)^{(1+2n)/n}}{1+2n} + \frac{n\tau_o^2(\tau_w - \tau_o)^{(1+n)/n}}{1+n} \right\} \tag{7}$$

$$Q_{exp} = Q_{theo} + Q_{slip} \tag{8}$$

Calculation results together with the experimental flow curves are given in figure 3. The theoretical flow rates calculated from equation (7) is found to be extremely small with respect to the experimental values, and the slip flow rates, to be in the same order of magnitude with the experimental flow rates in all pipes. All stages existing in the experimental flow curves in 3.5 and 5.6 mm pipes (figure 3(a) and (b)) and the slip flow from the beginning to the end of flow through 14.7 mm pipe (figure 3(c)) were verified by the calculation results. The study also indicated that only slip flow exists in flows through pipes with an inner diameter greater than 5.6 mm. The initial stages before the onset of slip flow are observed only in small pipes; therefore, the transformation to slip flow is not evident in literature based on observations in larger diameter pipes [21, 22, 24].

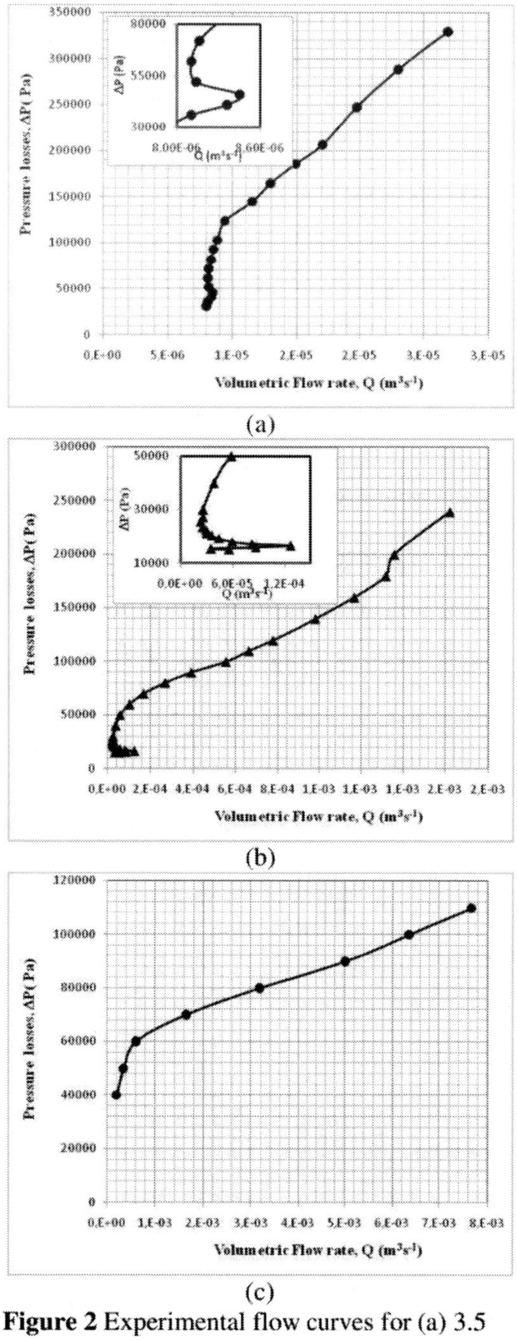


**Figure 2** Experimental flow curves for (a) 3.5 mm, (b) 5.6 mm and (c) 14.7 mm pipes

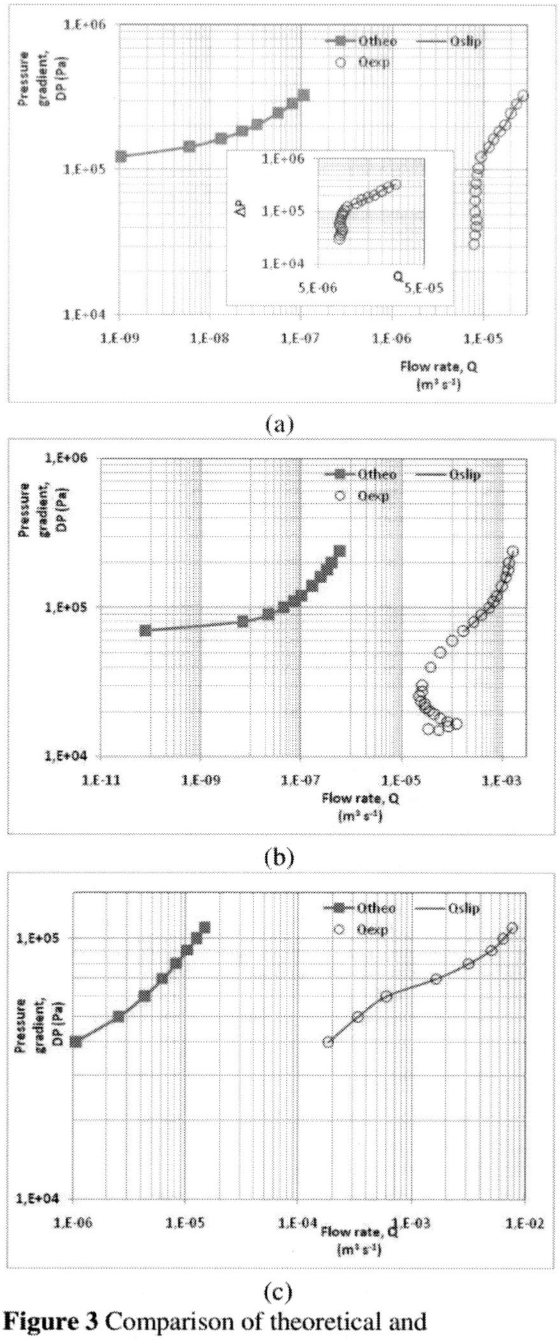


**Figure 3** Comparison of theoretical and experimental flow curves for (a) 3.5 mm, (b) 5.6 mm and (c) 14.7 mm pipes

*3.4 Flow of HIPRE through converging pipes*

Shear stresses exist along the walls of converging pipes. In addition normal stresses exist due to extensional (elongational) flow. Extensional flow is mainly responsible for drop deformation and consequent break-up [17]. As a result, variation in the drop size is expected, provided there is no slip at the walls. For this end, HIPREs at different prestirring rates were prepared and their initial drop size distributions are given in figure 4(a). Emulsions tend to be monodisperse at lower rates of prestirring, but the predominant (mode) drop diameters are larger.

The variations in the drop size distributions after passage through short (20 mm) and long (30 mm) converging pipes are plotted in figure 4(b) and (c) for emulsions prepared at 600 and 800 rpm prestirring rates: Drop sizes are reduced on initial application of shear: The sizes at 800 rpm prestirring rate are less than those at 600 rpm. If the number of drops is not large and thick liquid films with plateau borders exist, the drops are stretched and reformed as larger drops as observed at 600 rpm prestirring rate after flow through converging pipes. Formation of large drops increases with increasing length of the converging pipe. For small sizes of drops and thin films, stretching ruptures the drops leading to slip flow. Emulsions prestirred at 800 rpm ruptures under the effect of wall shear stress with resultant slip flow. Therefore, the variation in the drop size distribution is negligible.

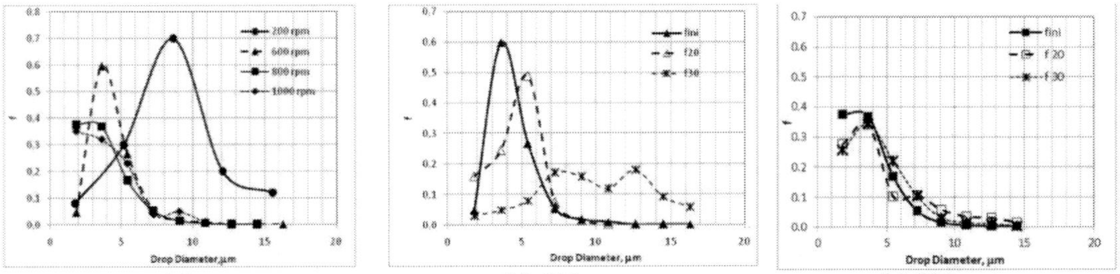


(a) (b) 600 rpm (c) 800 rpm

**Figure 4** Variations in drop size distribution before and after the flow through converging pipes as a function of prestirring rate: a) before the flow , b) emulsions prestirred at 600 rpm after the flow, c) emulsions prestirred at 800 rpm after the flow

## 4. Conclusion

In conclusion, the stability of drops in a thin film along the walls determines the onset of slip flow in pipes. Only HIPREs with sufficient external liquid phase in the plateau borders can elongate during passage through converging pipes. Greater size of drops is pre-requisite: In the absence of liquid stored in the plateau borders, the drops rupture along the high-shear pipe walls and slip flow takes place without elongation during passage through the converging pipe.

## References

[1] Cameron N R and Sherington D C. 1996 *Adv Polym Sci* **126** 163
[2] Maekawa H, Esquena J, Bishop S, Solans C and Chmelka BF 2003 *Adv Mater* **15** 591
[3] Espelt L, Clape´s P, Esquena J, Manich A and Solans C 2003 *Langmuir* **19** 1337
[4] Solans C, Esquena J and Azemar N 2003 *Current Opinion in Colloid and Interface Science* **8** 156
[5] Calderó G, Patti A, Llinàs M and García-Celma M J 2012 *Current Opinion in Colloid and Interface Science* **17** 255
[6] Llinàs M, Calderó G, García-Celma M J, Patti A and Solans C 2013 *Journal of Colloid and Interface Science* **394** 337
[7] Solans C, Pinazo A, Caldero´ G and Infante M R 2001 *Colloids and Surfaces A: Physicochemical and Engineering Aspects* **176** 101
[8] Princen H M 1983 *Journal of Colloid and Interface Science* **91** 160
[9] Princen H M, Aronson M P and Moser J C 1980 *Journal of Colloid and Interface Science* **75** 246

[10] Princen H M 1985 *Journal of Colloid and Interface Science* **105** 150
[11] Princen H M and Kiss A D 1989 *Journal of Colloid and Interface Science* **128** 176
[12] Capdevila M, Maestro A, Porras M and Gutiérrez J M 2010 *Journal of Colloid and Interface Science* **345** 27
[13] Metzner A B and Otto R E 1957 *AIChE J* **3** 3
[14] Bakker A and Gates L E 1995 *Chem. Eng. Prog.* 25
[15] Okufi S, Pérez de Ortiz E S and Sawistowski H 1990 *Can. J. Chem. Eng.* **68** 400
[16] Galindo S A, Puel F, Briançon S, Allémann E, Doelker and Fessi H 2005 *Eur. J. Pharm. Sci.* **25** 357
[17] Akay G 1998 *Chemical Engineering Science* **53** 203
[18] Pal R 2006 *Food Hydrocolloids* **20** 997
[19] Becu L, Grondin P, Colin A and Manneville S 2004 *Colloids and Surfaces A: Physicochem. Eng. Aspects* **263** 146
[20] Masalova I,.Malkin A Y, Slatter P, Wilson K 2003 *J. Non-Newtonian Fluid Mech.* **112** 101
[21] Masalova I and .Malkin A.Y 2013 *Chemical Engineering Research and Design* **91** 204
[22] Lee H-B, Yeo I W, Lee K-K 2013 *Advances in Water Resources* **53** 242
[23] Du L, Bodiguel H, Cottin C and Colin A 2013 *Chemical Engineering and Processing* **68** 3
[24] Al-Yaari M., Al-Sarkhi A., Hussein I A,Chang F and Abbad M 2013 *Chemical Engineering Research and Design* (http://dx.doi.org/10.1016/j.cherd.2013.09.001) in press

Journal of Physics: Conference Series **602** (2015) 012033 doi:10.1088/1742-6596/602/1/012033

# Magnetorheological fluids with two-component dispersed phase

**E V Korobko[*], Z A Novikova and M A Zhurauski**

Heat and Mass Transfer Institute of NASB, 220072, 15 P. Brovka str., Minsk, Belarus

E-mail: evkorobko@gmail.com

**Abstract**. In the paper the results of investigations of rheological properties of magnetorheological fluids with two-component dispersed phase are presented. Normal and tangential components of the mechanical stress in fluids in the magnetic field are determined. It is showed that using of carbonyl iron and chromium oxide as dispersed phase components allows to achieve essential structural response on magnetic influence

## 1. Introduction

The currently used methods protection from vibrations, performing important functions, however, due to increasing demands on the protection of technical equipment and precision electronic devices, especially for transport, require upgrades and new solutions for the implementation of a more efficient and directional damping effects of various vibration types [1].

One of the promising methods for targeted reduction of vibration exposure on the object is its vibroisolation, in particular, with the use of viscous damping. This method finds new control features through the use of modern intellectual damping fluids with magnetosensitive nanofillers [2]. In the magnetorheological fluids (MRF) containing non-colloidal particles as the dispersed phase, under the influence of external fields a strong structuring of particles is observed, which is defined substantially by their magnetic properties – magnetic permeability, saturation magnetization, remnant magnetization, coercive force. Thanks to it, MRF exhibits elastic and viscoplastic properties according to the level of external influence. This allows one to control its damping characteristics specifically and reversibly (response time less than $10^{-3}$ s). It is known, that MRF are used as fluids for flow hydraulic systems and fluid dampers, squeeze dampers [3–6], including ground transportation systems and precision instruments. In the present time there is a very urgent task of creating hydraulic multi-purpose damping elements on the base of MRF, having improved characteristics. The main difficulties in development of MRFs of new generation are securing the aggregation and sedimentation stability, also increasing of sensitivity of their elastoviscoplastic characteristics to the magnetic field. The stability of such systems can be provided by prevention of spontaneous irreversible coagulation of magnetic owing to the adding of colloid-type substances that form structural-mechanical barriers between particles.

Usually the classical MRF contains a magnetic dispersed phase of the magnetosoft or magnetohard type[7,8]. In the case of magnetosoft many-domain particles, for example, carbonyl iron the MRF structure is distorted even at small shear rate after the field has been turned off. MRF of single-domain

[*]Corresponding author.

particles "store" the viscosity change due to the magnetic field even after the field has been turned off. Adding an additional high-dispersed magnetohard powder can be one way for improving the classical MRF consumer characteristics. Its particles form a spatial structure in the carrying medium due to the intrinsic magnetic moment in the absence of the magnetic field, i. e. they serve as an additional stabilizer.

## 2. Materials and equipment

The aim of this study is to investigate the rheological and damping response to stationary and dynamic influences of different types on MRF containing two-component dispersed phase particles of soft magnetic material – carbonyl iron, as the second – particles of hard magnetic material (chrome dioxide $CrO_2$ or iron gamma-oxide $\gamma$-$Fe_2O_3$) or non-magnetic particles of aerosil $SiO_2$, under the influence of the external magnetic field. Mineral oil was dispersion media.

Rheological measurements were performed by means of rheometer Physica MCR 301 by Anton Paar with using of a measuring cell of parallel plate type with diameter of 20 mm and the clearance between them is 0.7 mm. The upper plate can perform a rotary tangential (horizontal) shear. Furthermore, it can be moved in the vertical direction at a constant speed. The values of the induction of the constant magnetic field have been varied up to the value of $B$ = 1 T, the range values of shear stress values is 0–128 kPa.

Experiments were conducted in the following modes of deformations on MRM:

1) tangential shear mode of the top plate at a constant shear rate in the range of 0.01 – 1000 $s^{-1}$;

2) continuous vertical shift mode of the upper plate, wherein the clearance $d$ during 50 s decreased from 0.7 mm to 0.65 mm with a corresponding relative deformation of the material $\varepsilon = \Delta d / d$ in the range of values 0 – 0.07;

3) continuous vertical movement mode of the upper plate (like to case 2) with a simultaneous tangential shear deformation at a constant shear rate of 10 $s^{-1}$;

4) linear shear stress growth from zero to given value slightly exceeding yield stress of MRF.

Magnetic properties of MRF are defined by the standard technique of measurement of magnetization with two Hall sensors. Magnetization curves are received in the range of magnetic field strength up to 450 kA/m.

In all modes of deformation the values of resistance force in a material to the shift of the upper plate $F$ (normal $F_N$ and tangential $F_T$ components) were recorded. They were determined by pressure sensors on the top measuring plate of the experimental cell at material's shear. The dependencies are presented in the form of specific values of force $F_{N,T} / S$, where $S$ – area of the plate, i.e., normal $\tau_N$ and tangential $\tau_T$ values of shear stress on the magnetic field at various modes of deformation.

## 3. Results and discussion

Firstly, magnetorheological fluids with high volume concentration of the dispersed phase (42 vol. % of carbonyl iron – the first phase – and 2.4 vol. % of the second phase) were studied. Compositions of MRF and their characteristics without magnetic field are listed in Table 1.

**Table 1.** Compositions of high-concentrated MRF containing complex dispersed phase.

| MRF | The second phase | $\tau_0$ at $B$=0 | $\tau_T$ at $B$=0, 30 $s^{-1}$ |
|---|---|---|---|
| 1 | $CrO_2$ | 2.78 | 309 |
| 2 | $\gamma$-$Fe_2O_3$ | 1.41 | 291 |
| 3 | $SiO_2$ | 0.76 | 141 |

Magnetization curves of these MRF are presented in figure 1. Magnetization of MRF-3 containing nonmagnetic silica is smaller than MRFs 1 and 2 containing only magnetosensitive powders. Because of saturation magnetization values of $CrO_2$ and $\gamma$-$Fe_2O_3$ are very close (480 and 440 kA/m correspondingly) their magnetization curves are similar. At the magnetic field strength more than 300 kA/m growth of magnetization reduces, the saturation begins.

Dependences of static yield stress and shear stress at shear rate 10 $s^{-1}$ on magnetic field induction are presented in figures 2 and 3. As particles of magnetohard powders have smaller sizes (long axis is less than 0.7 μm) than ones of carbonyl iron (3–4 μm) they form shells around carbonyl iron particles. In the external magnetic field all ferromagnetic phase are magnetized, at this needle-like particles of $\gamma$-$Fe_2O_3$ or $CrO_2$ aspire to align to field with long axis forming on the surface of carbonyl iron particles peculiar brushes. Interactions of such rough formations lead to additional energy dissipation in shear flow.

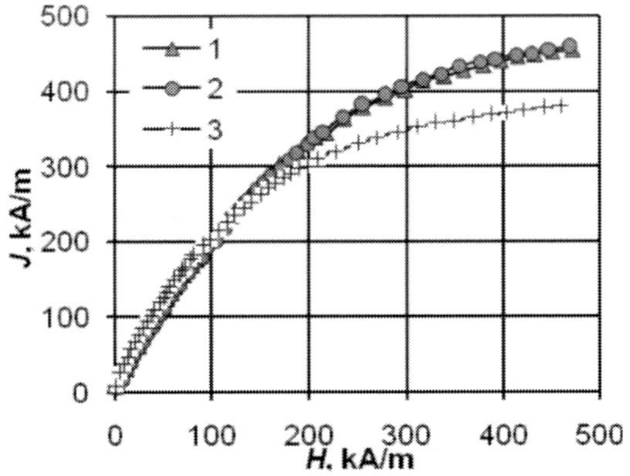


**Figure 1.** Magnetization curves of concentrated MRF.

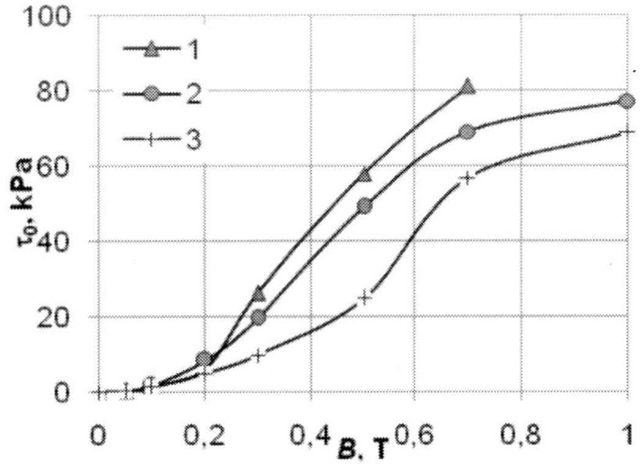


**Figure 2.** Dependence of static yield stress of MRFs (mode 4) on magnetic field induction.

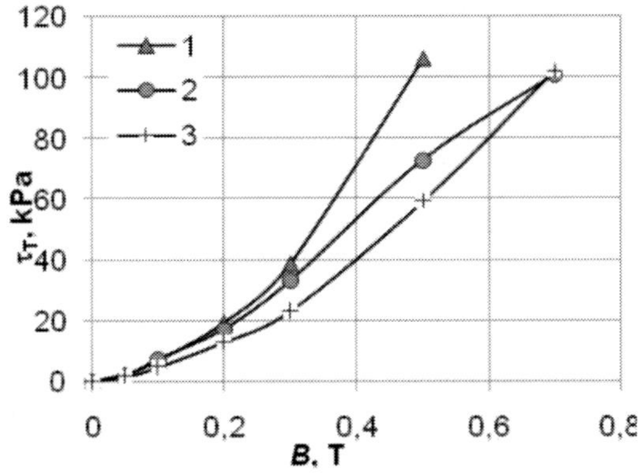


**Figure 3.** Dependence of shear stresses of MRFs on magnetic field induction (mode 1), shear rate – 10 $s^{-1}$.

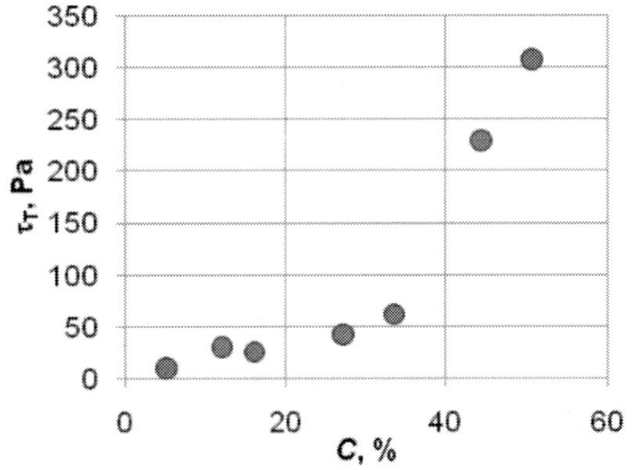


**Figure 4.** Dependence of shear stresses of MRFs on volume concentration of dispersed phase without magnetic field, shear rate – 11.5 $s^{-1}$.

As it is seen from figures 2 and 3, the MRF-1 shows the greater values of shear stresses in the magnetic field, what can be explained by bigger elongation of chromium dioxide particles as against $\gamma$-$Fe_2O_3$ ones and their higher magnetization.

The experiments showed, that MRFs containing general volume concentration of dispersed phase no more than 30% have optimal combination for vibroprotecting devices of low viscosity without magnetic field and essential increment of energy dissipation in the field. For example, in figure 4 dependence of shear stresses of some composition MRFs on volume concentration of dispersed phase without magnetic field is showed.

Therefore for further experiments we have chosen magnetorheological fluid which contains insulating oil and 15.9% of complex filler from iron, chromium oxide $CrO_2$ and silicon oxide $SiO_2$ was investigated in different regimes. Dependences of the shear stress $\tau_T$ vs. shear rate for this fluid presented in figure 5 allow us to determine that with increasing the induction of the external magnetic field all the elastic and viscoplastic characteristics of the MRF increase sharply. First of all, limiting shear stresses, increasing which the plastic shear of the material appears, followed by a flow in the gap between the plates. It is characterized by the small range of relative deformations (less than 1%) and the growth of values $\tau_T$ in 3000 times in comparison with those without the field.

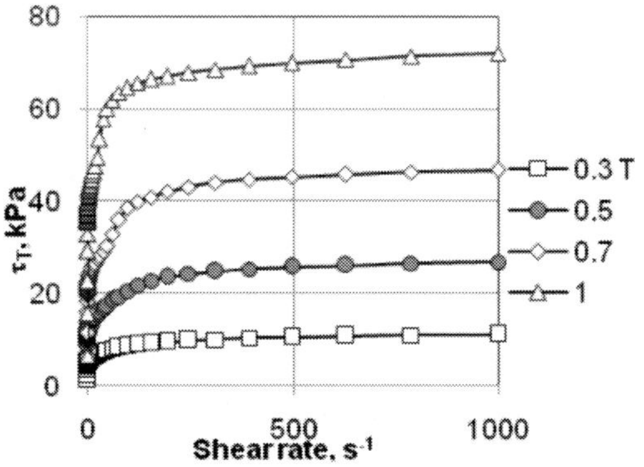


**Figure 5.** Dependence of shear stress of MRE on shear rate in mode 1 at different magnetic field induction.

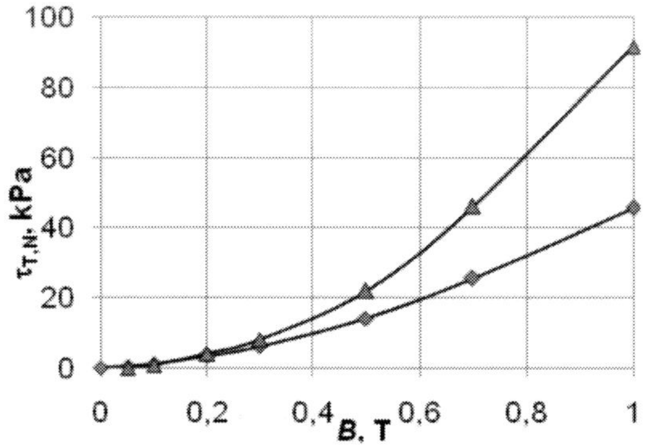


**Figure 6.** Dependence of normal and shear stresses of MRE on magnetic field induction in mode 1, shear rate – 11.5 s$^{-1}$; rhombuses – $\tau_T$, triangles – $\tau_N$.

Normal shear stresses (figure 6) in the continuous deformation mode are practically independent of the shear rate in the range of shear rates about 30 s$^{-1}$. At high shear rates, they increase in the magnetic field with induction of more than 0.1 T. The normal stresses are approximately equal to the shear stresses in the range of the magnetic field up to 0.25 T. At higher fields the normal stresses exceed the shear stresses, reaching values of 120 kPa.

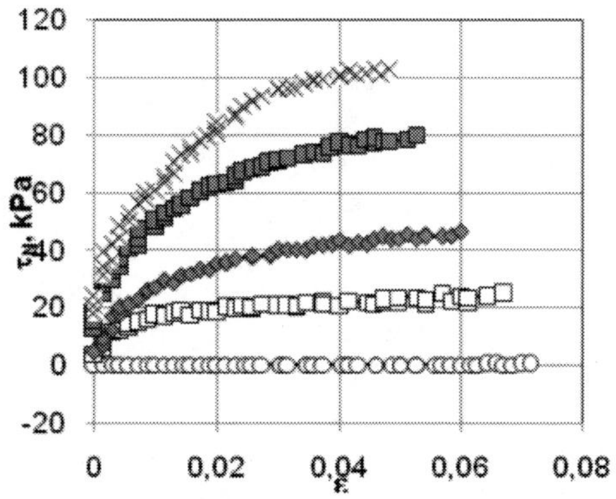


**Figure 7.** Dependence of $\tau_N$ on compressive deformation of MRE (mode 2) at different magnetic field induction (○ – $B$=0, □ – 0.2 T, ♦ – 0.3, ■ – 0.5, × – 1).

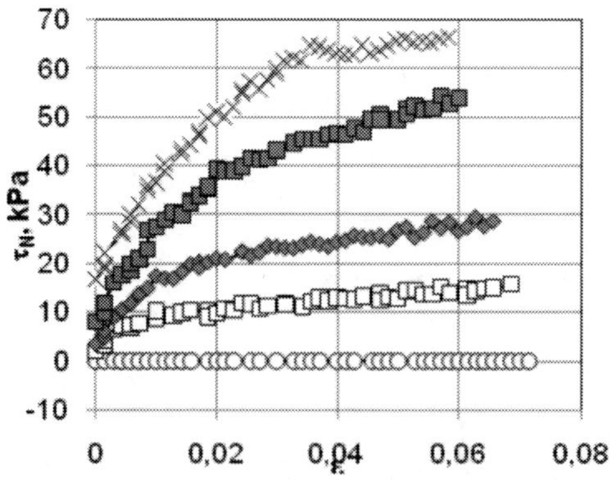


**Figure 8.** Dependence of normal stress on compressive deformation of MRM (mode 3) at shear rate 10 s$^{-1}$ at different magnetic field induction (○ – $B$=0, □ – 0.2 T, ♦ – 0.3, ■ – 0.5, × – 1).

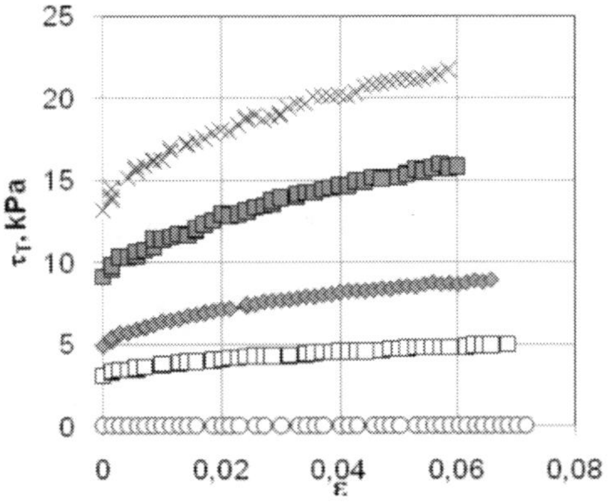


**Figure 9.** Dependence of shear stress $\tau_T$ on compressive deformation at continuous shear deformation with shear rate 10 $s^{-1}$ and vertical compression (mode 3) at different magnetic field induction (○ – $B$=0, □ – 0.2 T, ♦ – 0.3, ■ – 0.5, × – 1).

The dependence of the normal stresses on the strain under compression of MRM according to the mode 2 is shown in figure 7. It is seen that the increase in the normal stresses is the most significant at the beginning of the compression. With the increase in the compressive strain the growth of $\tau_N$ slows down. The higher the intensity of the magnetic field, the smaller compression of the material can be achieved and the higher its resistance to the compression is. Similar relationships are observed for the normal stresses in the vertical compression mode with a continuous simultaneous tangential shear deformation, but $\tau_N$ magnitude is smaller (figure 8). The increase of the shear stress $\tau_T$ at compression is smaller than the increase of $\tau_N$ (figure 9). The shear stresses grow linearly with deformation; they increase at the compression by about 60–70% (mode 3).

## 4. Conclusion

The rheological properties of magnetorheological fluids with complex dispersed phase in the magnetic field depend not only on magnetic properties, carbonyl iron concentration, but and on material of second component of dispersed phase. It has been experimentally shown that for certain modes of the magnetic field impact on the damping magnetosensitive material one can effectively increase the resistance to the external vibration influence both in the tangential and the normal directions. The use of magnetorheological fluids possessing higher structural response on magnetic influence for these purposes will allow to implement tuning parameters of the damping unit to multidirectional vibrations of broad frequency and amplitude range, arising as a result of the effects of internal and external sources.

## References

[1] Qiao Y et al 2007 *Smart Materials and Structures* **16** pp 332–339
[2] Wang X and Gordaninejad F 2001 *Proceedings of SPIE Conference on Smart Materials and Structures* **4331** pp 82–91.
[3] Han Y-M, Seong M-S, Choi S-B and Wereley N M 2010 *Proceedings of the Institution of Mechanical Engineers, Part C: J. Mechanical Engineering Science* **224** pp 293–304
[4] Zhu X, Jing X and Cheng L 2012 *J. Intel. Mater. Syst. and Struct.* **23** pp 839–873
[5] Wang J, Meng G, Feng N and Hahn E 2005 *Smart Materials and Structures* **14** pp 529–539
[6] Choi S B, Nam M H and Lee B K 2002 *J. Intel.Mater.Syst. and Struct.* **11** pp 936-944
[7] Carlson D 2002 *J. Intell.Mater.Sys.Struct.* **13** pp 431-435
[8] Sunkara S R, Root T W, Ulicny J C and Klindenberg D J 2009 *J Phys.: Conf.Ser.* **149** 012081

# Magnetorheology of core-shell carbonyl iron/ZnO rod-like particle silicone oil suspensions under oscillatory shear

**M Mrlik[1, 2*], M Machovsky[1, 2], V Pavlinek[1, 2] and I Kuritka [1, 2]**

[1] Centre of Polymer Systems, University Institute, Tomas Bata University in Zlin, Nad Ovcirnou 3685, 760 01 Zlin, Czech Republic
[2] Polymer Centre, Faculty of Technology, Tomas Bata University in Zlin, nam. T. G. Masaryka 275, 762 72 Zlin, Czech Republic

*E-mail: mrlik@ft.utb.cz

**Abstract**. The aim of this study is a preparation and application of inorganic coating on the surface of carbonyl iron particles. The two step solvothermal synthesis provides core-shell CI/ZnO rod-like morphology. Compact coating of particles has a slightly negative impact on their magnetic properties (measured for magnetic field strength in the range from 0 to 213 mT); however, there is a suitable magnetorheological performance investigated under oscillatory shear, suitable to be applied in real applications.

## 1. Introduction

Suspensions consisting of the solid magnetic particles dispersed in the liquid medium usually silicone oil, changing their basic rheological quantities (viscosity, viscoelastic moduli) upon external magnetic fields are called generally Magnetorheological (MR) [1-9]. Nowadays, material exhibiting this behaviour repeatedly can find various applications as shock absorbers [5] or also in the medical application in the local cancer hyperthermia treatment [9].

In order to obtain the highest MR efficiency of the suspensions usually ferromagnetic particles with low value of coercivity dispersed in medium with negligible magnetic properties are employed [11, 12]. Furthermore, sedimentation stability is another factor significantly influences the MR efficiency. MR suspensions exhibit poor stability, especially when the carbonyl iron is employed as a dispersed phase. Therefore, various techniques were utilized to obtained sufficient sedimentation properties such as bidispersed or dimorphic MR fluids [11, 13-15]. Promising approach consists of preparation of the core-shell based magnetic particles, where a magnetic core is coated with a suitable layer decreasing the overall density of the core-shell particles and enhancing interactions between particles and liquid medium [16, 17].

In the preceding study [18] another approach based on inorganic-inorganic CI/ZnO core-shell particle silicone oil suspension was developed. The sedimentation stability as well as thermo-oxidation one were significantly enhanced and also improved MR efficiency under steady state conditions were confirmed.

In this study, the viscoelastic behaviour of the CI/ZnO was discovered in detail. While the results from steady state conditions provide mostly information about the MR efficiency, the viscoelastic ones describing the behaviour close to the real-life applications.

## 2. Experimental

*2.1 Materials*

Carbonyl iron microparticles (SL grade, α -iron content >99.5 %, BASF, Germany), zinc acetate dihydrate Zn(CH3COO)2.2H2O (ZAD), zinc nitrate hexahydrate Zn(NO3)2.6H2O and hexamethylenetetramine (CH2)6N4 (HMTA) were all purchased from PENTA (Czech Republic) and used as received without further purification. Demineralized water with conductivity about 10-7 Scm-1 and ethanol were used throughout experiments.

*2.2 Synthesis of core-shell particles*

ZnO/CI particles were synthesized via two step reaction. In the first step the seeds of the ZnO particles were created on the surface of the CI. In this case the 0.005 M solution of ZAD and ethanol was prepared and sonicated with CI particles. Then the mixture was heated up to 80°C for 2 hours, filtered and dried at 60°C under vacuum. The second step is creation of the ZnO-rodlike structures on the surface of the CI. Here the mixture of aqueous solution of zinc nitrate hexahydrate (0.05 M) and HMTA (0.05 M) was carried out at 80 °C for 2 h. Finally, the synthesized CI/ZnO particles were dried under vacuum at 60°C.

*2.3 Characterization of the prepared particles*

Crystalline phase of particles was characterized by the powder X-ray diffractometer X´Pert PRO X-ray (PANalytical, The Netherlands) with a Cu-Kα X-ray source (λ = 1.5418 Å) in the diffraction angle range 5-85° 2θ.

The magnetic properties were studied using a vibrating sample magnetometer VSM 7400 (Lake Shore, United States).

*2.4 Suspension preparation*

ZnO/CI core-shell particles were suspended in silicone oil (Lukosiol M 200, viscosity $\eta_c$ = 194 mPa s, density $d_c$ = 0.970 g cm$^{-3}$, relative permittivity $\varepsilon'$ = 2.89, loss factor tan $\delta$ = 0.0001, Chemical Works Kolín, Czech Republic) with 20, 40, and 60 wt. % particle concentrations. The suspensions were mechanically stirred before each measurement. The rheological properties under an external magnetic fields in the range 0–300 mT were investigated using a rotational rheometer Physica MCR502 (Anton Paar GmbH, Austria) equipped with a Physica MRD 170/1T magneto-cell. The true magnetic flux density was measured using a Hall probe.

## 3. Results and Discussion

*3.1 Particles morphology and structure characterization*

As can be seen from the SEM images bare CI particles (Fig. 1a) are mostly spherical including the two main fractions. First and smaller particles are about two microns in diameter. Second and bigger ones are nearly 5 microns in diameter. After the first step of the synthesis the CI particles were coated with ZnO small particles (Fig. 1b), which in the second step promote the radial crystal growth on the ZnO. After the second step, the CI/ZnO rod-like particles (Fig. 1c) can be seen as a result of the solvothermal synthesis.

From the XRD patterns of bare CI, CI with small ZnO particles, and CI/ZnO rod-like particles are can be seen various peaks at $2\theta$ = 44.6°, 65°, and 82.3° correspond to (110), (200), and (211) reflections of iron with the cubic structure (ICDD PDF-2 entry 01-087-0722). The diffraction patterns of CI with small ZnO particles seem to remain unchanged comparing to diffractogram of bare CI, while no more peaks corresponding to ZnO crystal phase were detected. On the other hand the presence of the small ZnO particles on the CI surface was further confirmed by SEM. Diffraction patterns of CI/ZnO rod-like particles contain peaks related to iron cubic structure peaks at $2\theta$= 44.6°, 65°, and 82.3° and as expected also peaks located at $2\theta$= 31.7°, 34.3°, 36.1°, 47.4°, 56.5°, 62.8°, 67.8°

and 68.9° which correspond well to hexagonal wurtzite crystal structure of ZnO (ICDD PDF-2 entry 01-079-0207).

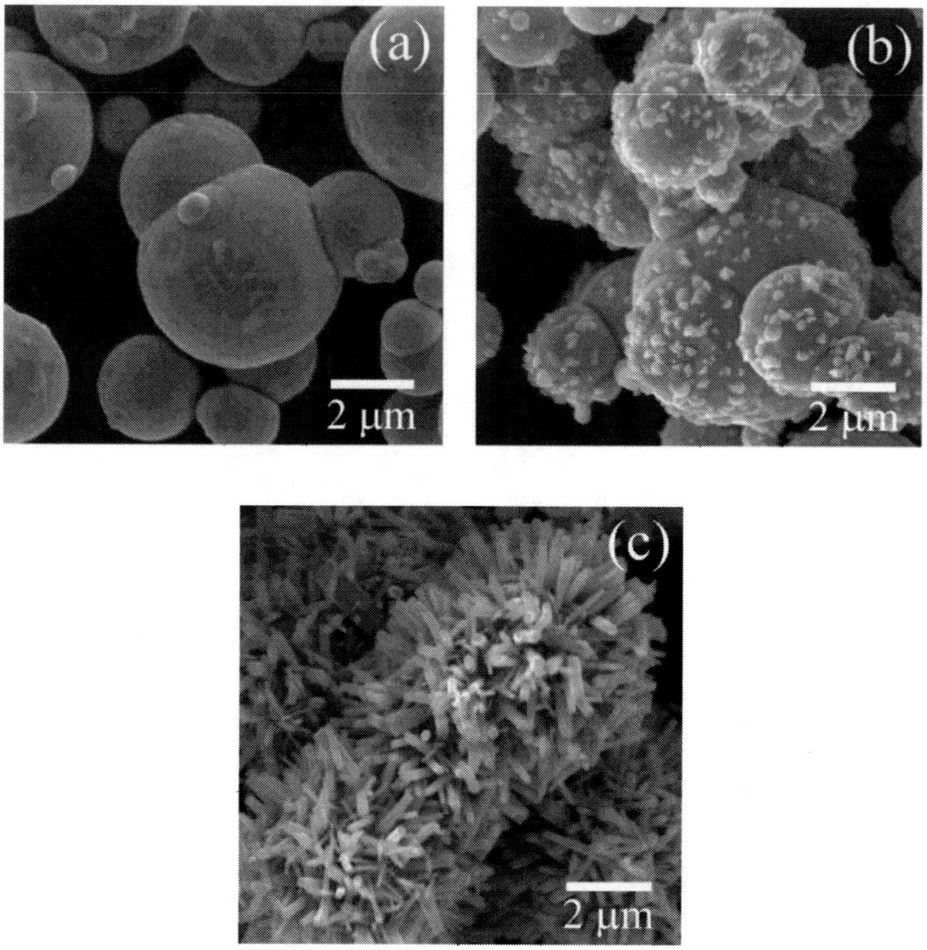


**Figure 1.** SEM images of (a) bare CI particles, (b) CI/ZnO seeded particles, (c) CI/ZnO rod-like particles.

*3.2 Magnetic properties*

The magnetic properties of the synthesized particles were measured due to predict the ability of them to respond on the magnetic field application. All samples exhibit soft magnetic behaviour with negligible remanence and coercivity. The highest magnetization saturation was observed for bare CI particles around 183 emu $g^{-1}$. After the coating of the CI particles with small ZnO ones, magnetization saturation decrease to the 160 emu $g^{-1}$. In the second step after the ZnO rod-like structures were synthesized on the surface of CI the magnetization saturation decrease to 127 emu $g^{-1}$, firstly due to the environmental presence during the synthesis and also due to substantial coating of CI with ZnO rod-like structures.

*3.3 Magnetorheology of particle silicone oil suspensions.*

As can be clearly visible from the viscoelastic investigation of the prepared suspensions, in the absence of the external magnetic field (Fig. 2a), particles are randomly dispersed in the system, because viscous modulus exhibit higher values than that of elastic one. However, after the application of the external magnetic field particle are magnetized and oriented in the streamlines of the external field and form internal structures. Nevertheless, such structures are not strong enough, because with increasing oscillation material start to be more liquid-like (Fig. 2b). With further increasing the magnetic field strength (Fig. 2c and Fig. 2d) magnetostatic forces dominates over hydrodynamic ones and elastic modulus overcame the viscous one in the whole measured frequency range confirming appropriate stiffness of created internal structures.

Furthermore, in order to elucidate the influence of the particle wt. % fraction on the viscoelastic properties of prepared system, the elastic modulus obtained at angular frequency 1 rad $s^{-1}$, was plotted against magnetic flux density. Here can be clearly seen that with increasing particle fraction also increases the stiffness of the created internal structures and is highest for 60 wt. % of CI/ZnO rod-like particles in the suspension. Moreover, such values of the dynamic behaviour of this system can be very promising for their potential applications.

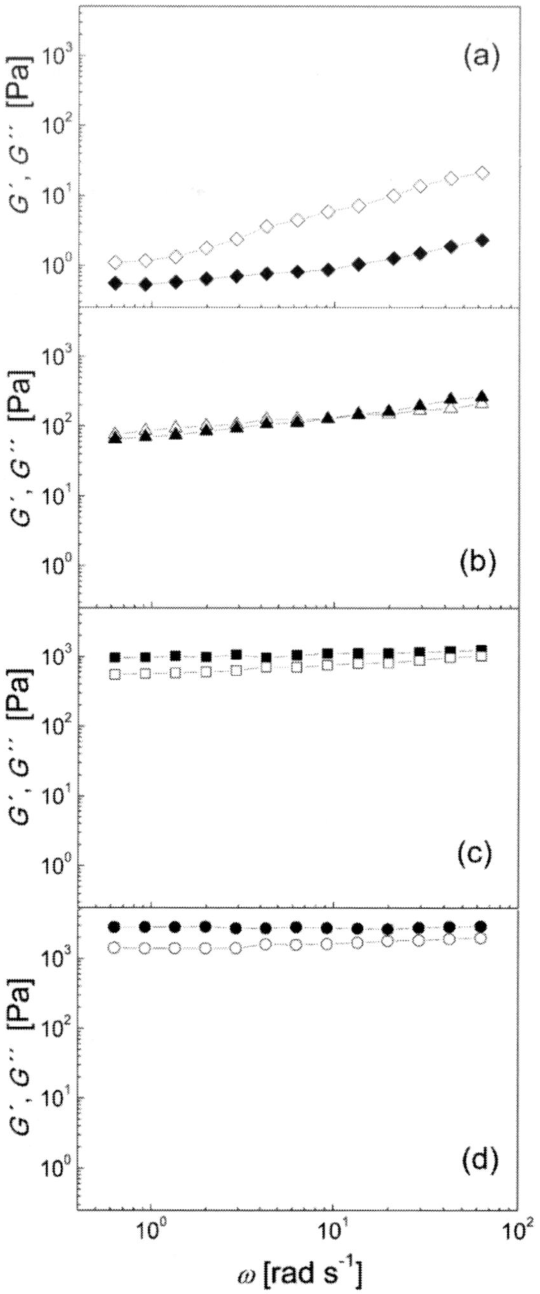


**Figure 2.** Dependence of elastic moduli , $G'$, (solid symbols) and loss moduli, $G''$, (open symbols) on the angular frequency, $\omega$, for 20 wt.% of CI/ZnO rod-like particles silicone oil suspensions under various magnetic flux densities (mT): (a) 0, (b) 44, (c) 126, (d) 213.

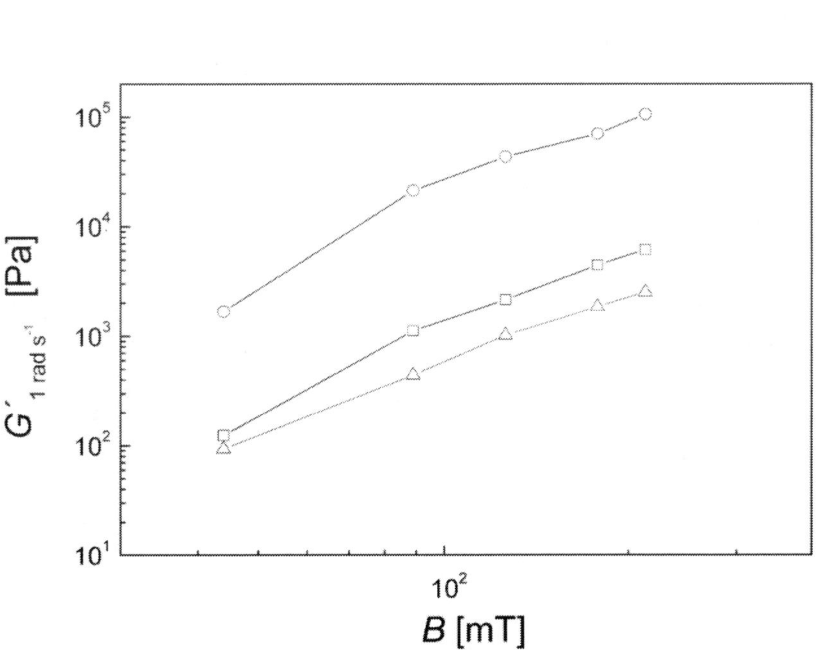


**Figure 2.** Dependence of the elastic modulus G′, at $\omega = 1$ rad s$^{-1}$, on the magnetic flux density for silicon oil suspensions with various concentrations of CI/ZnO rod-like particles (wt. %): (△) 20, (□) 40 and (○) 60.

## 4. Conclusion

In the present study, the simple preparation of the CI/ZnO rod-like particles was introduced. The coating of CI particles was realized via solvothermal synthesis. Magnetic properties of prepared particles are lower in comparison with the bare ones due to coating of CI particles with inorganic ZnO layer. Viscoelastic investigation elucidated the suitable dynamic behaviour of the presented system. Moreover, with increasing particle wt. % fraction the MR performance of such system considerably enhanced and thus can be utilized in their potential applications.

## Acknowledgement

The authors wish to thank the Grant Agency of the Czech Republic for the financial support of Grant No. 202/09/1626.

This contribution was written with support of Operational Program Research and Development for Innovations co-funded by the European Regional Development Fund (ERDF) and national budget of Czech Republic, within the framework of project Centre of Polymer Systems (reg. number: CZ.1.05/2.1.00/03.0111).

## References

[1] de Vicente J, Klingenberg D J and Hidalgo-Alvarez R 2011 *Soft Matter* **7** 3701
[2] Bica I 2006 *J. Ind. Eng. Chem.* **12** 501
[3] Bossis G, Lacis S, Meunier A and Volkova O 2002 *J. Magn. Magn. Mater.* **252** 224
[4] Park B J, Fang, F F and Choi H J 2010 *Soft Matter* **6** 5246
[5] Wang D H and Liao W H 2011 *Smart Mater. Struct.* **20** 023001
[6] Sedlacik M, Pavlinek V, Lehocky M, Mracek A, Grulich O, Svrcinova P, Filip P and Vesel A 2011 *Colloid Surf. A-Physicochem. Eng. Asp.* **387** 99
[7] Mrlik M, Sedlacik M, Pavlinek V, Bazant P, Saha P, Peer P and Filip P 2013 *J. Appl. Polym. Sci.* **128** 2977

[8] Mrlik M, Ilcikova M, Pavlinek V, Mosnacek J, Peer P and Filip P 2013 *J. Colloid Interface Sci.* **396** 146
[9] Sedlacik M, Pavlinek V, Saha P, Svrcinova P and Filip P 2012 *Mod. Phys. Lett. B* **26** 6.
[10] Sedlacik M, Moucka R, Kozakova Z, Kazantseva NE, Pavlinek V, Kuritka I, Kaman O and Peer P 2013 *J. Magn. Magn. Mater.* **326** 7
[11] Lopez-Lopez M T, de Vicente J, Gonzalez-Caballero F and Duran J D G 2005 *Colloid Surf. A-Physicochem. Eng. Asp.* **264** 75
[12] Sedlacik M, Pavlinek V, Saha P, Svrcinova P, Filip P and Stejskal J 2010 *Smart Mater. Struct.* **19** 115008
[13] Fang F F, Choi H J and Jhon M S 2009 *Colloid Surf. A-Physicochem. Eng. Asp.* **351** 46
[14] Wereley N M, Chaudhuri A, Yoo J H, John S, Kotha S, Suggs A, Radhakrishnan R, Loven B J and Sudarshan T S 2006 *J. Intell. Mater. Syst. Struct.* **17** 393
[15] Sedlacik M, Pavlinek V, Vyroubal R, Peer P ,Filip P 2013 *Smart Mater. Struct.* **22** 035011
[16] Fang F F, Choi H J and Seo Y 2010 *ACS Appl. Mater. Interfaces* **2** 54
[17] Fang F F, Choi H J and Choi W S 2010 *Colloid Polym. Sci.* **288** 359
[18] Machovsky M, Mrlik M, Kuritka I, Pavlinek V and Babayan V 2014 *RSC Adv.* DOI:10.1039/C3RA44982C.

# Plasticizer Effect on Rheological Behaviour of Screen Printing Pastes Based on Barium Titanate Nanopowder

**I Dulina[1], S Umerova[2], A Ragulya**

Frantsevich Institute for Problems of Materials Science of National Academy of Science of Ukraine, 3, Krzhyzhanovsky St., Kyiv, 03142, Ukraine

E-mail: i_risha@online.ua and saideshashine@mail.ru

**Abstract**. The dependence of rheological behaviour of pastes based on $BaTiO_3$ nanopowder vs. plasticizer content has been investigated. All pastes prepared for research can be divided into groups by structure types and viscosity. Such a grouping has been explained by different interaction between nanoparticles and binder in the pastes. Particles with molecules of binder form clusters – the representative units in the volume of paste where particles are uniformly distributed. Plasticizer adding effects on binder molecule conformation and change clusters size. Bond strength between clusters can be specified with rheopexy in the area of low shear stress and low strain rates. Rheopexy degree increasing authenticates interaction intensification between clusters. Rheopexy structure destruction leads to separate clusters formation and initiation of the pseudoplastic flow stage. The end of pseudoplastic flow corresponds to structure with clusters assembled into separated layers. Further shear stress increasing leads to inter-clusters bonds appear which can be deformed elastically and the temporary local linkage is possible. Such a phenomenon fully discloses the features of thixotropic structure destruction in plasticized pastes.

## 1. Introduction

The miniaturization of electronic devices such as multilayer ceramic capacitors supports the global trend of downsizing their dimensions and gradual application of nanosize particulate components. The specific capacity of the multilayer ceramic capacitors is reversely proportional to square of dielectric layer thickness and directly proportional to number of electrodes. Thickness of metallic and dielectric layers can be decreased to 100-200 nm by using of nanosize particles in combination with new techniques for layer manufacturing. Screen printing technology is one of the promising one for thin and dense dielectric layers obtaining, high precision, rapidness, low cost [1]. The method allows printing with print's thickness in the range 100 nm – 100 μm [2] complex geometry and high thin line resolution [3]. But such properties as roughness, evenness and film thickness, print geometry strongly depend on viscosity and rheological behavior of pastes and screen printing process parameters [1, 3, 4]. In accordance with [2-8], the main requirements to rheological behavior of pastes are pseudoplastic or pseudoplastic-thixotropic flow type and viscosity value must be low enough for paste retention on screen and free smoothing of pattern surface after deposition.

Accordingly to our previous investigation [9-10], ethyl cellulose has considerable affinity to $BaTiO_3$ nanopowder. This leads to sizeable paste densification, rheological behavior change and solid content decrease to 10 – 15 % wt of nanopowder as compared to typical amount for submicron powders of 50 – 70 % wt. One more important factor directly effecting on rheology is plasticizer content. Thus, present paper is aimed at investigation of dependence of rheological behavior of screen printing pastes based on $BaTiO_3$ nanopowder from plasticizer content.

## 2. Materials and Methods

Screen printing pastes have been prepared by using of $BaTiO_3$ nanopowder with mean particles size of 20 nm, ethyl cellulose with viscosity of 5 wt. % toluene-ethanol solution 10 mPa·s (Merck) as a binder, terpineol (Merck) as a solvent and dibutyl phthalate (Merck) as a plasticizer. Solid and binder content in pastes were invariable during experiments of 10 and 3 wt. % respectively.

Pastes have been prepared in planetary ball mill Pulverisette 6 (Fritsch) with rotating velocity of 450 rpm. Pastes have been characterized with rheological viscosity analysis in shear stress range of 1 – 1000 Pa using rotary viscometer Rheotest RN 4.1 (Rheotest Messgerate Medingen GmbH) with cylinder measuring system.

## 3. Results and Discussion

The flow curves in the Fig. 1 – 3, Table 1 and Fig. 4 show that the plasticizer content increasing determines the viscosity and rheological properties indirectly because the molecule of ethyl cellulose changes conformation with addition of plasticizer. First, the change in molecule conformation directly effects on adsorption of polymer molecules on nanoparticle. In general, several adsorption types may be realized in this systems, for example polymer adsorption on the nanoparticles surfaces, nanoparticles adsorption on polymer molecule, polymer molecule bonding to nanoparticle surface through one part or whole molecule. Adsorption type determines possible structuration type and viscosity of the system.

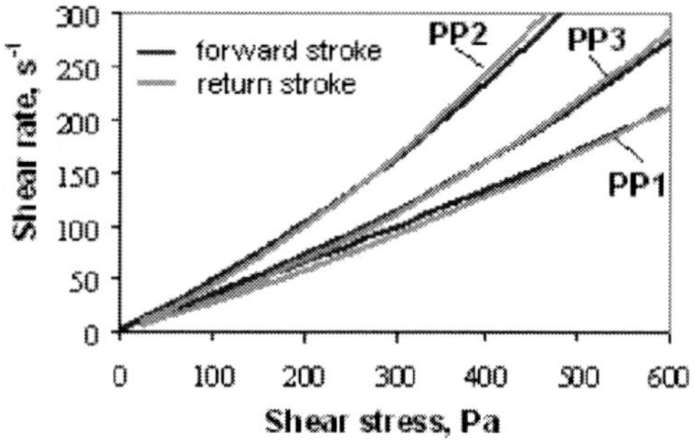


**Figure 1.** Flow curves of pastes PP1 – PP3

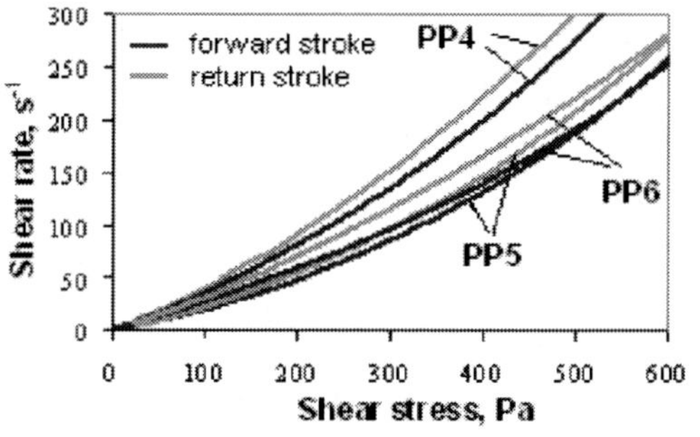


**Figure 2.** Flow curves of pastes PP4 – PP6

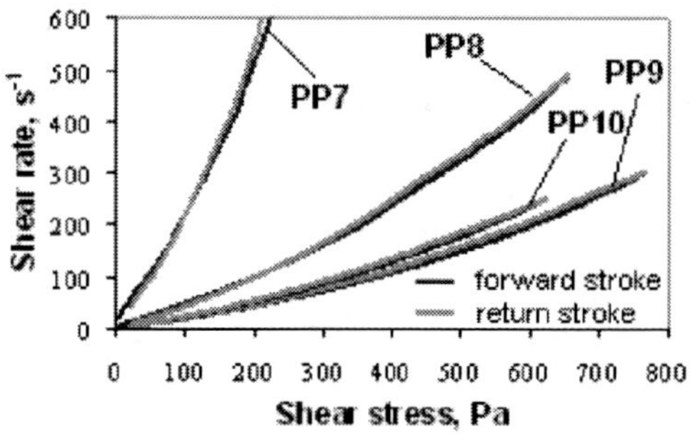


**Figure 3.** Flow curves of pastes PP7 – PP10

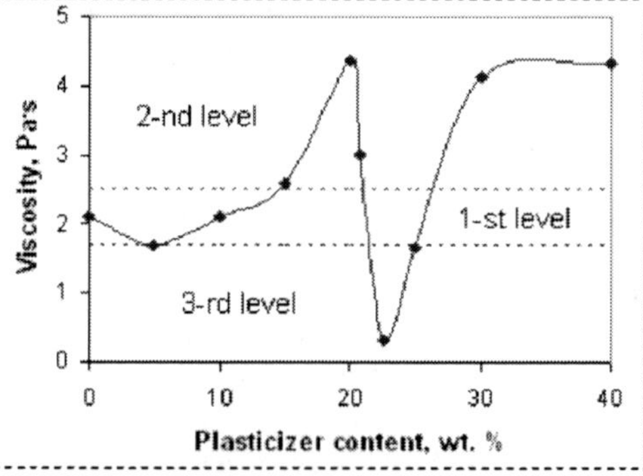


**Figure 4.** Viscosity dependence on plasticizer content

**Table 1.** Composition and rheological properties of pastes based on $BaTiO_3$ nanopowder and plasticizer (ethyl cellulose 10 mPa·s)

| Paste | Plasticizer content, wt. % | Yield stress, Pa | Dilatant flow finishing, $s^{-1}$ | Maximum viscosity of dilatancy, Pa·s | Thixotropy degree, MPa/s | Dilatancy degree, a.u. | Rheopexy degree, Pa·s | Degree of equilibrium structure estruction | Pseudoplastic flow start, Pa |
|---|---|---|---|---|---|---|---|---|---|
| **PP1** | 0.0 | 185.28 | 100.90 | 3.04 | 0.0002 | 0.4408 | 0.0033 | 2.1628 | 584.40 |
| **PP2** | 5.0 | 262.15 | 31.78 | 2.04 | 0.0041 | 0.2000 | 0.0009 | 0.5000 | 248.60 |
| **PP3** | 10.0 | 254.72 | 47.69 | 2.76 | 0.0012 | 0.3019 | 0.0014 | 0.8101 | 362.10 |
| **PP4** | 15.0 | 243.57 | 23.60 | 2.87 | 0.0100 | 0.1124 | 0.0000 | 0.2164 | 0.00 |
| **PP5** | 20.0 | 320.00 | 14.40 | 4.91 | 0.0081 | 0.1261 | 0.0000 | 0.1864 | 25.33 |
| **PP6** | 20.8 | 294.09 | 32.23 | 3.54 | 0.0130 | 0.1761 | 0.0000 | 0.3557 | 23.64 |
| **PP7** | 22.5 | 156.99 | 137.00 | 0.48 | 0.0080 | 0.3988 | 0.0008 | 0.7351 | 0.00 |
| **PP8** | 25.0 | 259.64 | 33.46 | 2.05 | 0.0039 | 0.2202 | 0.0008 | 0.5211 | 212.90 |
| **PP9** | 30.0 | 307.57 | 23.68 | 4.95 | 0.0041 | 0.1985 | 0.0001 | 0.3402 | 140.60 |
| **PP10** | 40.0 | 230.88 | 16.16 | 4.62 | 0.0026 | 0.0670 | 0.0000 | 0.1349 | 20.13 |

In general, all pastes of Table 1 can be divided into 4 groups in concordance with structuration types. For the pastes PP1, PP2, PP3, PP8 and PP9, the rheopexic pseudoplastic flow regime succeeded by pseudoplastic and thixotropic pseoudoplactic regimes. In the PP5, PP6, PP10, the pseudoplastic flow regime succeeded by thixotropic pseoudoplactic one. The PP4 has demonstrated thixotropic pseoudoplactic flowing type only and in the PP7, the rheopexic pseudoplastic flow regime succeeded by thixotropic pseoudoplactic one. In addition, three viscosity levels have been considered (Fig. 4). For the pastes PP1, PP2, PP3, PP8, an unbreakable structure viscosity was 1.7 – 2.5 Pa·s (1-st viscosity level), for the pastes PP4, PP5, PP6, PP9, PP10, unbreakable structure viscosity was over 2.5 Pa·s (2-nd level) and for the paste PP7, the viscosity was found less than 1.7 Pa·s (3-rd level). Pastes have been classified by different structural types and by viscosity levels in accordance with the interaction between structural particles and organic components in the pastes.

Existence of the dilatant flow interval on the curves for all pastes is considered the best evidence for 3D-net structure formation. At low shear stresses the structural bonds of such net elongate and viscosity increases. In this case, the dilatancy degree characterizes dimension rising and initial structure strength. Rheopexic or pseudoplastic flow is the evidence of 3D-net disintegration as a result of critical shear stress reaching. Existence of rheopexy is explainable with strong interaction between elementary structural clusters consisting of nanoparticles with polymers tailored into net. Breaking of these bonds requires significant energy to allow destruction dominate over structure reduction processes under shear stress applied (Fig. 5). In substitution pseudoplactic flow interval indicates the bounds braking between clusters as result of lattice volume increasing because of solvent molecules penetration in structure. According to adsorption theory, free functional groups in ethyl cellulose molecules can be observed in rheopexy structures. For the pastes PP1 – PP3, this may be explained with insufficient plasticizer quantity and for the PP8, PP9, with the presence of such groups in the separate molecule parts. In the case of pseudoplactic structuration, the polymer remains bonded with $BaTiO_3$ nanoparticles in that conformation, where all free functional groups are used. Such an adsorption type strongly increases the viscosity of pastes as result of more uniform cluster distribution in the volume of pastes (2-nd viscosity level). This behavior corresponds to decreasing of dilatancy degree (Fig. 6) and dilatant flow interval on flow curves (Fig. 7).

The destruction of rheopexic structure leads to separate clusters formation. This corresponds to initiation of the pseudoplastic flow stage. The rheopexy degree increase shifts the beginning of pseudoplactic stage to higher stress range (Fig. 8) due to reinforcement of bonds between clusters in the initial 3D-net. The end of pseudoplastic flow corresponds to structure with clusters assembled into separated layers. Further increasing of shear stress leads to difference between strain rates of adjacent cluster layers. Thus, the arrears of intercluster bonds are observed clearly. In this case, the inter-clusters bonds appear which can be deformed elastically and the temporary local linkage is possible.

Such a phenomenon fully discloses the features of thixotropic structure destruction in plasticized pastes and determines the reduction degree for the structure right after unloading.

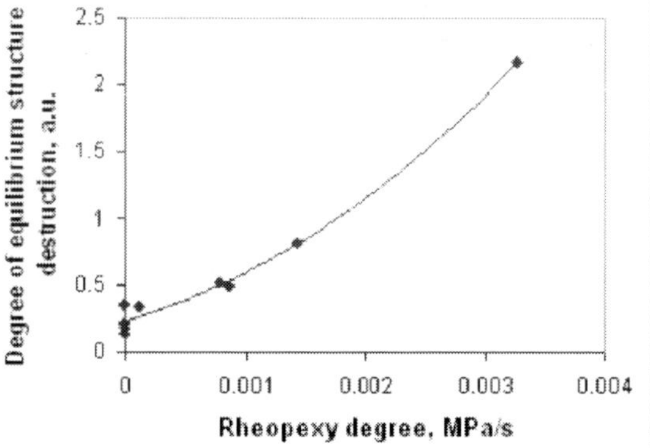


**Figure 5.** Dependence of degree of equilibrium structure destruction from rheopexy degree for plasticized pastes

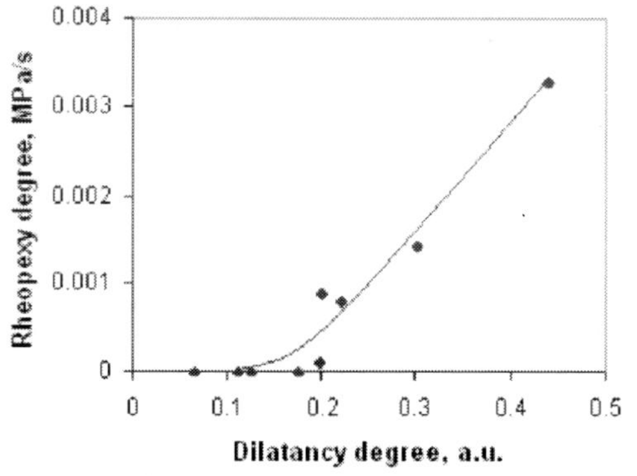


**Figure 6.** Rheopexy degree dependence on dilatancy degree for plasticized pastes

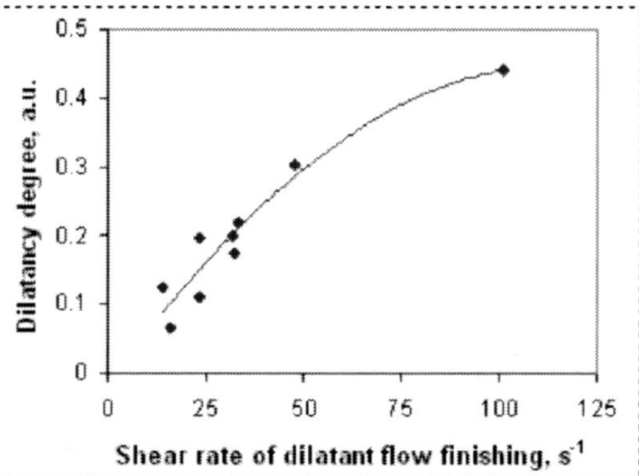


**Figure 7.** Dilatancy degree dependence on dilatant flow finishing

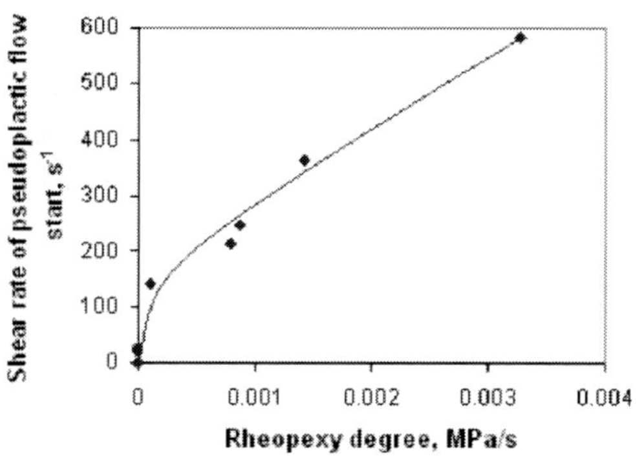


**Figure 8.** Dependence of pseudoplastic flow start from rheopexy degree

## 4. Conclusions

The dependence of rheological behavior of pastes based on $BaTiO_3$ nanopowder vs. plasticizer content has been investigated. Plasticizer addition leads to changing of molecule conformation and adsorption mechanism of ethyl cellulose on $BaTiO_3$. Interaction degree of polymer-plasticizer determines the viscosity for unbreakable structure and system stability under various conditions, which are specified for various flow types under adjusted shear stress range. Dilatant flow region has been observed for all nanopowder-polymer-plasticizer structures and related to the 3D-net in the pastes. Paste dilatancy degree determines the critical destruction stress for structure and structure strength. Interaction strength between system clusters is specified with existence of rheopexy at low shear stresses. The rheopexy degree increasing authenticates interaction between clusters. Rheopexy structure destruction leads to separate clusters formation and initiation of the pseudoplastic flow stage. The end of pseudoplastic flow corresponds to structure with clusters assembled into separated layers. Further shear stress increasing leads to inter-clusters bonds appear which can be deformed elastically and the temporary local linkage is possible. Such a phenomenon fully discloses the features of thixotropic structure destruction in plasticized pastes.

**Acknowledgments**
Authors are grateful to the State Agency on Science, Innovations and Informatization of Ukraine for a financial support for the project NN352-2012.

**References**
[1] Pan J, Tonkay G L, Quintero A 1998 *Proc. Int. Symp. on Microelectronic (San Diego)* p 264
[2] Beeby S, Ensell G, Kraft V, White N 2004 *MEMS Mechanical Sensors* (Boston-London.: Artech House, Inc.)
[3] Aegerter M A, Mennig M 2004 *Sol-gel technologies for glass producers and users* (New York: Springer)
[4] Tsai T N 2008 *Comp. Ind. Eng.* **54** 374
[5] Vijatovic M M, Bobic J D, Stojanovic B D, Malic B. 2010 *Process. Appl. Ceram* **4** 2 53
[6] Im D-H, Hyun S-H, Park S-Y, Lee B-Y, Kim Y-H 2006 *Mater. Chem. Phys.* **96** 228
[7] Leppavuori S, Uusimaki A, Hannula T 1981 *Thin Solid Films.* **86** 287
[8] Lin H-W, Chang C-P, Hwu W-H, Ger M-D 2008 *J. Mater. Process. Technol.* **197** 284
[9] Dulina I O, Umerova S O, Ragulya A V 2013 *Pros. Int. Conf. Nanomaterials: Application and Properties (Alushta)* **2** 1 p 01NTF10-1
[10] Dulina I, Umerova S, Ragulya A 2013 in L A Gömze *Book of Abstracts 1st Int. Conf. on Rheology and Modeling Materials (Miskolc-Lillafured, Hungary)* **ISBN: 978-963-08-7390-1** p. 128

# Effect of $BaTiO_3$ Nanopowder Concentration on Rheological Behaviour of Ceramic Inkjet Inks

**R Kyrpal[1], I Dulina, A Ragulya**

Frantsevich Institute for Problems of Materials Science of National Academy of Science of Ukraine, 3, Krzhyzhanovsky St., Kyiv, 03142, Ukraine

E-mail: k_r_o@ukr.net[1]

**Abstract**. The relationship between rheological properties of ceramic inkjet inks based on $BaTiO_3$ nanopowder and solid phase concentration has been investigated. In the ink volume takes place the formation periodic colloidal structures (PCS). The determining factor of structure formation is powder-dispersant ratio. Structural constitution of in the system with the low pigment concentration represented as $PCS_2$, that contains solid particles in deflocculated that stabilized by the presence of adsorption-solvate layers. Dilatant structure formation for such inks explained by constrained conditions of the interaction. Samples with high $BaTiO_3$ concentration have been classified as $PKS_1$. Dilatant properties of the $PKS_1$ resulted in particles rearrangement under the influence of the flow. In the region of some values powder-dispersant ratio take place conversation $PKS_2$ to $PKS_1$ and ink structure transformation from monodisperse to aggregate state.

## 1. Introduction

Current trend to decreasing portable devices leads to significant reduce the overall size of capacitors with a simultaneous increase in the specific capacity. It is known that the specific capacitance of a ceramic capacitor can be increased by thinning the dielectric layers and increasing of electrodes number. Thickness dielectric layers can be decreased to 100-200 nm by using of nanosize particles and new techniques of layer manufacturing. Inkjet printing is one of the most promising techniques for this application. The main advantages of inkjet printing, compared to other deposition methods one-step processing, low-cost and compact equipment, and applicability to various substrates [1].

Rheology of the ink is of great importance for its performance during jetting and spreading on the substrate. During the printing process high shear rates are exerted on the ink, so the shear dependent viscosity of an ink may strongly influence the inkjet pumping mechanism. In accordance with [1-3] the rheology of suspensions in general shows a shear dependency that depends on the volume fraction of solids, the particles size and shape, and the interparticle forces.

In accordance with our previous research [4-5], the most significant stabilization of the 5 wt. % pigment was observed for samples formulated n-butyl alcohol doped with 20 wt. % of dispersant. Rheological behavior has been changed on adding different amounts of dispersant. But the most important factor which effects on rheology directly has been solid fraction content. Thus, work is aimed at dependence of rheological behavior of ceramic inkjet inks based on $BaTiO_3$ nanopowder from powder content.

## 2. Materials and Methods

Inks have been formulated in the medium of n-butanol (Lab-Scan). BaTiO3 nanopowder with a mean particle size of 20 nm has been used as green material for the preparation of ceramic inks by the mechanical mixing method. Saturated aliphatic polyatomic alcohol with a symmetrical arrangement of hydroxyl groups has been used as a dispersant.

The ink for this application has been produced by two steps. At first one, the barium titanate nanopowder has been milled for one hour with liquid dispersant in planetary-type mill to obtain adsorbed layer on the surface of particles and then solution of the rest components has been added and milled for an hour again. Agate balls of 5 mm in diameter were used to pulverizing the soft agglomerated particles into primary particles. Formulated inks have been characterized by rheological viscosity analysis in shear stress range 1 – 1500 Pa at 25°C with using of rotary viscometer Rheotest RN 4.1 (Rheotest Messgerate Medingen GmbH) with cylinder measuring system.

## 3. Results and Discussion

The flow curves in the Fig. 1 – 4, Table 1 shows that all ceramic inks based on $BaTiO_3$ nanopowder demonstrate dilatant flow with an established rate of dilatant deformation. It means that all systems flow as completely deflocculated dispersions after attaining the established equilibrium level of strain rate [6]. In general, according to the changing of viscosity and rheological characteristics all formulated inks can be groped into newtonian-dilatancy and dilatant-thixotropic systems.

**Table 1.** Composition and rheological properties of investigated ceramic inks

| Ink | Composition, wt. % | | | Viscosity, mPa·s (22 Pa) | Dilatancy degree, a.u. | Boundary shear stress, Pa | Tixotropy degree, mPa/s | Equilibrium shear rate, $s^{-1}$ |
|---|---|---|---|---|---|---|---|---|
| | $BaTiO_3$ | dispersant | n-butanol | | | | | |
| **Ink 1** | 0 | 20.0 | 80.0 | 10.95 | 2.88 | 16.08 | – | 3205 |
| **Ink 2** | 0.5 | 19.9 | 79.6 | 9.49 | 4.74 | 8.84 | – | 3210 |
| **Ink 3** | 1 | 19.8 | 79.2 | 9.56 | 5.10 | 6.21 | – | 3235 |
| **Ink 4** | 2 | 19.6 | 78.4 | 9.79 | 5.68 | 2.45 | – | 3240 |
| **Ink 5** | 3 | 19.4 | 77.6 | 10.06 | 6.89 | 0.88 | -76.93* | 3080 |
| **Ink 6** | 4 | 19.2 | 76.8 | 10.72 | 13.68 | 1.84 | 1169.15 | 3130 |
| **Ink 7** | 5 | 19.0 | 76.0 | 11.52 | 15.15 | 3.19 | 1487.03 | 3140 |
| **Ink 8** | 10 | 18.0 | 72.0 | 18.22 | 18.56 | 4.60 | 2477.96 | 3150 |

*Note: The value modulo corresponds to the rheopexy degree

In general, such rheological behavior can be explained by physicochemical and molecular-kinetic interactions between components of the system. The determining factor of structure formation is powder-dispersant ratio which determines the composition and shape of the structural elements and their interaction conditions. Fig. 5 and Fig. 6 shows that in the absence of the solid phase (Ink_1), the structural elements formation occurs due to the fluctuation dipole interaction through the hydrogen bonds formation between hydroxyl groups of the dispersant and n-butanol. While the shear stress is low, hydrogen bonds have time to regroup, which leads to the longest length of newtonian initial region. At higher shear rates fluctuation contacts are destroyed, resulting in compressed deformation conditions [7].

Adding to system the small amount of powder (Ink_2), accompanied by a reducing amount of the free kinetic dispersant through the formation of adsorbed solvate shells on the highly developed surface of barium titanate nanopowder. In connection with the decreasing of kinetically free hydroxyl groups leveling couple fluctuation and the consequent degeneration of the ability of the structure to the applied shear stress relaxation is take place. This causes a sharp decreasing of the viscosity value of the Ink_1 - Ink_2, and duration initial Newtonian segment.

Further increasing $BaTiO_3$ concentration Ink_2 - Ink_5 resulted in the growth of specific effective volume of the dispersed phase and the expansion of adsorption layers. In turn, vacancy volume reducing results in more confined structure deformation conditions that makes difficult relaxation of the applied stress. This causes displacement of the beginning dilatant area flow curves in the region of smaller shear rates and higher pressures and also the gradual reduction newtonian flow range. In addition, there is a gradual increase in viscosity and degree of dilatancy values. Ink_5 components ratio corresponds to the formation of highly filled system.

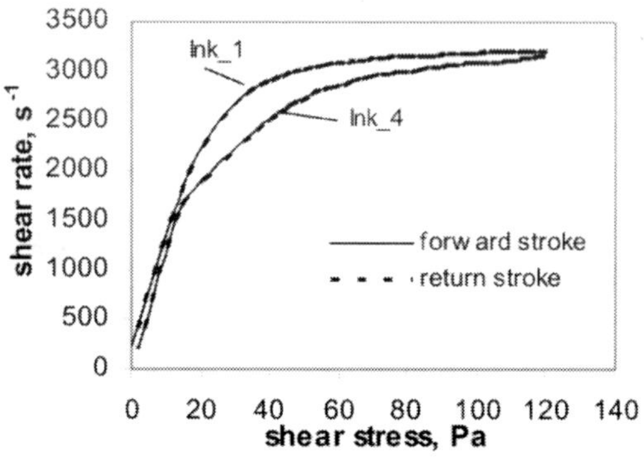


**Figure 1.** Flow curves of Ink_1 and Ink_4

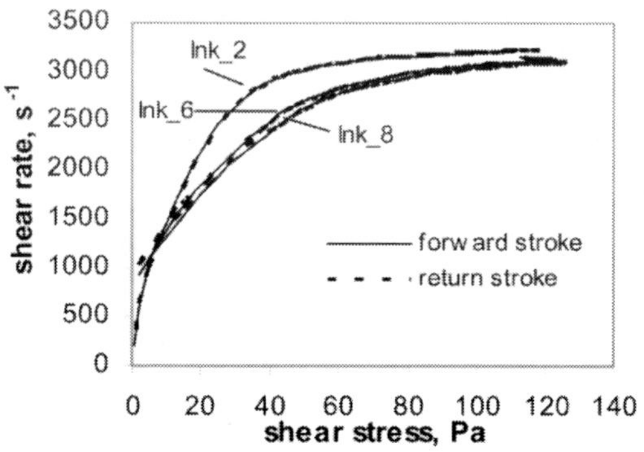


**Figure 2.** Flow curves of Ink_2, Ink_6 and Ink_8

Probably, Ink_5 structure is characterized by thinning the ink adsorption solvate shells due to the increasing specific surface area particles of the barium titanate volume fraction, resulting in a more compressed condition interaction. This assumption is confirmed by complete degradation newtonian plot on the corresponded flow curve and appearance the rheopectic-thixotropic hysteresis loop (fig.4).

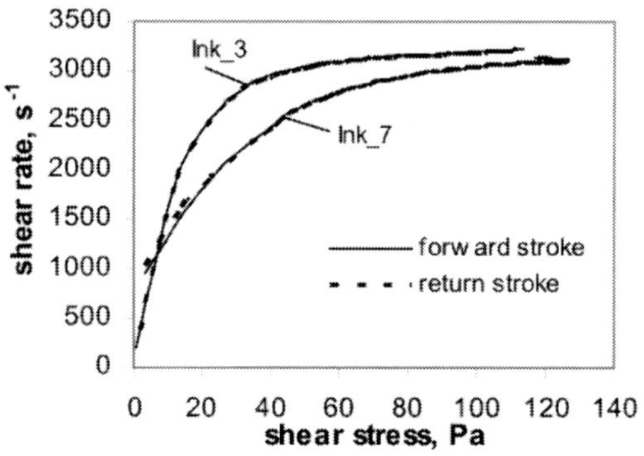


**Figure 3.** Flow curves of Ink_3 and Ink_7

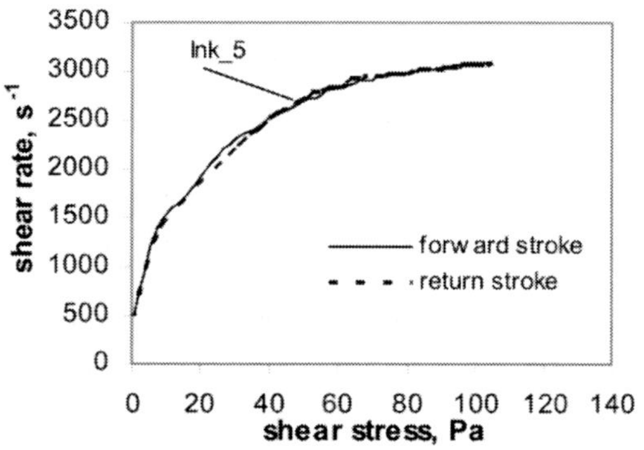


**Figure 4.** Flow curves of Ink_5

With further reduction of the dispersant concentration on the interval Ink_5 - Ink_8 thinning solvent protective layer take place that enhances the processes of aggregation and coagulation. This is expressed in sharp difference in the concentration dependences and is accompanied by a display of the dilatant flow with thixotropic viscosity hysteresis on the initial period of curves Ink_6 - Ink_9 that caused by the strengthening of colloidal structure under the influence of directional flow due to destruction of the original compact aggregates. Besides, there is take place a destruction this aggregates with their simultaneous peptization under the influence of hydrodynamic forces. As a result, structural elements formation with more compact aggregates occurred during system relaxation. A typical hysteresis loop rupture between the forward and reverse curves indicates that the structural change of the system proceeds irreversibly [8].

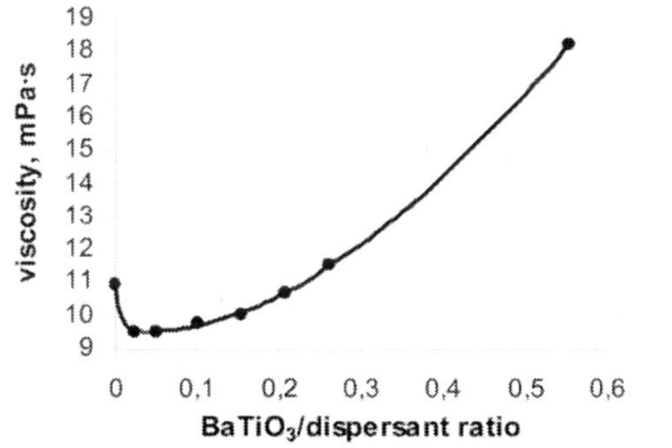


**Figure 5.** Dependence of viscosity from $BaTiO_3$–dispersant ratio

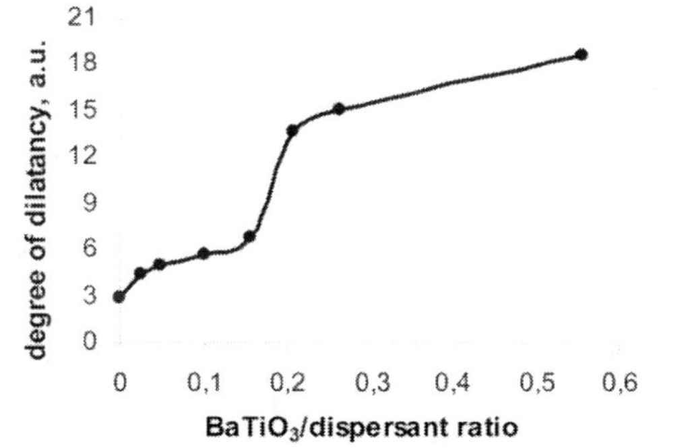


**Figure 6.** Dependence of dilatancy degree from BaTiO3–dispersant ratio

These results suggest that in the ink volume take place the formation periodic colloidal structures (PCS). In this case, structural constitution of the Ink_1 - Ink_4 in the system represented as $PCS_2$, that contains solid particles in deflocculated that stabilized by the presence of adsorption-solvate layers. Dilatant structure formation for such inks explained by constrained conditions of the interaction. At the same time, despite the demonstration of the rheological hysteresis effects, structure of the Ink_5 also belong to ordered $PCS_2$. This is explained by the fact that the system is more rheopectic than thixotropic, indicates that the dominance of repulsive forces under the forces of attraction between the structural elements. Therefore, one could argue about the overwhelming action of disjoining pressure $\pi_2 > 0$, that typical for the second kind of colloidal structures. Most probably that the formation of relaxation effects on the Ink_5 flow curve related to thinning of the solvation shells due to increasing value of the $BaTiO_3$–dispersant ratio.

In turn, the structure nature of samples Ink_6 – Ink_8 has been classified as $PKS_1$. This assumption evidenced by the formation of structurally defining hysteresis and anomalous effects such as initial loop breaking and disrupts the flow. Dilatant properties of the $PKS_1$ lead to stabilization of the colloidal structure as a result of its rearrangement under the influence of the flow. The end of tixotropic flow corresponds to formation completely deflocculated structure. The initial structure of the Ink_6, Ink_7 and - Ink_8 presented in the form of the compact aggregates [7].

Thus, during increasing of $BaTiO_3$-dispersant ratio to the values that lies between Ink_5 and Ink_6 (0.10 – 0.15) the formation systems with inverse properties take place. In other words, there is conversion $PKS_2$ to $PKS_1$. And summing all of the above, we can conclude that the dispersant content of Ink_5 (19.4 wt. %) corresponds to the critical concentration stabilization system.

## 4. Conclusions

The relationship between rheological properties of ceramic inkjet inks based on $BaTiO_3$ nanopowder and solid phase concentration has been investigated. Formation periodic colloidal structures (PCS) take place in the ink volume. The determining factor of structure formation is ratio powder-dispersant. At low powder concentrations system behaves as a dilatant periodic colloidal structure of the second type with a limited volume of the medium ($PCS_2$) in which a positive disjoining pressure $\pi_2 > 0$ operates. In turn, the nature of the highly loaded samples refers to $PKS_1$, as evidenced by the formation of hysteresis, and manifestation anomalous effects such as break the loop and disrupt the flow. In the region of values powder-dispersant ratio of about 0.10 – 0.15 take place conversion $PKS_2$ to $PKS_1$ and ink structure transform to destabilize monodisperse to aggregate state. Notably, critical solvation shell volume reached at the dispersant concentration 6.5 wt. %. All formulated inks demonstrate dilatant flow with an established rate of dilatant deformation. It follows that at high shear rates samples as completely deflocculated $PKS_2$.

## 5. Acknowledgments
Authors are grateful to the State Agency on Science, Innovations and Informatization of Ukraine for a financial support for the project NN352-2012.

## References
[1] Magdassi S 2010 *The Chemistry of Inkjet Inks* (Singapore.: World Scientific Publishing)
[2] Korvink J G, Smith P J., Shin D.-Y. 2012 *Inkjet-based Micromanufacturing* (Weinheim.: Wiley)
[3] Woo K, Jang D, Kim Y, Moon J. 2013 *Ceram. Int.* **39** 7015
[4] Kyrpal R O, Dulina I O, Umerova S O, Nikulin A G, Ragulya A V 2013 *Pros. Int. Conf. Nanomaterials: Application and Properties (Alushta)* **2** 1 p 01NTF35-1
[5] Kyrpal R, Nikulin A, Dulina I, Ivanchenko S, Ragulya A 2013 *Proc. 1-st Int. Conf. on Rheology and Modeling Materials (Miskolc-Lillafured)* p 134
[6] Pivinskii Yu E 1997 Refract. Ind. Ceram **38** 7015
[7] Efremov I F 1982 Russ. Chem. Rev. **51** 160
[8] Malkin A Ya, Isayev A I 2012 *Rheology: concepts, methods, and applications* (Toronto.: Chemtec Publishing)

# Determination of the liquid steel viscosity curves using a high temperature rheometer

**M. Korolczuk-Hejnak**[1], **P. Migas**[1], **W. Ślęzak**[1]
AGH-University of Science and Technology, 30-059 Krakow, 30 Mickiewicza Av., Poland

E-Mail: mkorolcz@agh.edu.pl

**Abstract.** Rheological measurements of high melting temperature materials require specialist equipment, which allow to obtain high temperature inside. This situation take place in case of iron alloys. In literature there is lack of information about rheological properties of liquid iron solutions. This paper is intended to supplement the data in this regard; included in the study of iron solutions rheological properties: five chemical composition with carbon amount in the range between 0.39 – 0.89 %, while used shear rate in the range between 40-180 $s^{-1}$. Obtained viscosity values are set in the range between 0.002-0.018 Pa·s. The obtained results could be used for verification of the model which calculate the parameters of liquid iron alloys during casting, in which the viscosity is one of the mane parameter.

## 1. Introduction

Rheology is a branch of science that has evolved from a branch of physics. Today, however, it is an independent area of knowledge that dates back over 70 years. The rheological behaviour of a material is described by the relationships between stresses, strains, strain rates, and the time in which the material has been subjected to such a strain. Such relationships are called rheological equations of state of the material, or for short - rheological equations [1]. The main task of rheology is to develop models to describe the behaviours of bodies that have been subjected to a force impact. Descriptions of the behaviour of ideal solid and liquid bodies have been formulated in the form of mathematical models that have taken into account the relationships between the stress, strain, strain rate and the stress growth rate (rheological function [2]). However, it should be emphasised that the following rheological equations of state (1-3) describe idealised bodies - ideal bodies that do not exist in reality. The concept of the ideal body simplifies theoretical considerations and enables mathematical methods to be applied to analyse any interesting phenomena. Ideal bodies are only extreme cases to which real bodies resemble to a lesser or higher degree. In specific conditions rheological properties of real bodies are approximated by mathematical rheological models; these are combinations of three basic rheological models of ideal bodies, such as [3]:
Hooke ideal elastic body,
St. Venant ideal plastic body,
Newtonian ideal viscous fluid.

Rheological tests of materials characterised by high melting temperatures require specialist equipment; these enable samples to be heated to high temperatures. This is the case for, among others, ferrous alloys: however, subject literature is short of data containing information on rheological

[1] To whom any correspondence should be addressed. mkorolcz@agh.edu.pl

properties of liquid ferrous solutions. This study aims at supplementing data in this area. It contains rheological tests of ferrous solutions, that is, five chemical compositions at applied shear rates from 40 to 180 s-1. The obtained values of dynamic viscosity coefficient are within the range of 0.002 - 0.018 Pa s. These results may be useful for the verification of models for calculating the parameters of liquid ferrous alloys, e.g. during the casting of steel, where viscosity is one of the basic parameters.

In his study [4] S. Sridhar emphasised the difficulties as regards access to high-temperature, thermophysical data that is needed for computer process modelling and for the understanding and developing processes from the point of science and technology, along with improving process control and product quality. Generating reliable measurements of physical properties of materials is time consuming and requires substantial experimental knowledge. In industrial conditions experimental data is often needed very quickly in order to verify and optimise a specific process. Therefore, there is substantial demand for mathematical models that describe the physical and thermophysical material parameters in their solid, liquid and semi-solid states.

Viscosity is a measure of "friction existing in a fluid", or "resistance" that occurs in liquid or gas during flow. It is defined for a laminar kind of flow which, in a simplified way, describes the flow of layers that do not undergo mixing – such a flow like this occurs at low velocities [2]. The viscosity effect in metallurgical processes is amongst the most important of effects. It influences the behaviour of reacting phases (metallic, slag and gaseous) with regard to the kinetics of mass exchange or chemical reactions [5]. Viscosity plays a significant role in all metallurgical processes - ironmaking, steelmaking, ladle refining, steel casting, and processes related to solidification [6,7]. It is very important from the standpoint of flow of liquid phases that are continuously moving during those processes, as well as due to the internal structure of the metallic or slag phase and related possibilities of contamination absorption, or even the ability to deform in their semi-solid states [8, 9, 10, 11, 12].

Issues of viscosity and rheological properties of liquid ferrous solutions are important from the perspective of modelling, and also the control of actual production processes related to the manufacturing of metals, including iron and steel. Conducted analysis within subject literature indicates that there are many theoretical considerations concerning the effect of viscosity of liquid metal solutions. The vast majority of models constitute a group of theoretical or semi-empirical equations, where thermodynamic parameters of solutions, or some parameters determined by experimental methods, are used for calculations of the dynamic viscosity coefficient.

## 2. Research on Liquid Ferrous Solutions

Figure 1 presents a schematic diagram of a high-temperature rheometer as used for rheological tests of liquid metals and slags [13, 14, 15, 16].

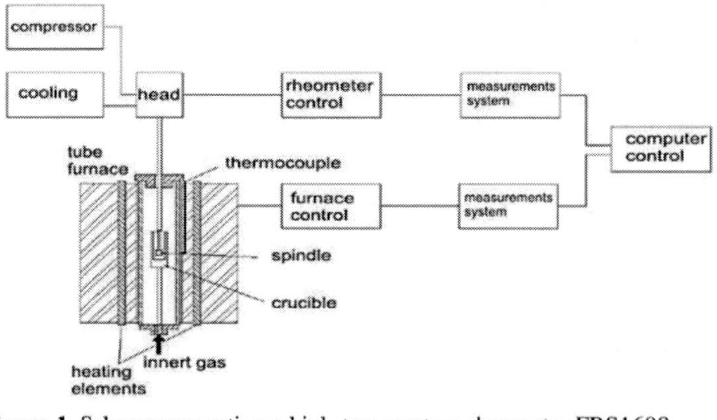


**Figure 1.** Scheme presenting a high-temperature rheometer FRS1600

The main part of the FRS1600 rheometer is its head, along with its cooling system. The head, like the furnace, is computer controlled. A thermocouple for temperature measurement is placed in the furnace, and the tubular resistance furnace with a mullite tube is controlled by a Eurotherm controller. An inert gas - argon with purity of 5.0 - is fed into the furnace tube to enable a protective atmosphere to be maintained during rheological measurements that last several hours. Figure 3 presents measurement tools used for rheometric tests.

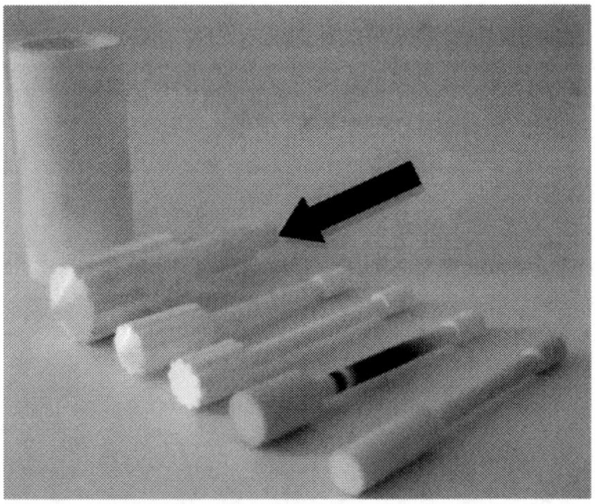

**Figure 2.** Measuring sistems

In this study, the measurement methodology in a system of coaxial Searle type cylinders was used for conducting high-temperature rheometric tests of steel. The cylindrical measurement method is based on a system consisting of an inner cylinder (bob) and an outer cylinder (cup).
In the study, the following ferrous solutions were examined.

*Tab. 1. Examined ferrous solutions [%]*

| Steel grade | C | Mn | Si | Cr | Ni | Mo | V |
|---|---|---|---|---|---|---|---|
| 90CrV6 | 0,89 | 0,26 | 0,19 | 1,43 | 0,44 | 0,06 | 0,106 |
| 34CrNiMo | 0,39 | 0,62 | 0,24 | 1,57 | 1,67 | 0,26 | 0,074 |
| DHQ3 | 0,80 | 0,26 | 0,69 | 2,93 | 0,13 | 0,54 | 0,011 |
| 42CrMo4 | 0,43 | 0,83 | 0,23 | 1,13 | 0,29 | 0,28 | 0,0068 |
| 45 | 0,46 | 0,74 | 0,30 | 0,17 | 0,24 | 0,06 | 0,052 |

For each of the above grades the values of solidus and liquidus temperatures of the above solutions (Table 2) [16] were calculated with the CompuTherm LLC thermodynamic databases, supplied together with the ProCast software package. The calculated temperature values are necessary to determine the measurement pattern, assuming that the steel tested is to be in the fully liquid state.

*Tab. 2. Liqidus and solidus temperaturę values valculated for examined steel grades[°C]*

| Gatunek stali | Temperatura likwidus | Temperatura solidus |
|---|---|---|
| 90CrV6 | 1463 | 1349 |
| 34CrNiMo | 1465 | 1343 |
| DHQ3 | 1461 | 1345 |
| 42CrMo4 | 1492 | 1422 |
| 45 | 1493 | 1418 |

Journal of Physics: Conference Series **602** (2015) 012037 doi:10.1088/1742-6596/602/1/012037

The rheological tests were conducted for five steel grades with different chemical compositions (Table 1). The tests were carried out at their liquidus temperatures (Table 2) and at temperatures up to 20 Celsius above and below the calculated liquidus temperatures. The measurements were taken every 10 degrees.

Figure 3 present the flow curves of five steel grades differing with chemical compositions. The objective of this presentation of results is to picture the differences between the values of dynamic viscosity coefficient and tangential shear for a given chemical composition at various temperatures.

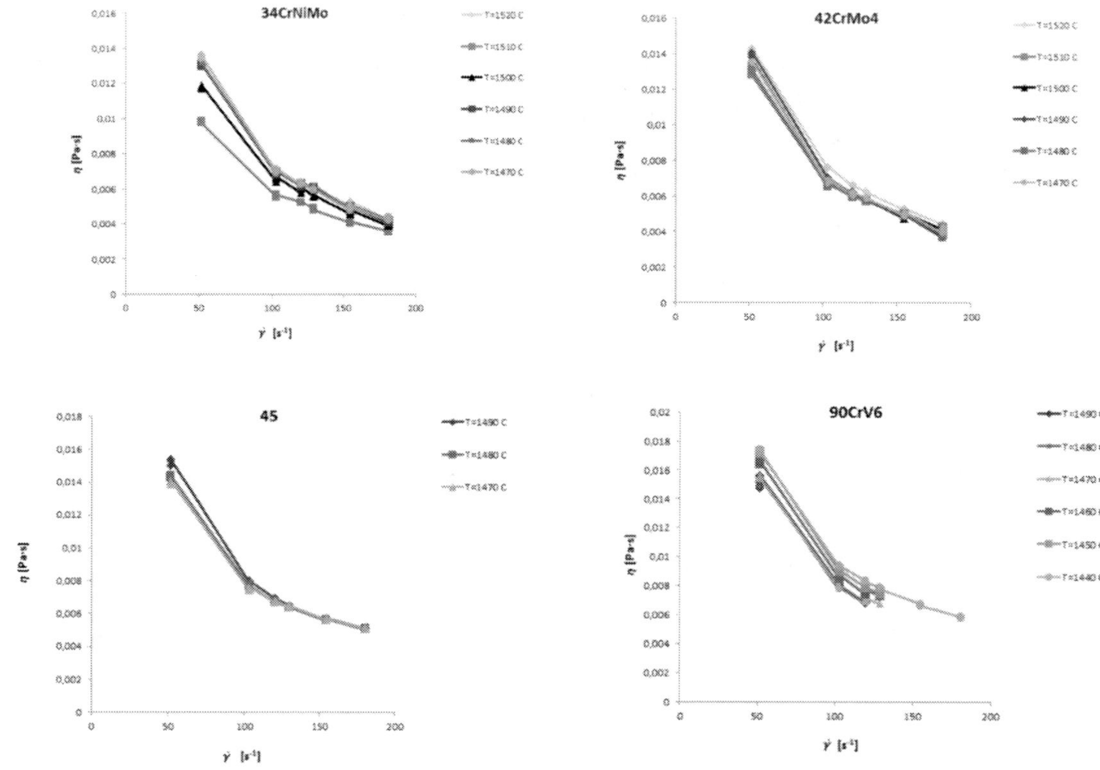


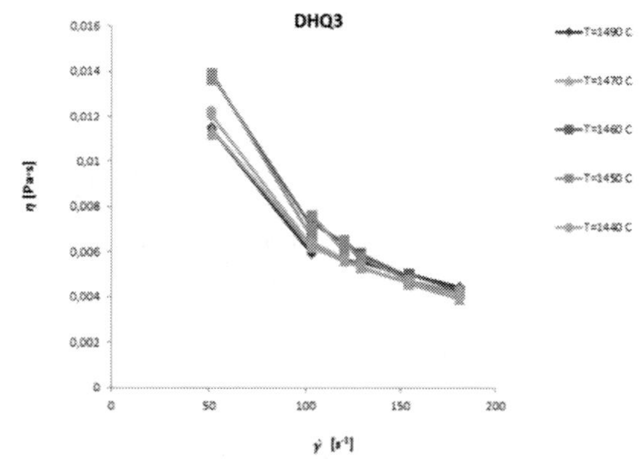


**Figure 3.** Viscosity curves for different steel grades

Based on the analysis of the above graphs (Fig. 50-54) it was found that the dynamic viscosity coefficient values range from 0.014 to 0.003; in addition for the 90CrV6 steel, the top value of its dynamic viscosity coefficient achieves 0.018 Pa s, and for 42CrMo4 the bottom value is 0.003 Pa s. In addition, it was observed that when the chromium and molybdenum content increases, the range of viscosity values extends. It is particularly visible for 34CrNiMO, which contains relatively high amounts of chromium and molybdenum, and has the highest content of nickel out of the steel grades tested. However, some differences in the top values of viscosity coefficient at various temperatures in the DHQ3 steel may be observed. This particular steel is characterised by the highest chromium and molybdenum content out of those steels analysed, and it has a relatively high carbon and low nickel content in its chemical composition. Therefore we can suppose that it is just the chromium and molybdenum content that creates favourable conditions for extending the range of dynamic viscosity coefficient of liquid steel. It was also observed that when the manganese content increases in the chemical composition of a ferrous solution, the range of the dynamic viscosity coefficient values contracts. 42CrMo4 has the highest manganese content, and for this steel the viscosity curves plotted for various temperatures virtually coincide.

## 3. Conclusion

The problem of viscosity and rheological properties of liquid ferrous solutions is important from the perspective of modelling, and also control of the actual production processes related to manufacturing of metals, including iron and steel. The conducted analysis of literature indicates that there are many theoretical considerations concerning rheological effects, while experimental results are scarce.

In this study, the measurement methodology in a system of coaxial Searle type cylinders was used for conducting high-temperature rheometric measurements of steel. During preparations for the tests, the shapes of measurement tools were designed, and tools made of various ceramic materials were tested. In terms of availability and strength, the optimal material for the measurement systems was selected.

Rheological tests of liquid steel solutions were then carried out, and five various chemical compositions were analysed. During these experiments the solution temperature was step changed every 10°C, from 20°C above the calculated liquidus temperature to 20°C below this temperature. In addition, during measurements the shear rate value was changed from 40 to 180 s-1. During these measurements, the time in which the sample tested was subjected to the same force was also modified. Finally, it was revealed that time was a rheological parameter, and which for measurements of liquid ferrous solutions (in constant process conditions), did not change the rheological nature of the medium examined.

As a result of these conducted experiments the values of liquid steel viscosity coefficient were obtained. These ranged from 0.002 to 0.018 Pa s.

For systems in their fully liquid state (at the theoretical liquidus temperature and above), the conducted rheological analysis allows us to understand that the liquid steel behaves like a Newtonian liquid, albeit with some departures within the tested range of shear rate which may be related - among others - to the occurrence of disturbances to laminar flow in the sample.

**Acknowledgements**

Financial support of National Science Center for research project "Developing an empirical model of the rheological properties of liquid metals on the example of iron solutions" no. 2011/01/N/ST8/07368 is gratefully acknowledged.

The research was conducted with the financial support from the statutory project no. 11.11.110.225 AGH University of Science and Technology.

**References**

[1] Derski W, Ziemba S: 1968 *Analiza modeli reologicznych* Państwowe Wydawnictwo Naukowe, Warszawa (in Polish)

[2] Hersey M D 1932 *Future Problems of Theoretical Rheology* Jour. Rheology **3** 196
[3] Kembłowski Z 1973 *Reometria płynów nienewtonowskich* Wydawnictwo Naukowo-Techniczne Warszawa (in Polish)
[4] Sridhar S 2002 *Estimation models for molten slag and alloy viscosities* JOM pp 46–50
[5] Benesch R, Janowski J, Mamro K. 1979 *Metalurgia żelaza. Podstawy fizykochemiczne procesów* Wydawnictwo Śląsk Katowice (in Polish)
[6] Karbowniczek M. 1998 *Pienienie żużla w procesach stalowniczych* Uczelniane Wydawnictwo Naukowo-Dydaktyczne Kraków (in Polish)
[7] Migas P, Karbowniczek M *Interactions Between Liquid Slag And Graphite During The Reduction Of Metallic Oxides* Archives Of Metallurgy And Materials **55** 4 pp 1147–1157
[8] Sołek K Korolczuk-Hejnak M Karbowniczek M *An Analysis Of Steel Viscosity In The Solidification Temperature Range* Archives Of Metallurgy And Materials **56** 3 pp 593–598
[9] Rogal Ł Dutkiewicz J 2012 *Effect Of Annealing On Microstructure, Phase Composition And Mechanical Properties Of Thixo-Cast 100cr6 Steel* Materials Characterization **68** pp123–130
[10] Rogal Ł and others 2010 *Characteristics Of 100cr6 Bearing Steel After Thixoforming Process Performed With Prototype* China Transactions of Nonferrous Metals Society of China **20** pp 1033–1037
[11] Hirt G Kopp R 2009 *Thixoforming. Semi-Solid Metal Processing* WILEY-VCH GMBH AND CO. KGAA Weinheim
[12] Modigell M Pape L Maier H R. 2006 *Rheology Of Semi-Solid Steel Alloys At Temperatures Up To 1500°C Solid State Phenomena* **116-117** pp 606–609
[13] Korolczuk-Hejnak M Migas P 2012 *Analysis Of Selected Liquid Steel. Archives Of Metallurgy And Materials* **57** No. 4 pp 963–969
[14] Korolczuk-Hejnak M Migas P 2012 *Selected Grades Of Steel As Rheologically Defined Liquid Bodies. Archives Of Metallurgy And Materials* **57** No. 2 pp 583–591
[15] Sołek K Korolczuk-Hejnak M, Ślęzak W 2012 *Viscosity Measurements For Modeling Of Continuous Steel Casting* Archives Of Metallurgy And Materials **57** No. 1 pp 325–333
[16] Miłkowska-Piszczek K Korolczuk – Hejnak M 2013 *An Analysis Of The Influence Of Viscosity On The Numerical Simulation Of Temperature Distribution As Demonstrated By The CCProcess* Archives Of Metallurgy And Materials **58** .4

# Examination of first and second order thermodynamic realignments of metals and metal-alloys, with application of DMTA equipment

**Béres Gábor, Dugár Zsolt, Kis Dávid, Hansághy Pál**
Kecskemét College - Faculty of Mechanical Engineering and Automation, Hungary

E-mail: gabberix@freemail.hu

**Abstract**
The researcher work pushes the borders of the application of a high efficiency equipment, in the matter of metals. With this equipment, which already testified in the compound industry, a flasher query of some metal-related technological parameters could be easily available. Towards the searching, we shaped different compound copper-alloy samples, in different states, and accordingly we monitored the realignments. Following the shaping, the DMTA detected such microscopic transformations (as we expected), according to heating rate, whereby we could determine the specialities of the transformations. In order to monitor the effect of the work hardening, we applied two different shaping grade: 50% and 75%. The heat treatment already took place in the DMTA, using 3 K/min heating rate. The specimens were loaded by inflection in the 2-point bending support with constant frequency and amplitude.
Our current object was the monitoring of the recrystallization, and the investigation of the influential factors of this process, but other transitions were also regarded. Those measurements' results, that the DMTA presented, had been compared with DSC and hardness analysis, whereby we try to conclude to the utility of DMTA, in matter of metal alloys.
Keywords: DMTA, recrystallization, DSC, hardness analyze, tin-bronzes, brasses, viscoelastic

## 1. Introduction
The researcher team examines both the industrial metals (principally) as well as the high-tech materials and special metals (for example shape memory alloys) with DMTA equipment. Our main object is the monitoring of the recrystallization, in case of copper alloys and the investigation of the influential factors of this process. The measurements' results, that the DMTA presented, had been compared with DSC and hardness analyses.

The literature research added up to that, we are not the first to examine metals with DMTA. But other (particularly Asian) researchers usually take into consideration the copper as alloying element or as compound, not as all-metal. For example Tang et al. examined the structure of copper electropolyurushiol film. The EPU-$Cu^{2+}$ film was characterized by dynamic mechanical thermal analysis, but the $Cu^{2+}$ content in the coordination polymer is only 8.63 %, as compound [5]. There are still many events, when the copper gives the base for surface treatment. Loo et al. investigated N,N-dimethylthioacetamide adsorbed on Ag and Cu electrode surfaces. The comparison was made by vibration analysis of DMTA [6].

In contrast with them, we would like to write up the measurement method, whereby the properties of the industrial material itself, would be easier defined.

The DSC detected the convection, due to energy release or energy attenuation. The DMTA is able to load mechanical stress, next to heating [1]. Therefore it can determine the changes of the mechanical properties, otherwise than DSC. Furthermore, it is a very precise instrument. The exact temperature monitoring with hardness analyses are difficult because of the variation of the furnaces' constant temperature [2].

In case of a visco-elastic material the σ bending stress is delayed to the deformation with delta angle (figure 1). In case of clearly viscous materials, like the water, the delay is 90°. Because the copper alloys are elastic materials, the σ bending stress and the deformation should have be in the same phase [1].

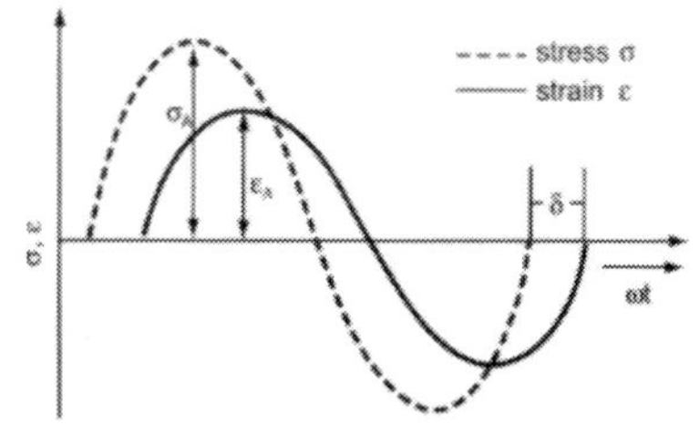


**Figure 1.** DMTA curve

In the curves of the DMTA, the loss factor appeared. It means that in certain cases, the investigated copper alloys also can show similar properties than visco-elastic materials.

## 2. Method

### *2.1. Samples*

The baseline was four different compound copper alloy ingots. The rolling was started from 15 mm thickness, on 500°C. This relatively low temperature ensured the avoidance of the oxidation. Samples were manufactured from all of the ingots on 2 mm and on 4 mm thickness. Next to four different compositions, it means eight pieces of samples. All of the samples were softened one hour.

After that, all of the samples were rolled to 1 mm thickness, on room temperature. Based on this I made two different shaping grades. In case of the initial 4 mm thickness samples, the deformation grade became 75%. In case of the initial 2 mm thickness samples, the deformation grade became 50%. It is well known, that the temperature of the recrystalliaztion is depend on the structure, and the shaping grade of the material [4]. In this way I tried to monitor both the effect of the shaping and the composition during the searching.

I selected such copper alloys, which processing is done by forming. It was necessary because of the recrystallization. The mark of the samples originated from University of Miskolc. The determination of the exact composition was done by spectrometer:

- sample 31 (sample 10): 58,8 % Cu 40,9 % Zn
- sample 24: 61,3 % Cu 26,8 % Zn > 5 % Ni
- sample 21: 91,5 % Cu 6,19 % Sn
- sample 23: 95,8 % Cu 3,70 % Sn

At first sight, the samples can be partition into two groups:

- the first two samples are brasses: (31, 24)
- the last two samples are tin-bronzes: (21, 23)

Examining brasses, in case of the sample 24 the average Zn content is lower than 30%, and the average Ni content is more than 5%. Based on this, the sample 24 belongs to formed brasses. The sample 31 may seem rigid alloy because of the high Zn content. However, in the lack of other elements, it also can be organized to the group of formed brasses [3].

In case of the tin-bronzes, the average Sn content is between 2% and 14%, and there are no casting type elements. So these samples belong to the group of formed tin-bronzes [3].

### *2.2. Preparation*

I applied two point bending method. The measurement frequency was 5 Hz, the amplitude of the elongation was 10 μm. The measurements were carried out in air. The DMTA provided the heating of the samples with 3 K/min heating rate. The DSC samples were manufactured by cutting. The cutting made the samples' structure formed, therefore some DSC measurements failed. Furthermore I had to clear the samples in alcohol. The hardness analyses were done by Instron Wolpert 2100B Vickers hardness tester. The samples were heated from room temperature to 500°C. The frequency of the measures was increased between 400°C and 500°C, for more precise results. Five parallel measurements were done in all temperatures. The edge values were left and the middle three values' average was taken.

## 3. Results

In case of copper alloys, the peak of the loss factor is around at 0,1 and 0,2 on the DMTA curves. The value is smaller than polymers by one decimal digit. The value of the loss factor is usually about at 1,5 at polymers.

### *3.1. Sample 21 (tin-bronze)*

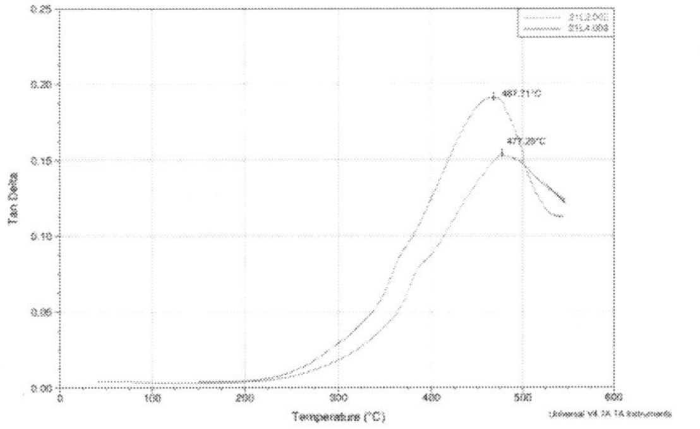


**Figure 2.** DMTA curve of sample 21, tanδ highlighted: 50% (green) and 75% (blue) shaped

The transitions are well visible in both diagrams. The blue curve shows the 75%, the green curve shows the 50% deformed sample. The main peaks around 470°C, refer to the β-β' transitions. The temperature of β-β' transition is as lower as the deformation higher. It is also true for the recrystallization. The bigger deformation grade gives the bigger area under the curve. The recrystallization starts at around 400°C, according to the hardness analyze (figure 3). The smaller, tanδ peaks would refer to it (figure 2).

Journal of Physics: Conference Series **602** (2015) 012038 doi:10.1088/1742-6596/602/1/012038

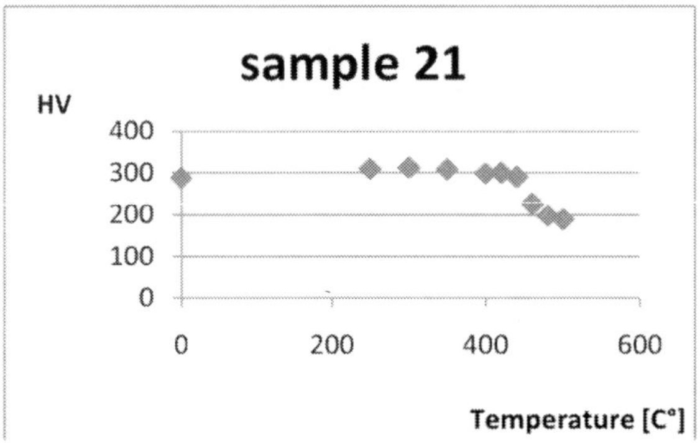


**Figure 3.** Hardness analyze result of sample 21

*3.2. Sample 23 (tin-bronze)*

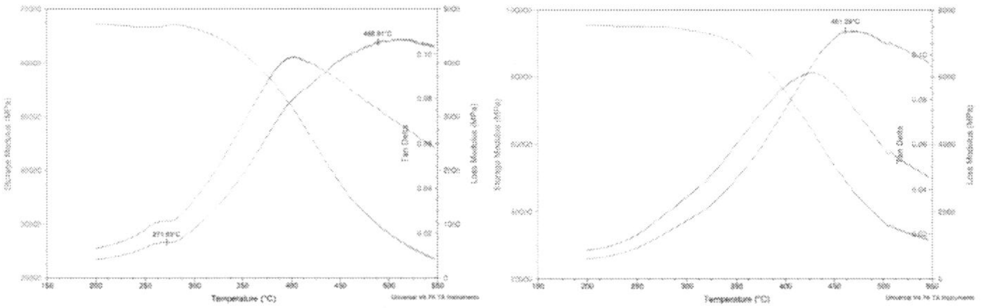

**Figure 4.** DMTA curve of sample 23, 50% (left) and 75% (right) shaped

The β-β' transitions are clear in both cases. Unfortunately the recrystallization is not visible in the curves. On the left side of figure 4, a little peak can be seen roughly on 400°C, which refer to it. It is justified by the DSC curve on figure 5. The recycling is also clear, in lower temperature range, at 271°C. The curve of the 75% deformed sample (right) is flat until it reaches the β-β' transformation point. The reason for this believed, that too small grains grown up during to the recrystallization, which can't display the transition. The temperature of β-β' transition is also as lower as the deformation higher again.

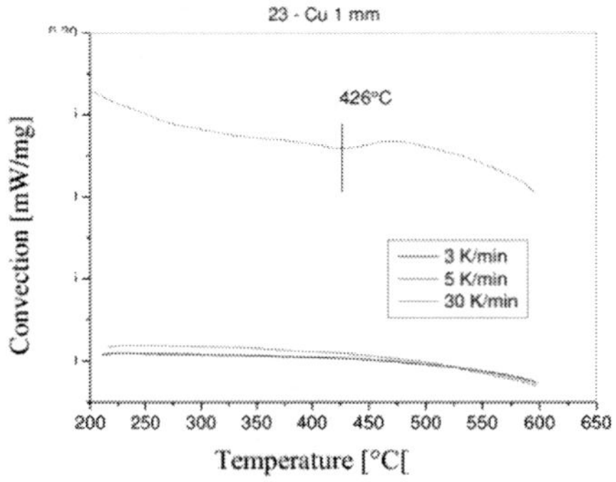


**Figure 5.** DSC curve of sample 23

### *3.3. Sample 10 (brass)*

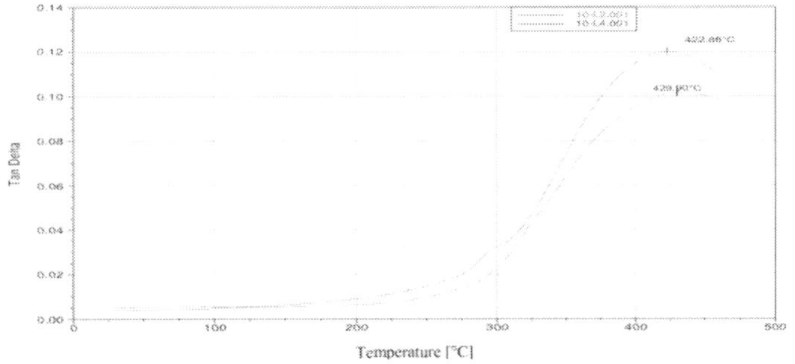


**Figure 6.** DMTA curves of sample 10,
tanδ highlighted: 50% (green) and 75% (blue) shaped

Based on the results, I conclude that the recrystallization takes place at around 300°C, lower than tin-bronzes. It can be clearly seen in case of 75% deformed samples, but no peaks appeared on 50% shaped sample's curve. The β-β' transitions are also visible, as in the previous cases.
The diagram clarifies well the discrepancy between the effects of the shaping grade. The biggest area under the curve, and the lower temperature of β-β' transition belongs to the biggest deformation, which is shown by the blue curve.

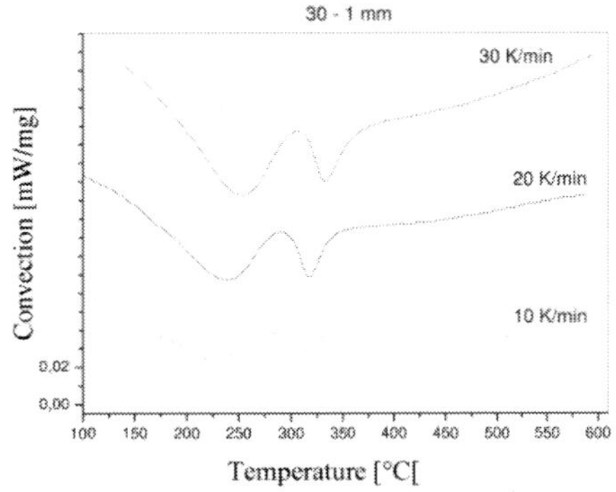


**Figure 7.** DSC curve of sample 10

The place of the DSC peaks was changed by the heating rate. But it still can be seen, that the recrystallization happened around 300°C as the DMTA and hardness analyze (figure 8) show.

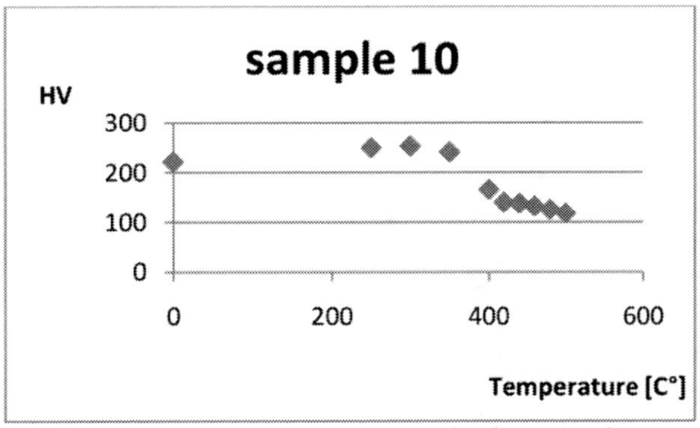


**Figure 8.** Hardness analyze result of sample 10

### *3.4. Sample 24 (brass)*

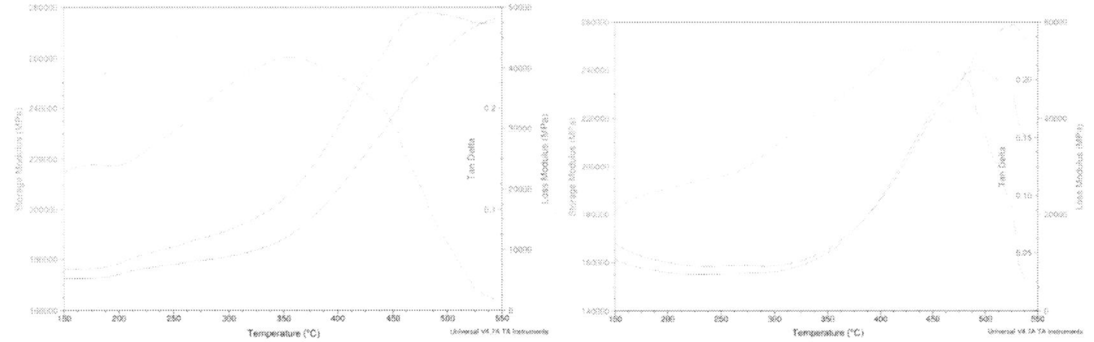


**Figure 9.** DMTA curve of sample 24, 75% (left) and 50% shaped

In case of 75% deformed sample, the β-β' transition took place such high temperature, that was outside the boundaries of the measuring. There is an uncertainty at around 450°C, which is maybe not enough to declare the recrystallization, but according to the DSC (figure 10 left) and the hardness analyze (figure 10 right) it should be there. In lower temperature range, between 200°C and 250°C, the recycling appeared.

In case of the 50% deformed samples, only the β-β' transformation can be seen. This is already the second event of brasses, when the recrystallization does not appear at lower deformation grade. Furthermore it should be mentioned that the highest loss factor values are shown by the DMTA curve of sample 24. Both curves' peaks transcend the 0,2 tanδ value, the more deformed sample better, the less deformed sample less. Based on this it is observed, that this sample is the most elastic.

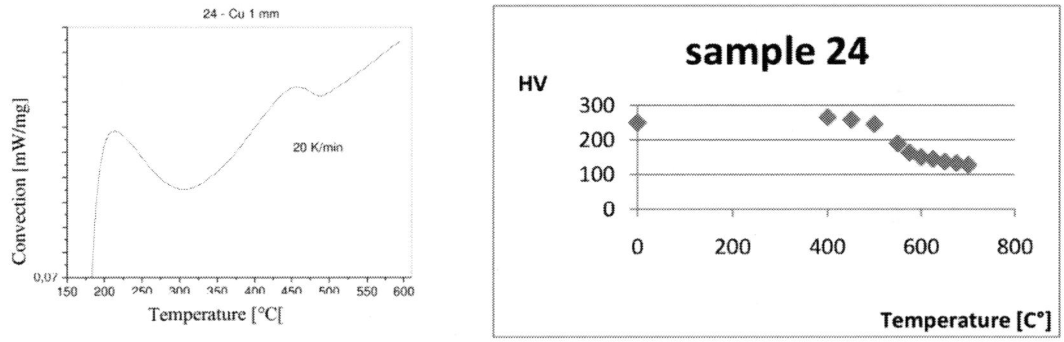


**Figure 10.** DSC curve (left) and hardness analyze result (right) of sample 24

Left, on the DSC figure, the first peak refers to the recycling, the next one refers to the recrystallization. The end of the process is not in the figure, as in the case of DMTA.

## 4. Conclusion

Realignments of copper alloys were investigated in different circumstances. To reveal the transformations we applied DMTA, DSC equipments, and hardness analysis. Transitions have been demonstrated in several cases. β-β' transition is always visible, expect only in one or two times. In case of brasses, the presence of the recrystallization depends on the shaping grade. It is never appeared under 75% deformation. The recycling is visible in more cases in lower temperature ranges.

Based on the results, we succeeded in demonstrating the appearance of the loss factor, with this new measurement process. Furthermore, it is observed, that the thermo-dynamic measurement process is a suitable method for monitoring realignments.

**Acknowledgement**
The research work presented in this paper/study/etc. was carried out as part of the TÁMOP-4.2.2.A-11/1/KONV-2012-0029 project in the framework of the New Széchenyi Plan. The realization of this project is supported by the European Union, and co-financed by the European Social Fund.

**References**

[1] Joel R. Fried 2013 Polymer Science and Technology (2nd Edition), Prentice Hall
[2] Raymond A. Higgins: 1993 Engineering Metallurgy, Applied Physical Metallurgy, Hodder Headline Group
[3] ASM Metals Handbook Desk Edition 2001. Volume 2: Properties and Selection: Nonferrous Alloys andSpecial-Purpose Materials
[4] William F. Hosford:, Robert M. Caddel 2007 Metal Forming, Mehanics and Metallurgy, Cambridge University Press.
[5] Tang, JY, Zhang, WG, Shen, GC and Gao, F 2001 Coordination structure of copper electropolyurushiol film and its catalytic performance for methyl methacrylate, Chinese Journal of Catalysis **22** (2) 119
[6] Loo BH, Lee YG and Kato T 1991 Raman-Spectra os N,N-Dimethylthioacetamide Adsorbed on Silver and Copper Electrode Surfaces, Bulletion of the Chemical Society of Japan **64** (10) 3126

# Optical Coherence Tomography Velocimetry with Complex Fluids

**A. Malm, T.A. Waigh*, S. Jaradat, R. Tomlin**

Biological Physics, School of Physics and Astronomy, University of Manchester, Oxford Road, Manchester, M13 9PL, UK

Photon Science Institue, University of Manchester, Oxford Road, Manchester, M13 9PL, UK

E-mail: *t.a.waigh@manchester.ac.uk

**Abstract.** We present recent results obtained with an Optical Coherence Tomography Velocimetry technique. An optical interferometer measures the velocity of a sheared fluid at specific depths of the sample using the coherence length of the light source. The technique allows the dynamics of 3 pico liter volumes to be probed inside opaque complex fluids. In a study of opaque starch suspensions, classical bulk rheology experiments show non-linear shear thickening, whereas observations of the velocity profiles as a function of distance across the gap show Newtonian behavior. The ability of the technique to measure velocity fluctuations is also discussed for the case of polyacrylamide samples which were observed to display shear banding behavior. A relationship between the viscoelasticity of the sample and the size of the apparent fluctuations is observed.

## 1. Introduction

When attempting to understand complex fluid behavior rheometers are only capable of telling us information about the bulk of the fluid, when in reality a range of phenomena can be occurring on a hierachy of different length scales. More detailed information is available if microrheology measurements are performed on a specimen [1]. A particularly useful microrheology measurement is to calculate the velocity profiles of a sample under flow (a variety of non-linear microrheology). Particle imaging velocimetry (PIV) is the most commonly used technique for measuring velocity profiles, but this technique has its limitations, mainly in terms of the thickness and concentration of the samples that can be probed [2]. Optical Coherence Tomography (OCT) Velocimetry is a relatively new technique and has been previously demonstrated to be a powerful tool for measuring complex flows and rheological behavior in a wide range of materials [3]. OCT is a new type of photonics technique that adds to the growing list of available rheo-optical methods [4].

## 2. Overview of OCT Velocimetry

OCT is a technique that was originally developed for imaging in opthalmology, since it allows the retina to be imaged relatively independently of the shape of the cornea. OCT has seen a wide range of applications and developments [5, 6]. In OCT Velocimetry a Doppler signal is measured using an interferometer for light scattered off a sheared fluid. High resolution measurement with this technique is achieved by using a light source with a small coherence length, which means that the back scattered interference signal can be associated with a thin layer of the sample at a specific depth [3]. The infra-red light source currently used has a coherence length and therefore an axial resolution of 9 $\mu m$, which factoring in the spot size of the focused beam, gives a probe volume of 3.4 pico liters. The Doppler frequency is measured in the form of a peak in the Power Spectrum of the light intensity recorded on a fibre coupled detector, and this in turn can be converted into a velocity, given the wavelength and angle of incidence of the light source. OCT has a penetration depth of the order of a few millimeters in opaque materials (in the case of our OCT Velocimetry system the depth of focus is 1.2 mm, so that the entire rheometer gap is in focus and no refocusing is required during an experiment), which is greater than in most conventional microscopy techniques, and is facilitated by the use of an infra-red super luminescent diode light source, with a wavelength of 1.3 $\mu m$.

Figure 1 gives a list of velocimetry methods that can be used to probe the velocity of complex fluids in a rheometer. OCT Velocimetry is competitive in terms of temporal resolution, spatial resolution and price. Our current realization of the technique requires samples to have relatively low flow birefringence. However this restriction can be lifted if a polarization sensitive detector is used [7].

**Figure 1.** Comparison of different commonly used velocimetry techniques to probe velocity profiles in rheometers. Adapted from [2]

| Technique | Spatial resolution | Temporal resolution | System requirements | Advantages | Disadvantages |
|---|---|---|---|---|---|
| *OCT shear rheometer* | 9 µm slices, 3 pL volume | 4 µs-1000 s dynamics, ~1 s per point | Source of speckle | High spatial and temporal resolution, Opaque materials | Would require polarization sensitivity to probe liquid crystals[1] |
| *Standard photon correlation spectroscopy (PCS) velocimetry* | 100 µm slices | 1 s per point | Weak scattering, No turbidity | Reasonable spatial and temporal resolution | No opaque materials |
| *Ultrasonic velocimetry (UV)* | 40 µm slices | 20 ms-2 s per point | Acoustic contrast | Reasonable spatial and temporal resolution, Opaque materials | Can require seeding |
| *NMR* | 50-100 µm slices | 1 s per profile | Correct nuclei | Opaque materials | Costly, Low resolution |
| *Particle tracking velocimetry* | 1-10 µm slices | 2.5 ms-0.1 s per profile | Tracer particles and optically transparent | Excellent spatial and temporal resolution | Requires seeding and transparent systems. Small depth of field. No opaque materials |

[1] Polarization sensitive OCT techniques with static samples have been demonstrated in the literature, and polarization sensitive detectors are commercially available [7].

## 3. Velocimetry of starch suspensions

Starch is a well studied colloidal material due to its importance as a food stuff and also as a raw material (e.g. paper production). It is an example of a non-Newtonian complex fluid due to its shear thickening properties [8]. In the case of starch suspensions, shear thickening is caused by jamming processes, and it is therefore reasonable to expect that non-linear velocity profiles across the rheometer gap might be observed for suitably high shear rates and concentrations [9]. Therefore aqueous corn starch suspensions of between 40 and 50 percent by weight were examined with a bulk rheometer and were found to display shear thickening behavior in viscosity measurements as shown in Figure 2.

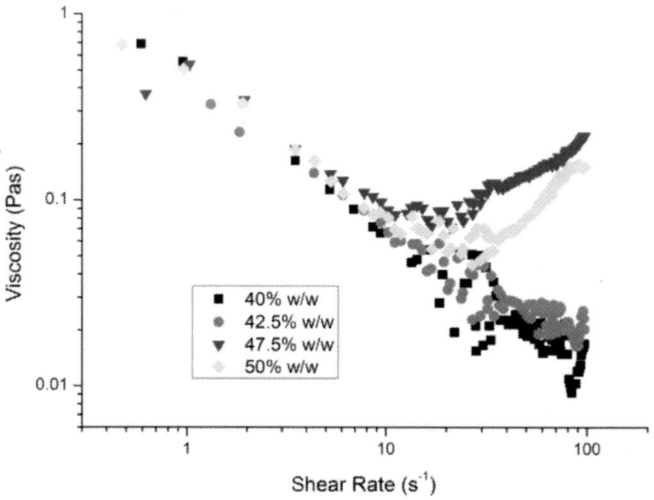


**Figure 2.** Viscosity plotted as a function of shear rate measured in a bulk rheometer for a range of concentrations of corn starch for a rheometer gap of 300 $\mu$m, showing shear thickening behavior at high shear rates (>10 $s^{-1}$) for concentrations above 45% $w/w$.

Figure 3 shows velocity profiles as a function of position in the rheometer gap at three different concentrations of corn starch, 40% $w/w$ which showed no shear thickening, and 45% $w/w$ and 50% $w/w$ which showed significant shear thickening [10, 11].

$$\dot{\gamma} = \frac{\sigma}{\eta} = \frac{dv}{dz} \tag{1}$$

Surprisingly, all three profiles in Figure 3 show linear Newtonian profiles, with no apparent slipping on either plate, conforming to Equation 1, where the shear rate of the fluid, $\dot{\gamma}$, is equal to the velocity gradient across the rheometer gap, $dv/dz$, and is related to the bulk viscosity of the sample, $\eta$, and stress applied to it, $\sigma$. A comparison between the velocity profiles for different shear rates for the 50% $w/w$ suspension, which showed the greatest amount of shear thickening, is shown in Figure 4. All of the velocity profiles are again linear.

## 4. Study of velocity fluctuations

Another useful aspect of the OCT Velocimetry method is its ability to quantify fluctuations in the flow velocity. This should prove to be useful in studying the onset of turbulence, since characteristic velocity fluctuations have been observed previously [12]. The velocity of the fluid can be calculated by fitting to a peak in the Power Spectrum, with a central frequency position

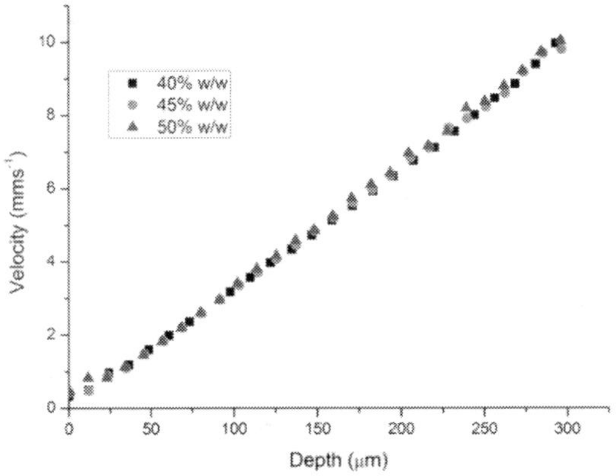


**Figure 3.** Velocity profile as a function of depth for opaque 40% $w/w$, 45% $w/w$ and 50% $w/w$ suspensions of corn starch with a parallel plate geometry and steady shear rate of 80 $s^{-1}$. Linear Newtonian behaviour is observed in all cases, Equation 1.

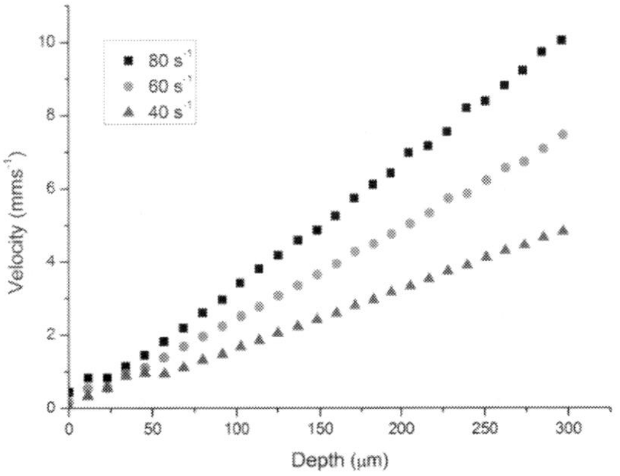


**Figure 4.** Velocity profiles as a function of depth for an opaque 50% $w/w$ corn starch sample at a range of shear rates, showing Newtonian behaviour in all cases as described by Equation 1.

$\omega$, which is then used to calculate the velocity at which complex fluids move within the 3 pico liter probe volume using Equation 2,

$$v = \frac{\lambda\omega}{2\sin\theta} \tag{2}$$

where $\lambda$ is the wavelength of the light source and $\theta$ is the angle at which the light is incident upon the sample. This peak is approximately Gaussian with a finite width ($\delta\omega$) which will be greater for volumes of fluid where the velocity is not constant. Also as the velocity of fluid within the probe sample varies over time, the position of the peak will also fluctuate over time. The size of velocity fluctuations can thus vary at specific depths within the sample ($\delta v$). An understanding of the velocity fluctuations can give greater insight into turbulent or unstable regions of flow. An example of different Power Spectra with varying peak widths is shown in Figure 5.

The OCT Velocimetry method has been previously used in a study of polyacrylamide samples, which displayed a range of flow phenomena with dynamic phase transitions that depended on the molecular weight and the concentration of the polyacrylamide, Figure 6 [13]. For the lowest molecular weight samples that were analyzed (1.5k and 10k MW), the velocity profiles across the gap were found to be always linear, regardless of the concentration of polymer in the sample. However for medium molecular weights (5M MW), shear banding occurred as the concentration of polymer was increased, and for high concentrations wall slip became more apparent. Similar results were obtained for higher molecular weights (18M MW), however the onset of shear banding and wall slip was observed for much lower concentrations. This shear banding behavior may be associated with the entanglement of the polymer chains and is also referred to as *elastic turbulence* in the literature [14, 15, 16].

By analyzing the degree of broadening of the peaks in the data that produced the velocity profiles, we can gain further insight into the role of polymer concentration in the determination of the flow behaviour. Figure 7 shows the reduced velocity fluctuations across the rheometer gap for a range of different concentrations of low molecular weight polyacrylamides and Figure 8 shows similar data for high molecular weight samples.

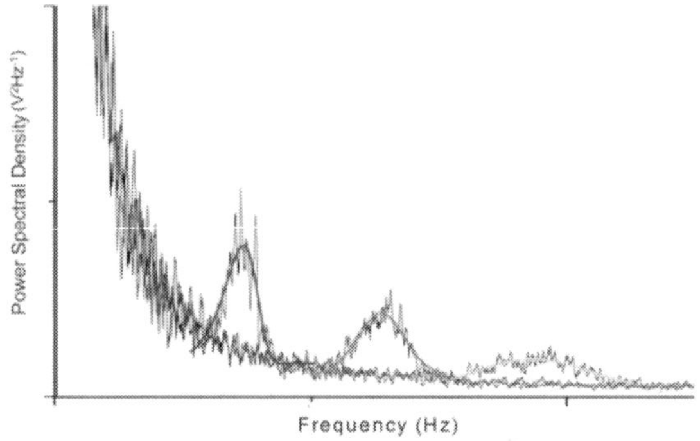


**Figure 5.** Three different Power Spectral Densities measured using OCT for different shear velocities of probe spheres embedded in a polyacrylamide solution [13]. The lines in bold show Gaussian fits. The fitted curves give shear velocities of $1.06 \pm 0.20$ mms$^{-1}$, $1.86 \pm 0.32$ mms$^{-1}$ and $2.73 \pm 0.77$ mms$^{-1}$ for the blue, red and green curves respectively

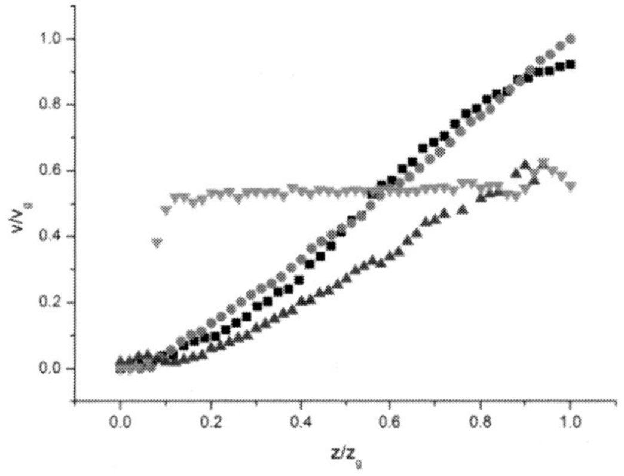


**Figure 6.** Examples of velocity profiles as a function of distance across the plate-plate geometry ($z/z_g$) for different concentrations and molecular weights of Polyacrylamide samples displaying Newtonian (red), Shear Banding (black) and Wall Slip behavior (blue and pink).

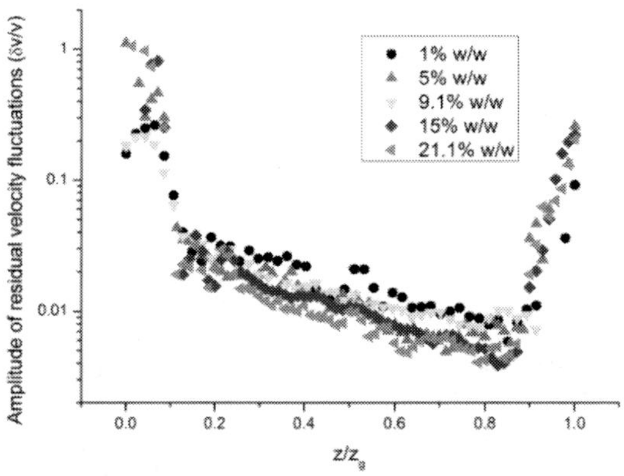


**Figure 7.** Velocity fluctuations ($\delta v/v$) as a function of distance across the plate-plate geometry ($z/z_g$) for low molecular weight (10k MW) polyacrylamide solutions at varying concentrations.

The reduced velocity fluctuations are plotted in order to compensate for the relative scaling of the fluctuations at higher shear velocities. In both cases the flow of the system is apparently unstable close to both of the plates of the rheometer, however we are currently unsure of the origin of this instability. It could be due to irregularity in the control loop of the rheometer (although the instability is also near the stationary plate), broadening due to the roughness/asymmetry of the plates (although the lower plate is highly polished glass), an artifact resulting from the limited ability of the instrument to measure low shear velocities or a real intrinsic hydrodynamic effect. Further away from the surface of the plates, the change in fluctuation size for different concentrations is small and most noticeable when plotted logarithmically. For the low molecular weight samples, the velocity fluctuations decrease for higher concentrations of polymer. This makes logical sense as the viscosity of the sample increases with concentration, and this effectively dampens any instability more than a less viscous sample. The opposite however seems to be the case for high molecular weights as the fluctuations increase with concentration, Figure 8. The velocity fluctuations in this case may be driven by the non-linear elastic instability that drives the non-Newtonian velocity profiles, Figure 6[17]. The elasticity of the samples at higher concentrations (they are reptating and above the entanglement concentration) may help to drive the instability.

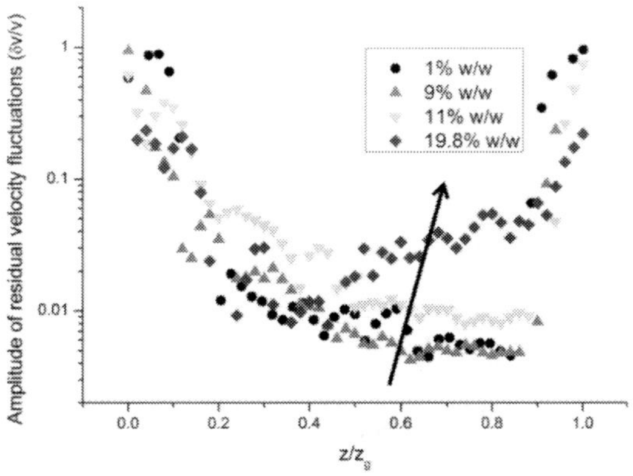


**Figure 8.** Velocity fluctuations as a function of distance across the plate-plate geometry ($z/z_g$) for high molecular weight (18M MW) polyacrylamide solutions at varying concentrations. The arrow indicates the increase in the velocity fluctuations that we associate with elastic turbulence and the onset of wall slip.

## 5. Future directions

There are a wide range of potential samples that can be investigated using OCT Velocimetry that are known to display complex flow behavior including; live bacteria, blood cells, food stuffs such as biopolymers and milk products, petroleum based products, surfactants and colloidal suspensions. The OCT Velocimetry method should in principal function with other types of rheometer such as concentric cylinders (a Couette cell), channel flow and also polymer melt rheometers, which will allow a wider range of materials to be studied. We are in the process of developing a new OCT Velocimetry design using an Electro Optical Modulator that we expect will provide improved resolution on the velocity fluctuations.

## 6. Conclusions

OCT Velocimetry allows quantitative analysis of complex flow systems both in terms of velocity profiles and velocity fluctuations. This has been demonstrated in a study of opaque starch suspensions which have been shown to display Newtonian behavior across the plate/plate geometry despite displaying shear thickening in conventional bulk rheometry measurements. Also polyacrylamide samples are examined in which the velocity fluctuations are a sensitive function of polymer concentration. However the trends observed for the velocity fluctuations differ depending on the molecular weights of the polymer chains in question. For the shear banding (elastic turbulence) specimens, velocity fluctuations increase with polymer concentration, whereas the opposite is true for specimens with Newtonian profiles.

## Acknowledgments

The authors would like to thank Malvern Instruments for the gift of a fluids rheometer and Gian Fulgoni for funding and also Andrew Harrison and Matthew Harvey for useful conversations.

This technique is patented [18], but for further information, including detailed descriptions on how to build a non-commercial OCT Velocimeter, please go to `http://www.personalpages.manchester.ac.uk/alex.malm` or contact Alex Malm directly: `alex.malm@manchester.ac.uk`

## References

[1] T A Waigh. Microrheology of complex fluids. *Reports on Progress in Physics*, 68(3):685, 2005.
[2] S. Manneville. Recent experimental probes of shear banding. *Rheologica Acta*, 3:301–318, 2008.
[3] M. Harvey and T. A. Waigh. Optical coherence tomography velocimetry in controlled shear flow. *Phys. Rev. E*, 83:031502, Mar 2011.
[4] G.G. Fuller. *Optical Rheometry of Complex Fluids*. OUP, 1995.
[5] M Hee, J Izatt, and J Fujimoto. Optical coherence tomography for ophthalmic imaging. *IEEE Engineering in Medicine and Biology Magazine*, 14(1), 1995.
[6] W. Drexler and J.G. Fujimoto. *Optical Coherence Tomography- Technology and Applications*. Springer, 2008.
[7] D.P. Dave and T.E. Milner. Optical low-coherence reflectometer for differemtial phase measurement. *Optics Letters*, 4:222–229, 2000.
[8] J. Mewis and N. Wagner. *Colloidal Suspension Rheology*. CUP, 2011.
[9] A. Fall. Shear thickening of corn starch suspensions. *Journal of Rheology*, 3:575–591, 2012.
[10] H. Barnes. Shear-thickening (dilatency) in suspensions of nonaggregating solid particles dispersed in newtonian liquids. *Journal of Rheology*, 2:329–366, 1989.
[11] M.E. Cates. Jamming, force chains, and fragile matter. *Physical Review Letters*, 9:1841–1844, 1998.
[12] P.A. Davidson. *Turbulence, and Introduction for Scientists and Engineers*. OUP, 2004.
[13] S. Jaradat, M. Harvey, and T.A. Waigh. Shear-banding in polyacrylamide solutions revealed via optical coherence tomography velocimetry. *Soft Matter*, 8:11677–11686, 2012.
[14] B. Schiamberg, L. Shereda, H. Hu, and R. Larson. Transitional pathway to elastic turbulence in torsional, parallel-plate flow of a polymer solution. *Journal of Fluid Mechanics*, 554:191–216, 2006.
[15] R. Larson. Instabilities in viscoelastic flows. *Rheologica Acta*, 31:213–263, 1992.
[16] M. Smieszek, O. Crummeyrolle, I. Mutabazi, and C. Egbers. Instabilities with polyacrylamide solution in small and large aspect ratios taylor-couette systems. *Journal of Physics- Conference Series*, 137, 2008.
[17] A. Morozow and W. van Saarloos. An introductory essay on subcritical instabilities and the transition to turbulence in viscoelastic parallel shear flows. *Physics Reports*, pages 112–143, 2007.
[18] T.A. Waigh and M. Harvey. Oct rheometer patent. Number- 1013945.9 August 24, 2010.

Journal of Physics: Conference Series **602** (2015) 012040 doi:10.1088/1742-6596/602/1/012040

# Optical microrheology measurements for determination of sol-gel transitions of chitosan hydrogels

**P Domagalski, M Orczykowska, P Owczarz and M Dziubinski**

Faculty of Process and Environmental Engineering, Lodz Technical University, ul. Wolczanska 213, 93-005 Lodz, Poland

**Abtract.** The combination of medicinal substance and an appropriate polymer matrix with specially adjusted permeability has became a basic form of medicine in the controlled drag delivery. One of main solutions used recently in pharmacology is the application of polymer matrices which form colloidal systems. The sol-gel phase transition is inhibited by a change of pH or temperature. Thermosensitive systems are produced from polymers whose sol-gel transition in water solutions depends on temperature. Thermosensitive hydrogel systems are currently obtained mainly from poly(N-isopropylacrylamide) (PNIPA), copolymer (polyethylene oxide – polypropylene oxide – polyethylene oxide), polyethylene glycol, lactic acid copolymer with glycolic acid and polyethylene glycol, and polysaccharides (xyloglucane, ethylcellulose, hydroxyethylcellulose and chitosan). At the temperature below 25°C these systems occur in the form of sols which at human body temperature are transformed into gels. Such systems belong to the so called intelligent ones and are a subject of special interest, particularly in view of their possible use as drug carriers and also as material for cell culture in the past.

***Keywords****: passive optical microrheology, chitosan, sol-gel transition*

## 1. Introduction

At present, when ecology is given much attention in industrial production, of special interest is the use of natural polymers in the formation of pH- and thermally-sensitive colloidal systems. Such polymers include chitin derivative, N-acetyl-D-glucosamine, called chitosan. Chitosan is a polymer insoluble in water and organic solvents. However, in the form of salt it shows high solubility. The most frequently used solvents are diluted organic acids, and the main solvent is acetic acid. Hydrochloric acid is also quite often applied as a solvent for chitosan.

Acid solutions of chitosan salts show the ability of sol-gel transition under environment (temperature or pH) change. Such ability allows for using the chitosan and its thermogels as polymer matrix carrier of active bio-compounds like medicines. However, for use of such carrier in vivo, the precise knowledge on lower critical solution temperature (LCST) is needed.

Position of the critical solution temperature results from an unstable equilibrium between hydrophilic and hydrophobic interaction of a polymer with water and depends on the content of hydrophilic and hydrophobic elements in a polymer molecule, their distribution, topology as well as molecular mass and presence of low-molecular substances in the solution. Below the phase transition temperature, a macromolecule is retained in the solution owing to the salvation zone. It is formed as a result of formation of hydrogen bonds between hydrophilic fragments of polymer chain and water molecules. A temperature increase weakens the hydrogen polymer-water bonds and promotes hydrophobic polymer-polymer interactions (dipole, hydrogen, van der Waals forces, etc.). Precipitation of the polymer from the water solution at increased temperature is accompanied by an increase of the system entropy. As a consequence, the chain structure changes into a ball, phase separation occurs, and in the case of a 3D polymer network a volumetric phase transition is observed.

Determination of the lowest critical solution temperature (LCST) proceeds by way of rheological measurements, i.e. determination of the point of intersection of G' and G" as the function of oscillation frequency ω or by determining a dramatic increase in viscosity of the system investigated during its heating.

Rheological measurements are dependent on volume of sample. In classical rheometry with coaxial cylinders or cone and plate measurement systems volume of sample is equal from several to several dozen millilitres. In pharmacological practice such a volume of samples is practically unattainable and simultaneously cost of their production is very high. During rheological measurement in rotational rheometer structure of investigated media is partially deformed (destroyed) according to acting on sample mechanical forces.

In addition, the dynamic nature of measurement, where the structure of investigated medium is deformed due to shear inducted by measurement's system rotor results in both temperature and mechanical effects responsible for destroying the building up structure.

The above measurement problems can be solved by use of passive optical microrheology.

### *1.1. Aim of work*

Due to huge amount of possible hydrogels and their derivatives they represent vast spectrum of properties. Rheology can quickly provide information on gel structure changes like sol-gel transition, therefore can act as primary, very effective characterization techniques. The main aim of presented study was determination of lower critical solution temperature LCST and transition kinetics of chitosan gel by high precision classical rotational rheometry and passive optical microrheology and comparison of results obtained by those two different rheological techniques.

## 2. Experimental

As an experimental media Shrimp chitosan -based matrix (average molecular weight Mw 862 kDa and degree of deacetylation 83,4 DD) was used. A method to form chitosan thermogels based on solutions of chitosan salts in 0.1 M hydrochloric acid using sodium β-glycerophospate as a neutralizing agent has been developed [1-5].

Rheological properties of chitosan gel were determined by Physica MCR 301 Anton Paar rotational rheometer using cone and plate system. The angle of 50 mm cone diameter was equal 1 deg. Investigations were carried out in four temperatures 25, 30, 35 and 40°C and range of oscillation frequency ω from 0.01 to 500 $s^{-1}$ with constant amplitude γ equal 1%.

The measurements were performed by placing the chitosan salt solution sample in measurement gap at temperature of 5 °C and increasing the temperature to desired lever of 25, 30, 35 or 40 °C with speed of 10 °C/s.

The passive microrheology techniques relies in general on tracking the Brownian motion of fluorescent tracer particles suspended in investigated medium. The rheological parameters are calculated based on mean square displacement MSD of tracer evolution with time and generalized Stokes-Einstein equation [6].

$$\tilde{G}(s) = \frac{k_B T}{\pi a s \langle \Delta \tilde{r}^2(s) \rangle} \qquad (1)$$

where $k_B$ is Boltzmann constant, T is temperature, a is the radius of the tracer particle, $\tilde{G}(s)$ is the Laplace transformed complex shear modulus, s is the Laplace frequency and $\Delta\tilde{r}^2(s)$ is Laplace transformed mean square displacement (MSD) defined as:

$$\langle MSD \rangle = \langle \Delta r^2(\tau) \rangle = \langle |r(t+\tau) - r(t)|^2 \rangle \qquad (2)$$

where r is particle position and τ is the lag time.

The passive microrheology measurements were carried out in temperature range between 39 and 45 °C on Leica DMI3000B inverted microscope equipped with HCX PL FLUOTAR 100x/1.30 oil immersion objective. The light was provided by Thorlabs Mounted High Power LED (470 nm, 830 mW) and filtered/splitted on Leica I3 (BP 450–490/510/LP 515) filtercube. The images were recorded by Lumenera Infinity 1 camera (1,3 MPix, ½ "CMOS). Polystyrene, fluorescent yellow-green tracers of 100 nm diameter were used (Sigma Aldrich, Poland). The temperature was controlled by Anton Paar Viscotherm V2 & PZO Poland temperature control system. Data processing was done in MATLAB (MathSoft, USA) using PolyParticleTracker script [7].

## 3. Results

Experimental values of storage G' and loss modulus G" for chitosan gels obtained in different temperatures in rheometrical measurements (rotational rheometer Anton Paar) are presented on fig.1.

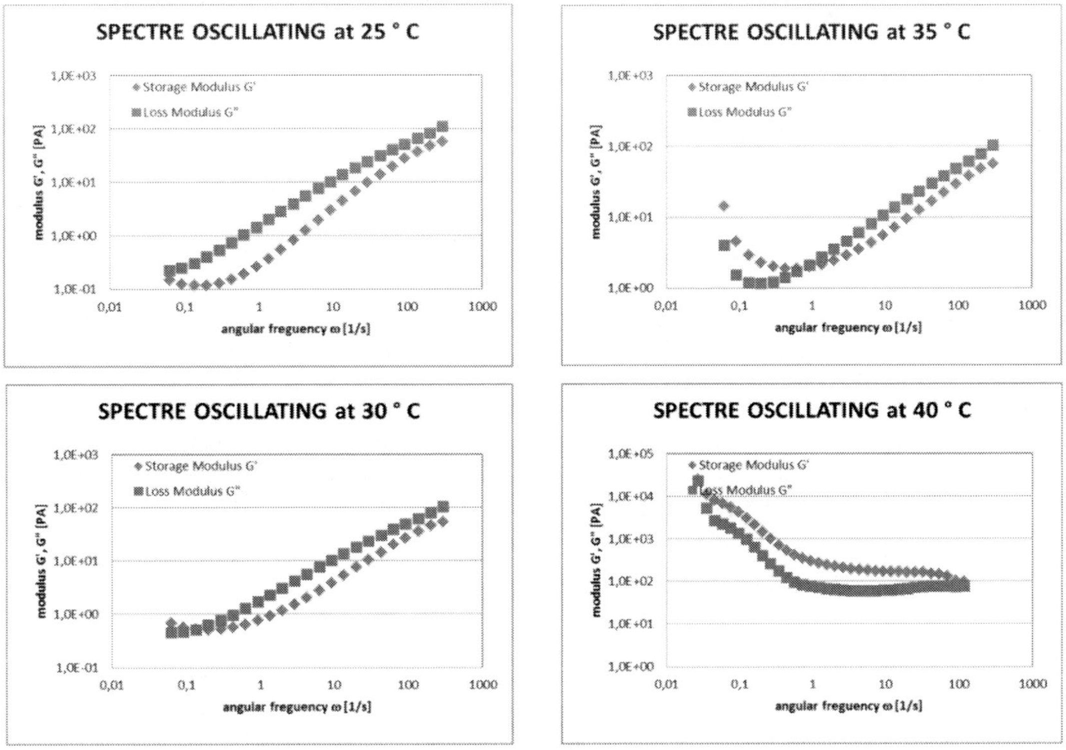


Fig. 1. The frequency-dependent storage G'(ω) and Loss Modulus G"(ω) for investigated hydrogel in different temperatures.

Presented on Fig. 1. graphs show that in 25-35°C the investigated chitosan salts solutions present non-Newtonian character with dominating viscous effects. Measurements performed at 35°C show however the sample viscoelastic properties, indicating the polymer network build up due to temperature rise.

The obtained values of complex viscosity and initial results of optical microrheology measurements are presented on fig. 2.

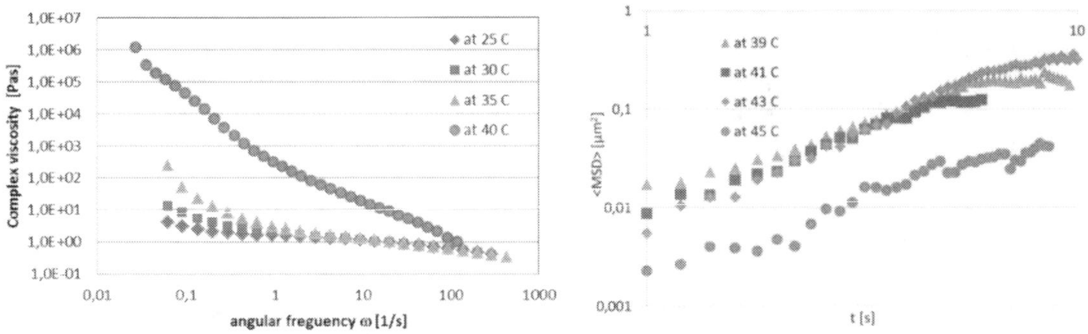


Fig. 2. Comparison of complex viscosity (Anton Paar) with optical microrheology results.

Visible at fig. 2 abrupt change in the complex viscosity in case of classical rotational rheometric measurements and mean squared displacement <MSD> in case of initial results of optical rheology measurements reflects the material mechanical parameters changes due to structural sol-gel transition.

## 4. Conclusions

As we show the passive optical microrheology can provide efficient way for lower critical solution temperature determination of sol-gel transition as the sample volume (order of μl) of expensive chitosan gel can be reduced by order of magnitude compared with classical rheological techniques, therefore the heat inertia and temperature control issues can be minimized. Another advantage of passive optical microrheology is easiness to maintaining sample sterility. This, together with low equipment cost, time of analysis of order of seconds and the potential of automated sample handling proves that optical microrheology is a very versatile and powerful rheological technique.

**Acknowledgements**
Work was supported by The National Science Centre (project N N209 764640)

**References**

1. Modrzejewska Z, Korus I, Owczarz P 2001 *J. Membrane Science* **181** 229
2. Modrzejewska Z, Owczarz P 2004 *Inżynieria Chemiczna i Procesowa* **25** 1417
3. Modrzejewska Z, Owczarz P 2011 *Thermosensitive chitosan system,* EKMA Warszawa, 203-224, ISBN: 978-83-61905-08-0
4. Cho J, Heuzey M-C, Begin A, Carreau P J 2005 *Biomacromolecules* **6** 3267
5. Cho J, Heuzey M-C, Hamidine M 2007 *Macromolecular Mater. and Eng.* **292** 571
6. Gardel M L, Valentin M T, Weitz D A 2005 *Microrheology, in Microscale Diagnostic Techniques* (ed. Breuer K.S.) Springer-Verlag Berlin Heidelberg 1-49
7. Rogers S S, Waigh T A, Zhao X, Lu L R. 2007 *Physical Biology* **4** (**3**) 220

# The influence of selected excipients on the rheological behaviour of chitosan based ocular pharmaceutical systems

**L Budai, E Szabadi, M Hajdú, M Budai, I Klebovich, I Antal**

Semmelweis University, Department of Pharmaceutics,
Hőgyes E. u. 7., Budapest, H-1092, Hungary
E-mail: budai.livia@pharma.semmelweis-univ.hu, antal.istvan@ pharma.semmelweis-univ.hu

**Abstract.** Aims: Chitosan, a modified natural carbohydrate polymer, has received great attention in diverse scientific fields including pharmaceutical and biomedical research areas. Besides its low toxicity, mucoadhesiveness and biodegradability its special favourable rheological feature makes it a unique gelling agent for the design of ocular systems. Chitosan based (2.0 w/v %) ocular systems containing selected excipients were formulated in order to investigate the rheological influence of applied auxiliary materials. Rotational and oscillatory rheological properties of propylene glycol (1.0-20.0 w/v %), glycerin (1.0-5.0 w/v %) and castor oil (1.0-5.0 w/v %) containing chitosan gels were evaluated. The rheological behaviour of formulated ocular gels were compared before and after steam sterilization.

Methods: Rotational and oscillatory rheological measurements were carried out with Kinexus Pro Rheometer. Comparison of flow curves and oscillatory frequency sweep measurements in the linear viscoelastic region made possible the evaluation of rheological effect of selected excipients.

Results: In the applied concentration range the effect of propylene glycol among the selected excipients presents the most significant impact on rheology of chitosan formulations. Steam sterilization results in reduced viscosity in most of chitosan gels. However, the presence of polyols appears to prevent the degradation of chitosan after steam sterilization.

## 1. Introduction

Chitosan is a widely used polysaccharide with outstanding rheological behaviour. The viscoelastic property of chitosan makes the polymer a unique gelling agent in many reseach areas including the pharmaceutical formulation development. Chitosan based pharmaceutical systems possess low toxicity, mucoadhesiveness and biodegradability. Furthermore, chitosan posesses viscoelastic behaviour, as human tear under physiologic conditions. This fact provides chitosan the ability to serve as an appropriate gelling agent in ocular dosage forms [1-2].

Glycerin, propylene glycol and castor oil are frequently used auxiliary materials in pharmaceutical technology. In ophtalmic formulations glycerin is mainly used as a solvent. Propylene glycol has become widely used as a solvent and preservative in a variety of pharmaceutical formulations. Application of castor oil in chitosan based ocular gels can be recommended in order to increase the solubility of lipophil active ingredients in the dosage forms. Required microbiological quality of the mentioned auxiliary materials can be achieved by steam sterilization [3].

Polyols such as glycerin and propylene glycol were used in this study because of the rheological protective effect attributed to polyols in chitosan formulations after steam sterilization [4-5].

Our aim was to prepare chitosan based ocular systems and to investigate the effect of selected pharmaceutical excipients on the rheological behaviour of formulated gels before and after steam sterilization.

## 2. Materials and methods

Chitosan was purchased from Sigma-Aldrich Ltd. (Budapest, Hungary). All the other excipients such as acetic acid, glycerin, propylene glycol, castor oil were purchased from Hungaropharma Ltd., (Budapest, Hungary). Distilled water was produced by MilliQ-water device (Millipore Ltd., Budapest, Hungary).

Chitosan formulations 2.0 % (w/v) were prepared by dissolving chitosan powder in acetic acid solution 0.5 % (v/v) under stirring for 1 h. If the preparations contained other excipients such as glycerin or propylene glycol, were the excipients first dissolved in acetic acid solution then was chitosan added to the preparation followed by an 1-hour-long stirring procedure. Besides the reference chitosan formulations - 0.5 % (v/v) acetic acid containing 2.0 % (w/v) chitosan gel - three sequences (containing one of the selected excipient above the components of reference sample) were prepared. Glycerin (1.0-5.0 w/v %), propylene glycol (1.0-20.0 w/v %) and castor oil (1.0-5.0 w/v %) containing chitosan gels - named as glycerin, propylene glycol and castor oil sequences - were examined.

Rheological measurements were carried out with Kinexus Pro rheometer (Malvern Instruments Ltd, UK). Measured data were registered with rSpace for Kinexus Pro 1.3 software. Preparations were measured using a cone and plate geometry. The gap between the cone and plate of sample placement was 0.15 mm. The temperature of the samples was controlled with an accuracy of ± 0.1 °C, by Peltier system of the rheometer. In all measurements a cylindrical cover made of stainless steel was placed over the samples, in order to create a closed, saturated volume round the sample and to prevent evaporation of the sample. Rotational measurements of all chitosan formulations were determined at 25 °C. Oscillatory experiments were carried out at 33 °C. Our temperature choice during rotational and oscillatory measurements can be explained with the following facts: the temperature of ophthalmic dosage forms during storage and application is about 25 °C, while the temperature of human tears on the cornea surface is approximately 33 °C. In case of each sample at least two parallels were analysed; mean values are demonstrated in present article. Boxer 300_75LR steam sterilizer was used for sterilization (20 mins, 121 °C). The pH of the samples was measured by Mettler Toledo AG (Switzerland).

## 3. Results

The compositions and pH data for formulated chitosan gels are summarized in Table 1. The pH values of chitosan formulations were between 5.65 - 5.90 prior to sterilization. After sterilization the pH values vary between 5.71 - 5.95. Any significant changes can be observed in pH values after sterilization.

**Table 1.** Compositions and pH of chitosan formulations

| Name of sequence | Concentration of chitosan % (w/v ) | Concentration of acetic acid % (w/v) | Concentration of excipient % (w/v) | pH before steam sterilization | pH after steam sterilization |
|---|---|---|---|---|---|
| "glycerin sequence" | 2.0 | 0.5 | 1.0 | 5.65 | 5.74 |
| | 2.0 | 0.5 | 2.0 | 5.66 | 5.76 |
| | 2.0 | 0.5 | 3.0 | 5.71 | 5.72 |
| | 2.0 | 0.5 | 4.0 | 5.76 | 5.85 |
| | 2.0 | 0.5 | 5.0 | 5.76 | 5.78 |
| | | | | | |
| "propylene glycol sequence" | 2.0 | 0.5 | 1.0 | 5.74 | 5.89 |
| | 2.0 | 0.5 | 5.0 | 5.73 | 5.93 |
| | 2.0 | 0.5 | 10.0 | 5.77 | 5.94 |
| | 2.0 | 0.5 | 15.0 | 5.80 | 5.93 |
| | 2.0 | 0.5 | 20.0 | 5.90 | 5.95 |
| | | | | | |
| "castor oil sequence" | 2.0 | 0.5 | 1.0 | 5.69 | 5.71 |
| | 2.0 | 0.5 | 2.0 | 5.63 | 5.71 |
| | 2.0 | 0.5 | 3.0 | 5.70 | 5.73 |
| | 2.0 | 0.5 | 4.0 | 5.80 | 5.79 |
| | 2.0 | 0.5 | 5.0 | 5.86 | 5.79 |

### *3.1. Choice of excipients*

The choice of **glycerin** can be explained by its frequent application in ophthalmic dosage forms. As it is known that glycerin decomposes on heating, with the evolution of toxic acrolein if the aqueous preparation contains more than 85 % glycerin. Thus glycerin containing samples can be sterilized by autoclaving if the preparation contains at least 20 % water and the conditions of steam sterilization follow the standard setting parameters (120 °C, 20 min). In ophthalmic preparations glycerin is commonly used in a concentration range of 0.5-3.0 %. In order to investigate the effect of glycerin on the rheological properties of chitosan gels, chitosan gels containing 1.0-5.0 % glycerin were prepared. The highest glycerin concentration was 5.0 %, regarding the applied concentrations in eye drops [3].

**Propylene glycol** is a frequently used excipient in pharmaceuticals. In parenteral formulations its concentration can reach 60 %, while in topical formulations 80 %. In topical preparations, propylene glycol is regarded as minimally irritant, although it is more irritant than glycerin. Propylene glycol can be sterilized by autoclaving. As in a concentration above 15 % propylene glycol possesses preservative effect, a further advantage of propylene glycol application in solutions and semisolid dosage forms is the preparation of pharmaceutical dosage forms without the traditionally used preservatives [3].

Based on the fact that chitosan containing gels 2.0 % (w/v) possess higher viscosity than aqueous solutions, **castor oil** in 1.0-5.0 % (w/v) can be dispersed homogeneously in chitosan gels. The highest concentration of castor oil used throughout our experiments was 5.0 % (w/v) in chitosan gels. Higher than 5.0 % (w/v) concentration of castor oil in gels would have required the application of emulsifiers which are regarded as irritant excipients in ocular systems.

### *3.2. Rotational measurements*

Rotational and oscillatory measurements of reference samples and sequences were compared (Figure 1). The viscosity was measured as a function of shear rate in a range of 0.1-10.0 $s^{-1}$. The applied shear rate spectrum describes the shear conditions of several applications. The shear rate values of about 0.1 $s^{-1}$ characterize the leveling, 1.0 $s^{-1}$ the pumping and sample application and 10.0 $s^{-1}$ the brushing, mixing or stirring. The viscosity values of samples vary between 0.1 and 10.0 Pa·s.

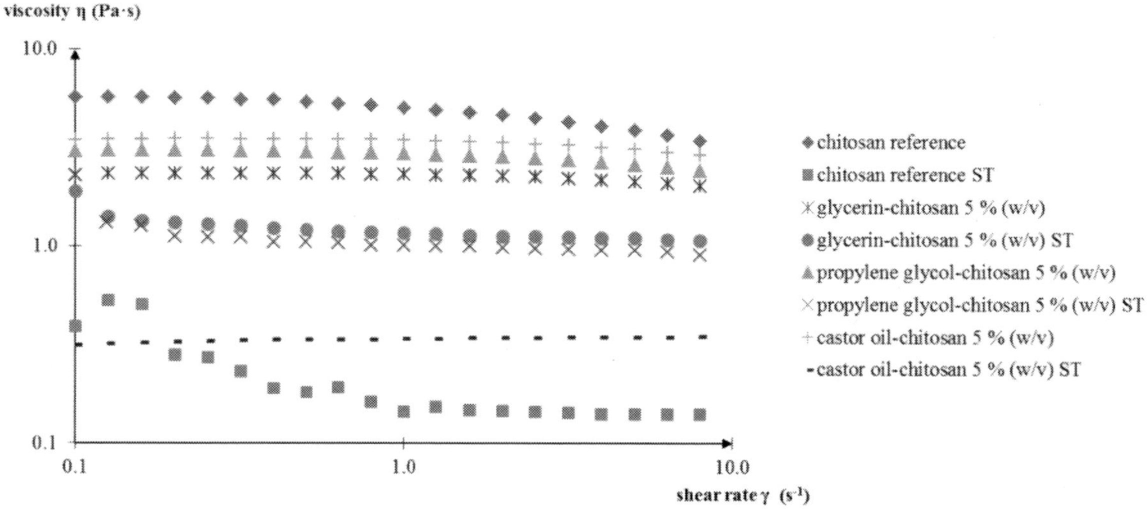


**Figure 1.** Comparison of rotational measurements. Rotational measurement results for 2.0 % (w/v) chitosan gels with various excipients. Concentration of excipient is 5.0 % (w/v) in each case. ST denotes steam sterilized samples.

The highest viscosity values belong to the reference chitosan sample. The viscosity loss is most expressed in case of steam sterilized reference and sterilized castor oil 5 % (w/v) containing samples whose viscosity is below 1.0 Pa·s. Almost all the excipients reduce the viscosity of gels comparing to reference chitosan sample. Further viscosity loss can be observed after steam sterilization. However, the viscosity of polyol containing chitosan gels is not reduced in such an extent after steam sterilization than that of reference sample. The 5 % (w/v) polyol-content in chitosan gels results in a viscosity of approximately 1.0 Pa·s after steam sterilization.

In the propylene glycol sequence propylene glycol content reached 20 % (w/v). Regarding the rotational viscosity results the higher propylene glycol concentrations leads to higher viscosity values (Figure 2). The viscosity of 20 % (w/v) propylene glycol containing sample exceeds the viscosity of reference sample in the shear rate range of 0.1-1.0 $s^{-1}$. The viscosity loss of sterilized samples is not so expressed in case of higher propylene glycol content as for samples with relatively low propylene glycol concentrations. The presence of castor oil reduces the viscosity of gels compared to the viscosity of reference before steam sterilization. However, increased viscosity values can be measured as a result of addition of castor oil for samples after steam sterilization.

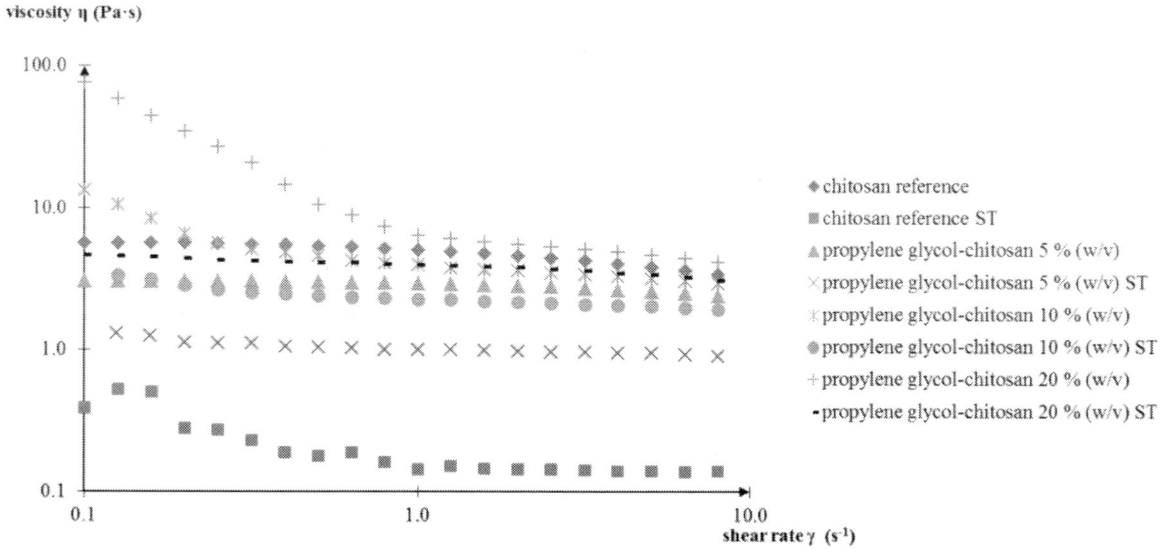


**Figure 2.** Rotational measurements of propylene glycol-chitosan gels. Rotational measurement results of propylene glycol-chitosan gels with various 5.0-20.0 % (w/v) propylene glycol concentrations. ST denotes steam sterilized samples.

### *3.3. Oscillatory measurements*

In the present work the complex viscosity was used for the comparison of oscillatory measurement. The complex viscosity ($\eta^*$) published and introduced by Cox and Merz 1958 serves an empirical relationship for polymeric systems between rotational and oscillatory viscosity data as a function of shear rate (G) and angular frequency ($\omega$) [6].

Complex viscosity can be represented by complex mathematics: $\eta^* = \eta' + i\,\eta''$. The terms $\eta'$ and $\eta''$ can be related to the G' and G'' through the following relationships: $G' = \eta''\omega$, $G' = \eta'\omega$. The complex viscosity can be defined as $\eta^* = G^* / i\,\omega$, where $G^*$ means complex modulus and $i$ is the so called imaginary unit considering the rules of complex mathematics [6].

The complex viscosity was measured in the angular frequency range of 1.0-100.0 (rad/s). The complex viscosity of samples varied between 0.1-3.5 Pa·s (Figure 3). Polyol (glycerin and propylene glycol) containing samples provide a protective effect against complex viscosity loss after steam sterilization.

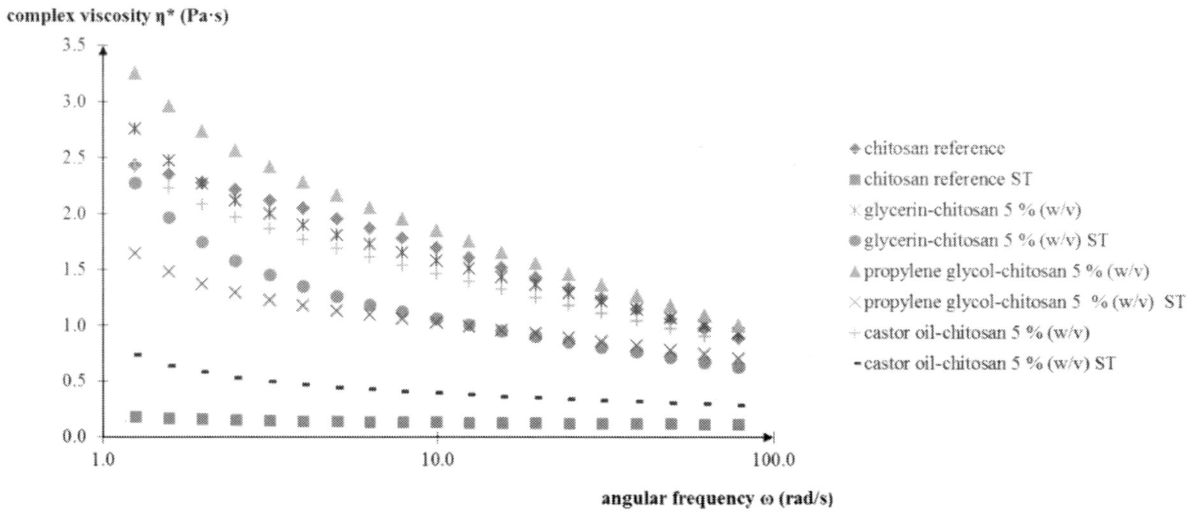


**Figure 3.** Comparison of oscillatory measurements. Oscillatory measurement results for 2.0 % (w/v) chitosan gels with various excipients. Concentration of excipient is 5.0 % (w/v) in each case. ST denotes steam sterilized samples.

Propylene glycol – on a concentration dependent manner - increased the complex viscosity of chitosan gels comparing to reference sample (Figure 4). The most expressed viscosity increase was observed for the 20.0 % w/v propylene glycol containing sample. Comparing samples with same propylene glycol concentrations before and after steam sterilization, after steam sterilization lower viscosity values can be registered.

The presence of castor oil reduces the complex viscosity of formulations compared to the viscosity of reference before steam sterilization. However, increased viscosity values can be measured as a result of addition of castor oil for samples after steam sterilization.

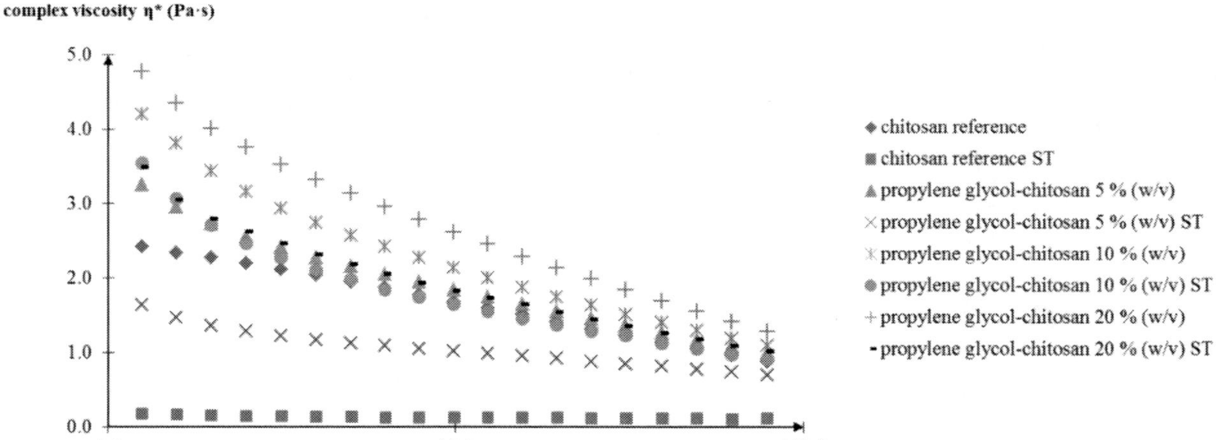


**Figure 4.** Oscillatory measurements of propylene glycol-chitosan gels. Oscillatory measurement results of propylene glycol-chitosan gels with various 5.0-20.0 % (w/v) propylene glycol concentrations. ST denotes steam sterilized samples.

## 4. Discussion

Most of the pharmaceutical excipients used in the present study reduce the viscosity of chitosan formulations. Steam sterilization also results in viscosity loss in the investigated samples. Both rotational and oscillatory measurements demonstrate that the applied excipients help to avoid the viscosity loss of chitosan gels after steam sterilization. The protective effect of polyols (glycerin and propylene glycol) is more significant than that of castor oil. In the applied propylene glycol concentration range (1.0-20.0 % w/v) the 20.0 % (w/v) possesses the most effective protective influence on rheology of chitosan systems. Furthermore, the application of propylene glycol is advantageous in pharmaceutical dosage forms as it serves as a preservative beyond its protective rheological feature. The propylene glycol provides preservative effect if its concentration reaches 15.0 % (w/v). Considering that the formulated gels are intended to apply as ocular systems, further investigations are needed concerning the safety and irritative potential of propylene glycol containing gels.

## 5. Conclusions

The rheological effects of selected excipients – glycerin, propylene glycol and castor oil – in chitosan ocular formulations were described before and after steam sterilization. Rotational measurement results correspond to oscillatory measurements results. Steam sterilization reduces the viscosity of chitosan 2.0 % (w/v) formulations. Polyol excipients, such as glycerin and propylene glycol protect the chitosan gels against viscosity loss after steam sterilization. The most expressed protective effect was observed in case of chitosan formulation with the highest propylene glycol concentration (20.0 % w/v).

## References

[1] Rodrigues S, Dionísio M, López C R and Grenha A 2012 *J. Funct. Biomater* **3** 615
[2] Tiffany M J 1991 *Int. Ophthalmol.* **15** 371
[3] Rowe R C, Scheskey PJ and Owen SC 2006 *Handbook of Pharmaceutical Excipients* 5th Edition Published by Pharmaceutical Press and American Pharmacists Association, London-Chicago
[4] Jarry C, Chaput C, Chenite A, Renaud M A, Buschmann M and Leroux J C 2001 *J.Biomed. Mat. Res.* **58** 127
[5] Jarry C, Leroux J C, Haeck J and Chaput C 2002 *Chem. Pharm. Bull.* **50** 1335
[6] Cox WP, Merz EH 1958 *J. Polym. Sci.* **28** 619

# Numerical Simulation of Rheological, Chemical and Hydromechanical Processes of Thrombolysis

**E Khramchenkov[1,2], M Khramchenkov[1]**
[1] Kazan Federal University – Institute of Mathematics and Mechanics– Kazan – Russia
[2] Joint Supercomputer Center – Kazan department – Kazan – Russia

ekhramch@gmail.com and mkhramch@gmail.com

**Abstract**. Mathematical model of clot lysis in blood vessels is developed on the basis of equations of convection-diffusion. Fibrin of the clot is considered stationary solid phase, and plasminogen, plasmin and plasminogen-activators - as dissolved fluid phases. As a result of numerical solution of the model predictions of lysis process are gained. Important influence of clot swelling on the process of lysis is revealed.

## 1. Introduction

Lysis of blood clots has been a subject of intensive investigation for a long time. Importance of such investigations is concluded in extreme significance of thrombosis and thrombolysis processes for human organism.

In critical situations caused by clot plugging of main blood vessels, necessity of clot lysis with special chemical agents arises. Such agents are tissue plasminogen-activator, urokinase plasminogen-activator and plasminogen. In mutual reaction plasminogen and activator produce plasmin that dissolves material of the clot (fibrin).

Rate of lysis process depends on several aspects: clot structure, clot disposition, volume of injected lytic agents. It is important to remark that the lysis strategy - volume and period of injection of lytic agents - are critical factors of thrombolysis process. In every case it is important to consider the necessity of fast clot lysis from one side, and the risk of system bleeding from another.

Structure of blood clot is very complicated and various. Mechanisms of clot formation are described in [1-3]. Physical and biochemical structures of a clot depends on such factors as: concentration of thrombin, concentration of fibrinogen, ionic forces in solution, hemodynamic force [2,4]. Platelets caught during the clot formation are able to compact fibrin down to the 0.1 of its initial volume, by pressing plasma from the clot. Compacted clots are especially resistant to lysis, perhaps because of minor fluid phase of plasminogen in clot. Clots in arteries are able to contain extremely dense fibrin net.

Clot consists of long dendrite fibrin fibers of approximately equal diameter [5]. It's possible to sort out fibrin structures, which were formed under low ionic strength (0.1 mM) and high ionic strength ( 0.3 mM).

First type are "coarse" structures - fibrin fibers have radius about 100-300 nm, permeability about 10-8 sm2, diameter of pores 3-10 um. Structures formed under high ionic strength are "fine" - fibrin fibers have radius about 8-100 nm, permeability about 10-10 cm2, diameter of pores 0.1-0.5 um [4-6]. The process of lysis is concluded in reaction between activator and plasminogen with production of active dissolving agent - plasmin. The latter reacts with fibrin fibers, destroying them gradually and moving into the clot material under the convection and diffusion. Tissue plasminogen-activator reacts with plasminogen on the surface of fibrin, urokinase reacts with plasminogen in fluid phase.

The problem is that transport of lytic agents arises from convection and diffusion. Convection in clot is hampered and diffusion takes a lot of time. The process of plasminogen transmutation to plasmin takes only few seconds, but it takes about an hour to destroy even average clot. Taking into account limits of therapeutic window it is necessary to have as much information about lysis dynamics and agents transport as possible.

The aim of this work was to develop clot lysis model with respect to peculiarities of its structure and properties as a porous deformable body with elastic and viscous-elastic rheology. On the basis of the model numerical scheme was established and processes taking place during the clot lysis were examined. Also, the aim was to investigate dynamic of lysis, transport of agents in both fluid and solid phases. Influence of clot characteristics on the lysis was examined as well.

## 2. Mathematical model

To obtain mathematical model of the process we have to build a system of filtration equations in deformable porous space with porous skeleton with variable mass. From the definition of the coefficient of dilatation [7]

$$\theta = \frac{V - V_0}{V_0}, \tag{1}$$

where $V$ and $V_0$ – current and initial values of the volume of the deformable porous media. Supposing values of $\theta$ to be small, we obtain the following equation

$$V = V_0 \exp\theta. \tag{2}$$

Then for the mass of the porous skeleton we have

$$M_s = (1-\varepsilon)\rho_s V_0 \exp\theta. \tag{3}$$

Consider $\rho_s$ constant and taking time derivative in the last equation, after some transformations, we receive:

$$\frac{\partial \varepsilon}{\partial t} = (1-\varepsilon)\frac{\partial \theta}{\partial t} - \frac{(1-\varepsilon)}{M_s}\frac{\partial M_s}{\partial t}. \tag{4}$$

In case of lysis, when clot consists of fibrin fibers of variable radius $R_s$, we can obtain the following equation:

$$\frac{1}{M_s}\frac{\partial M_s}{\partial t} = \frac{2}{R_s}\frac{\partial R_s}{\partial t} \tag{5}$$

Here $M_s$ – mass of the clot's solid phase. Equations of mass balance of the fluid phase, the dissolved fibrin and the fibrin of the porous skeleton:

$$\frac{\partial(\varepsilon\rho_f)}{\partial t} + \operatorname{div}(\varepsilon\rho_f \vec{V}_f) = 0 \tag{6}$$

$$\frac{\partial(\varepsilon c)}{\partial t} + \operatorname{div}(\varepsilon c \vec{V}_f) = j \tag{7}$$

$$\frac{\partial\left[(1-\varepsilon)\rho_s\right]}{\partial t} + \operatorname{div}\left[(1-\varepsilon)\rho_s \vec{V}_s\right] = -j \tag{8}$$

Here $\varepsilon$ is porosity; $\rho_s$ and $\rho_f$ are densities of the solid and fluid phases correspondingly; $\vec{V}_f$ is the velocity of the fluid phase; $\vec{V}_s$ is the velocity of the solid phase. Then we divide the equation (6) by $\rho_f$, equations (7) and (8) by $\rho_s$. Designating $c / \rho_s = C$, $\varepsilon(\vec{V}_f - \vec{V}_s) = \vec{v}$ and consider $\operatorname{div}\vec{V}_s = \partial\theta / \partial t$, we calculate sum of equations (6)-(8). Ignoring second-order infinitesimal, we obtain

$$\frac{\partial \varepsilon C}{\partial t} + \frac{\partial \theta}{\partial t} + \operatorname{div}\vec{v} + \operatorname{div}c\vec{v} + \varepsilon C \frac{\partial \theta}{\partial t} = 0. \tag{9}$$

Here $C$ is the dimensionless concentration of the dissolved fibrin; $c$ is the concentration of the dissolved fibrin. Adding (7) and (8), after dividing them by $\rho_s$, with respect to (4), one can obtain the following equation

$$\frac{\partial \varepsilon C}{\partial t} + \operatorname{div}C\vec{v} + \varepsilon C \frac{\partial \theta}{\partial t} = -\frac{2(1-\varepsilon)}{R_s}\frac{\partial R_s}{\partial t}. \tag{10}$$

Subtracting the last equation from (9)

$$\frac{\partial \theta}{\partial t} + \operatorname{div}\vec{v} = \frac{2(1-\varepsilon)}{R_s}\frac{\partial R_s}{\partial t}. \tag{11}$$

Equations (4), (10) and (11) and Darcy law describe filtration in deformable porous media with porous skeleton of variable mass. So we can now write obtained equations of general mathematical model. Equations of mass transport are:

$$\frac{\partial}{\partial t}\left(\varepsilon c_{tPA}^f + (1-\varepsilon)c_{tPA}^b\right) + \varepsilon c_{tPA}^f \frac{\partial \theta}{\partial t} = \operatorname{div}\left(\varepsilon D_L^{tPA}\nabla c_{tPA}^f - \vec{v}c_{tPA}^f\right) \quad , \tag{12}$$

$$\frac{\partial}{\partial t}\left(\varepsilon c_{uPA}^f\right) + \varepsilon c_{uPA}^f \frac{\partial \theta}{\partial t} = \operatorname{div}\left(\varepsilon D_L^{uPA}\nabla c_{uPA}^f - \vec{v}c_{uPA}^f\right), \tag{13}$$

$$\frac{\partial}{\partial t}\left(\varepsilon c_{plg}^f + (1-\varepsilon)c_{plg}^b\right) + \varepsilon c_{plg}^f \frac{\partial \theta}{\partial t} = \operatorname{div}\left(\varepsilon D_L^{plg}\nabla c_{plg}^f - \vec{v}c_{plg}^f\right) - \frac{k_2^{tPA}c_{tPA}^b c_{plg}^b}{K_m^{tPA} + c_{plg}^b}, \tag{14}$$

$$\frac{\partial}{\partial t}\left(\varepsilon c_{plm}^f + (1-\varepsilon)c_{plm}^b\right) + \varepsilon c_{plm}^f \frac{\partial \theta}{\partial t} = \operatorname{div}\left(\varepsilon D_L^{plm}\nabla c_{plm}^f - \vec{v}c_{plm}^f\right) + \frac{k_2^{tPA}c_{tPA}^b c_{plg}^b}{K_m^{tPA} + c_{plg}^b}. \tag{15}$$

Here $c_i^j$ is the concentration of the $i$-th component in the $j$-th phase; $D_L^i$ – coefficient of dispersion of the $i$-th component; $k_2$, $k_r$, $k_f$, $K_m$ are dynamic constants of the process(values have been taken from [8]). For upper indexes we have: b – concentrations which are bounded to the fibrin; f – fluid; L – value on the exit of the clot; plg – plasminogen; plm – plasmin; s – solid phase; tPA – tissue plasminogen-activator; uPA – urokinase plasminogen-activator; 0 – initial value. Equations (12)-(15) are gained in similar way as (10), with replacement of source in right part of (10) by corresponding source [8]. In one-dimensional case physical interpretation of coefficient of cubic widening $\theta$ (which can be both positive and negative, in latter case it is the measure of shrinkage) ) is concluded in accounting linear deformations of clot - part of clot is characterized by positive values of $\theta$, the other part - by negative values of $\theta$. The total length of the clot is considered to be constant.

Equations for components in solid phase are as follows [8]:

$$\partial c_{tPA}^b / \partial t = k_f^{tPA}c_{tPA}^f\left(\psi_{tPA} - c_{tPA}^b\right) - k_2^{tPA}c_{tPA}^b. \tag{16}$$

$$\partial c^b_{p\lg} / \partial t = k_f^{p\lg} c^f_{p\lg} \left( \psi_{p\lg/m} - c^b_{p\lg} - c^b_{plm} \right) - k_r^{p\lg} c^b_{p\lg} - \frac{k_2^{tPA} c^b_{tPA} c^b_{p\lg}}{K_m^{tPA} + c^b_{p\lg}}, \quad (17)$$

$$\partial c^b_{plm} / \partial t = k_f^{p\lg} c^f_{plm} \left( \psi_{p\lg/m} - c^b_{p\lg} - c^b_{plm} \right) - k_r^{plm} c^b_{plm} + \frac{k_2^{tPA} c^b_{tPA} c^b_{p\lg}}{K_m^{tPA} + c^b_{p\lg}}. \quad (18)$$

Rate of radius variation is described as [8]

$$\bar{R}_1 = \frac{k_2 c^b_{plm} \psi_{plm}}{K_m + \psi_{plm}}, \; \psi_{plm} = 4q\pi p_0 r_0 R \left\{ \frac{L_{\Delta V}}{\Delta V} \frac{2 \; fibrin \; monomers}{45 \; nm} \frac{1}{N_{AV}} \frac{\left(10^6 \; \mu mol / mol\right)}{\left(10^{-3} \; L / cm^3\right)} \right\}, \quad (19)$$

$\psi_{plm}$ is an auxiliary function; it helps to recalculate destroyed dimensional (45nm) fibrin monomers into the value with concentration dimension.

Degree of lysis is calculated from the following equation [8]

$$L = C_1 \int_0^t \bar{R}_1 dt, \; C_1 = \text{const}. \quad (20)$$

Equation for the radius of fibrin fiber, $L$ is the length [8]

$$R_s = \sqrt{R_0^2 - \frac{1}{\pi p_0} \frac{L}{\left\{ \frac{L_y}{Vol_T} \frac{2 \; fibrin \; monomers}{45 \; nm} \frac{1}{N_{AV}} \frac{\left(10^6 \; \mu mol / mol\right)}{\left(10^{-3} \; L / cm^3\right)} \right\}}}. \quad (21)$$

The definition of deformation tensor is:

$$e_{ij} = (1/2)(u_{i,j} + u_{j,i}), \; i, j = x, y, z; \quad (22)$$

Conditions of mechanical equilibrium are as follows:

$$\begin{aligned} &\frac{\partial \sigma_{xx}}{\partial x} + \frac{\partial \sigma_{xy}}{\partial y} + \frac{\partial \sigma_{xz}}{\partial z} = 0, \\ &\frac{\partial \sigma_{yx}}{\partial x} + \frac{\partial \sigma_{yy}}{\partial y} + \frac{\partial \sigma_{yz}}{\partial z} = 0 \\ &\frac{\partial \sigma_{zx}}{\partial x} + \frac{\partial \sigma_{zy}}{\partial y} + \frac{\partial \sigma_{zz}}{\partial z} + \frac{\partial p}{\partial z} = 0. \end{aligned} \quad (23)$$

Equations for elastic and viscous-elastic rheology of skeleton correspondently, assuming that the load on clot material is equal to zero (effective tensions are equal to the pressure), are as follows:

$$\sigma_{ij}^{(s)} = -(K - \frac{2}{3} G) \varepsilon \delta_{ij} - 2G \varepsilon_{ij}, \; \varepsilon = \sum_i \varepsilon_{ii} \quad (24)$$

$$\sigma_{ij} = 2Ge_{ij} + 2\eta \dot{e}_{ij}; i \neq j; \; \sigma_{ii} = 3Ke_{ii}; \; i, j = x, y, z; \quad (25)$$

Here $K, G$ are stress and shear elastic modules of the fibrin; $\eta$ – effective viscosity of the fibrin in the Kelvin-Voigt model. Variation of porosity is described by the equation:

$$\frac{\partial \varepsilon}{\partial t} = (1-\varepsilon)\frac{\partial \theta}{\partial t} - \frac{2(1-\varepsilon)R_t'}{R_s},\ R_t' = \frac{\partial R_s}{\partial t}, \tag{26}$$

Equation for shrinkage takes a form

$$\frac{\partial \theta}{\partial t} + \mathrm{div}\vec{v} = \frac{2(1-\varepsilon)R_t'}{R_s} \tag{27}$$

Darcy law (form of permeability coefficient is taken from [8])

$$\vec{v} = -\frac{k}{\mu}\nabla p,\ k = \frac{4R_s^2}{70(1-\varepsilon)^{3/2}\left\{1+52(1-\varepsilon)^{3/2}\right\}}, \tag{28}$$

where $k$ is the coefficient of the permeability of clot, $p$ is the pressure in fluid; $\mu$ is the viscosity of blood. Boundary conditions for concentration on the left border are as follows:

$$\vec{v}c_{inlet} = \left(\vec{v}c_i - D_L^i \nabla c_i\right)\Big|_{x,y,z=0},\ i = tPA,\ uPA,\ plg,\ plm. \tag{29}$$

$$c_{i,inlet} = \begin{Bmatrix} c_i = 0, t > t_{inlet} \\ c_i = c_{i,bolus}, 0 < t < t_{inlet} \end{Bmatrix}. \tag{30}$$

Condition (28) describes regime of medicament injection and dosage. In the specified period of time specified amount of lytic agents are injected (tPA, uPA, plasminogen).

Boundary conditions on the right border:

$$\frac{\partial c_i}{\partial \vec{n}} = 0\,|_{\Gamma}. \tag{31}$$

Boundary conditions for the pressure

$$p = p_0\,|_{x=0},\ p = p_L\,|_{x=L}. \tag{32}$$

Initial conditions are

$$c_i = c_{i,0}\,|_{t=0}. \tag{33}$$

Rheological constants $K$ and $G$ in (24) are taken from [9]. So, according to [9] $G = 0.8$ [cmHg]. We did not succeed in defining the value of $\eta$ from literature, so that is why we've taken the value $\eta = 80$ [cmHg][s], which corresponds to some superpolymers (natural caoutchoucs (raw rubbers, nitrocellulose [10]). It's supposed that values of $G$ and $\eta$ are constant during the lytic process. The initial value of porosity is $\varepsilon_0 = 0.9$. The length of the clot was set to 1 cm, the initial value of coefficient of cubic widening is zero. At first we run calculations of one-dimensional case with no account taken of rheological properties of the clot, fluid flow rate was considered constant (values were taken from [8]). Results of calculations were compared with results of [8]. Then calculations of the three-dimensional model with the elastic and viscous-elastic rheology were performed.

## 3. Results and Discussion

At the beginning of each time step we calculated values of concentrations in solid phase, and then with these new values we solved equations for concentrations in fluid phase. From obtained concentrations we calculated radius of fibrin. After that, on the basis of received data, we determined distribution of pressure and shrinkage. Then the program went on the new time iteration.

For calculation of lytic agents in fluid phase the explicit scheme was used because of complicated structure of correspondent equations (12)-(15). Space and time steps were determined according to the stability condition of the explicit scheme. This scheme has been found to be rather stable for variations of parameters of the model.

Stability of the scheme was increased, after usage of the common fourth-order Runge-Kutta method for calculation of concentrations in the solid phase.

Model was calculated in two variants. In first case parameters of the model were set in compliance with parameters from [8] and 1-dimensional case was investigated. The aim was to compare results with predecessors. Obtained coincidence is good enough; more detailed information can be found in [11].

In second case model was calculated for three-dimensional case with elastic and viscous-elastic rheology. PETSc solvers were used for calculations. Visualization of obtained results was performed in ParaView open-source software. (Contributors: National Center for Supercomputing Applications (NCSA) at the University of Illinois at Urbana-Champaign (UIUC), Lawrence Livermore National Laboratory (LLNL), Sandia National Laboratories (SNL), Los Alamos National Laboratory (LANL), Jean-Loup Gailly and Mark Adler (gzip library)).

Some of the results are provided below.

We performed three-dimensional numerical simulation of an injection of dissolving agent into homogeneous porous layer with vertical borehole as an example of the real physical problem. The carbonate rock was considered as the layer, water was considered as the dissolving agent. Data for speed of dissolution, solubility and other parameters were taken from [8].

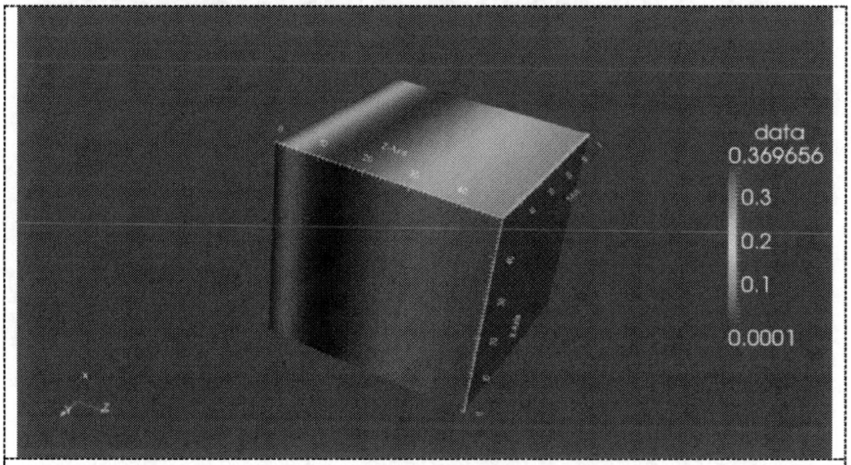


**Figure 1.** Distribution of tPA agent. Middle of the process.

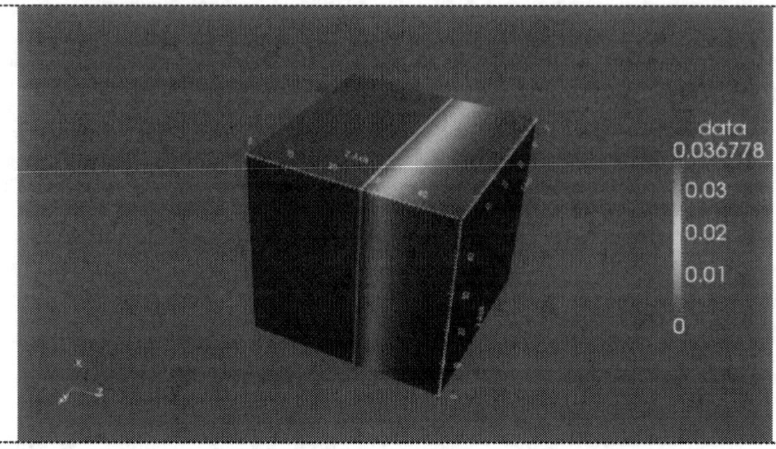


**Figure 2.** Plasmin wave through the clot. Middle of the process.

**Figure 3.** Pressure distribution. Middle of the process. Red color shows zone where pressure is equal to the blood pressure.

**Figure 4.** Fibrin distribution. Middle of the process. Red zone represents undamaged fibrin. Blue zone represents destroyed fibrin.

## 4. Conclusion

The main result of this work in our opinion is the clot lysis model. Clot is considered as porous body with skeleton which is dissolving in chemical reactions during the filtration. At first we developed general formulation of the problem, which describes the variation of the skeleton and porosity caused by the filtration of chemical active fluid. This formulation also takes in the account deformations of the skeleton under a load (in our case it happens because of pressure drop in the clot, with zero total load). Then the obtained model was specified for the case of reaction of fibrin with active agents of blood, which cause lysis of the clot. This model corresponds to the case of blood vessel plugging by the clot with the risk of thromboembolism propagation. The purpose was to investigate the influence of rheological properties and deformations of the clot on the process of lysis, and on the volumes and periods of injection of lytic agents. This interest was caused by anticipation of nonuniform deformation (and of variable porosity) zones appearance, in which lysis goes different from the case investigated in [8]. Computations for the 3-dimension model of flow and deformations show that zones of entrance and exit of the fluid in the clot also suffer significant deformations, which depend on rheological properties of the clot. In this article the simplest 1-dimension problem is investigated. The computations fully affirmed our predictions.

Obtained results show the significant influence of rheological properties of the porous matrix of the clot on the process of lysis. As it follows from computations the higher the clot material elasticity is, the faster lysis goes. The rate of lysis is even faster in case of elastic-viscous rheology of the fibrin. In the latter case lesser values of $\eta$ lead to higher rate of the lysis. The rate of clot destruction was non-constant, and the process was going faster in the ending phase. Model also has opportunities for future enhancements (complex form of the clot, non-linear properties of the material, anisotropic binding sites concentration).

**Acknowledgements**
Work is conducted under the RFBR project №12-07-00007 and ISTC 3868.

**References**

[1] Doolittle R. F. 1984 Fibrinogen and fibrin // Annu. Rev. Biochem. **53** 195
[2] Weisel J. W. and Nagaswami C. 1992 Computer modeling of fibrin polymerization kinetics correlated with electron microscope and turbidity observations: clot structure and assembly are kinetically controlled // Biophys. J. **63** 111

[3] Hunziker E. B., Straub P. W. and Haeberli A. A. 1990 New concept of fibrin formation based upon the linear growth of interlacing and branching polymers and molecular alignment into interlocking single-stranded segments // J. BioL Chem. **265** 7455
[4] Carr M. E., Shen L. L. and Hermans J. 1977 Mass-length ratio of fibrin fibers from gel permeation and light scattering // Biopolymers **16** 1
[5] Blomback B., Carlsson K., Hessel B., Liljeborg A., Procyk R. and Aslund N. 1989 Native fibrin gel networks observed by 3D microscopy permeation, and turbidity // Biochim. Biophys. Acta **997** 96
[6] Carr M. E. and Hardin C. L. 1987 Large fibrin fibers enhance urokinase-induced plasmin digestion of plasma clots // Blood **70** 400
[7] Sedov L.I. 1973 Mechanics of continuous media T.1. M.: Nauka 536 p. (in Russian)
[8] Diamond S. L. and Anand S. 1993 Inner Clot Diffusion and Permeation during Fibrinolysis // Biophysical Journal **65** 2622
[9] Diamond S.L. 1999 Engineering Design of Optimal Strategies for Blood Clot Dissolution // Annu. Rev. Biomed. Eng. **01** 427
[10] Ghlikman S.A. 1959 Introduction to physical chemistry of superpolymers. Saratov: Saratov university press, 379 p. (in Russian)
[11] Khramchenkov E.M. Khramchenkov M.G. 2011 Mathematical modeling of the lysis of clots in blood vessels Journal of Engineering Physics and Thermophysics **84** (5) 1

# Rheology and physical-chemical characteristics of the solutions of the medicines

**A Urakov[1, 2], N Urakova[2]**
[1] IM Ub RAS – Department deformed solid body – 426067 Izhevsk – Russia
[2] Izhevsk State Medical Academy – Department General and Clinical Pharmacology – 4260034 – Izhevsk – Russia

urakoval@live.ru, urakovanatal@mail,ru

**Abstract**. In the laboratory studied the dynamics of rheology of water solutions with plasma-inflammatory and antiseptic funds when mixing them with blood, plasma and pus under the influence of the following physical and chemical factors of local interaction: gravity, specific gravity, temperature, relative viscosity, internal pressure, sparkling water, total concentration of the ingredients, surface activity, volume of acid and osmotic activity of medicines. Found that the rheology of biological liquids improve hyperthermic, highly alkaline and highly carbonated solution medicines. For the dilution of pus, dense festering mass of sulfur plugs and tear stones invited to apply heated to +39 - +42°C with aqueous solution of 0.5 - 3% hydrogen peroxide and 0.5 - 10% sodium bicarbonate saturated with carbon dioxide to excess pressure 0.2 ATM.

## 1. Introduction

Currently in medicine and veterinary medicine are very widely used antiseptic and substitutes plasma, which shall be entered accordingly in the festering wounds (cavity) and in the blood [1]. Traditionally, nobody examines the processes of diffusion of solutions in these biological fluids, believing that these funds are equally bred blood and pus. The medicines of these groups are used in the form of water solutions entered in very large quantities (amounts), calculated in hundreds of milliliters. And still the essence of conventional technologies the use of these drugs, namely the essence of intravenous and intracavitary injections, is to introduce into the blood and pus cold substitutes plasma, antiseptic and even disinfectants solutions [2-6]. Such technology application drugs is so routine that its efficiency is not in doubt, processes of diffusion of drug in the blood and pus not investigated and no new technologies radically increase the efficiency of cultivation of blood (plasma) and dense festering mass [5,7,8]. At the same time, us has received a large amount of scientific evidence proving the existence of significant shortcomings of traditional technologies of drug cultivation of blood and pus, explaining the nature of the deficiencies identified and showing the way of effective modernization of conventional breeding techniques of blood and pus using famous plasma-inflammatory and antiseptic funds [9-15]. Our results show that to achieve a powerful clinical success (for the rapid and radical dissolution of blood clots, remove blood clots, dense purulent masses, ear wax and others) it is enough only to optimize processes rheology of drugs and dense and viscous colloidal fluids [16-21].

Initially, we examined the dynamics infiltrations postinjection medicines in the subcutaneous fat. It turned out that long-term local action solutions of medicinal products in the large concentration of subcutaneous fat causes devastating erosion , the appearance of which we explained excessive distinction physical and chemical characteristics of medicines and subcutaneous fat of each other [1]. Next, our study showed that the most significant destructive influence of such physical-chemical interaction factors as temperature, acidity (alkalinity) and the saturation gas solution [22].

In this regard, the aim of our study was to investigate the peculiarities of rheology and interstitial diffusion solutions of medicines in the thick of purulent mass changes of the temperature, pH, internal pressure and oxygen content and/or carbon dioxide.

## 2. Materials and methods

The study of the state of purulent masses conducted in laboratory conditions. In the experiments used purulent contents of the pleural cavity of patients TB pleural empyema, hospitalized in the surgery Department of TB clinic of the Udmurt Republic in 2007-2012, abdominal cavity patients with suppurative spilled peritonitis, hospitalized in the surgery Department clinics № 9 of the city of Izhevsk in 2010 - 2012, and conjunctival cavity of patients by a purulent conjunctivitis, received by the «Clinic Medlife» the city of Perm in 2010. Changing the state of isolated purulent masses and rheology of water solutions of medicines during pouring them into a festering mass of the volume ratio of 1:1 at +24, +37 +42°C degrees studied with the eye on the previously published method [22]. Temperature of purulent masses in their interaction with solutions of drugs in vitro is in the infrared range of a spectrum of radiation with the help of thermal imager NEC TH91XX (USA), with the subsequent processing of information by means of programmes Thermography Explorer and Image Processor. Microstructure of pus up and 15 minutes after the start of interaction with the solutions studied by his strokes, preparing and similarly stained smear of blood on standard laboratory methods of dyeing 0,5% paint Main-Grjunval cooked 96° ethyl alcohol, and paint Romanovsky-Giemsa. Osmotic activity of aqueous solutions defined cryoscopy using vapor-pressure osmometer OSMOMAT-030 RS production company ANSELMA Industries (Austria). The value of acidity (alkalinity) solutions and purulent masses identified with strips of the universal indicator paper company Lachema. Visualization of gas bubbles in carbonated solutions carried out using an ultrasound device «ALOKA SSD - ALPHA 10» with the use of sensor convex type with a frequency of 3 to 7 MHz.

## 3. Results and Discussion

Laboratory studies conducted on isolated fragments of thick purulent masses, who in the initial state at a temperature of +24, +37 +42°C with looked in the visible range of the spectrum of radiation is not transparent, have a consistence of the formless, thick, viscous and sticky mass of yellow-grey. While in the infrared range of the spectrum of radiation on thermovisor purulent masses looked solid colour to match the colour of inanimate objects when their temperature is +24, +37 +42° C (respectively). The microstructure of the pus was a relatively homogeneous and moderately transparent intercellular environment with rarely placed in it opaque cells, half of which was located in isolation from each other, and the other half are closely adjoined to each other, forming groups, including 2 - 9 cells.

15 minutes after the start infiltration festering mass at a temperature of +24 or +37° C water for injection, or solutions of 0.9% and 10% sodium chloride, 10% and 20% sodium sulfatsil, 0,02% furatsilin, 0,4% «Ekor-Forte», 3% «Alaminol», 0,25% «Lizophormine - 3000», «Ahdez 3000» (solution, ready to use), 4% double-main salt hypochlorite calcium, as well as a solution prepared by dissolving 1 effervescent tablet «Pjurgjavel» in 1 liter of water, rheology, appearance and microstructure of pus has not undergone significant changes.

Slight decrease in the viscosity of dense purulent mass were observed at infiltration it these same solutions, but heated to a temperature of +42° C. In this case, 15 minutes after the start of interaction with drugs purulent masses looked swollen and their viscosity decreased (increased turnover).In the infrared spectrum thermovisor purulent masses looked solid colour to match the colour of inanimate objects at a temperature of +42° C. Microstructure of pus after its interaction with water, solutions of 0.9% sodium chloride, furatsilin (at a dilution of 1:5000) or 1.5% hydrogen peroxide testified about the division of a homogeneous mass of pus translucent turbid liquid multiple «islands». In pockets of pus in the most part kept the microstructure in the form of a relatively homogeneous moderately transparent intercellular environment in which rarely were spread non-differential cellular elements. Islets of unchanged pus had a variety of irregular shape, irregular partially blurred edges and look isolated from each other due to the fact that around them unevenly and irregularly and were placed layers of translucent liquid. Layers of the liquid looked as many splitting and merging streams surrounding all sides of the fragments of thick pus from the original microstructure.

At the same time, in 15 minutes after the interaction with pus solutions «Ahdez 3000», 4% «Alaminol», 1,5% «Lizophormine 3000», 2% «Tephlex», 2% «Pjurgjavel», 0,4% «Ekor-Forte», or 4% double-main salt hypochlorite calcium in addition to the above changes in pus appear destroyed cellular elements and areas of the destroyed structure of the extracellular environment. The sites are identified uneven infiltration thick, purulent substrate influencing solutions in the form of clusters. We conducted identification of indicators of the acidity of the water and the solutions showed that all of them except solutions 10% and 20% sodium sulfatsil have a pH lower than 7.0. In particular, water and ready-to-use solution «Ahdez 3000» has a pH about 5.0, solutions 4% «Alaminol» and 1.5% «Lizophormine 3000» have a pH of 6.5, solution 2% «Pjurgjavel» has a pH of about 6.8, and the solution of 0.4% «Ekor-Forte has a pH of approximately 7.0.

In the next series of experiments in vitro we studied viscosity and microstructure of thick pus after 15 minutes of his infiltration water and water solutions of medicines at room temperature (+24° C) in the artificial lowering of the pH to 2.0 by introducing hydrochloric acid or when raising the pH to 12.0 through the introduction of sodium hydroxide. It is established that acidification not allows, and alkalization allows to decrease the viscosity of dense purulent masses. So, after 15 minutes of interaction thick festering mass of water or solution containing one of the following medicines: solution 0,9% and 10% sodium chloride, 10% and 20% sulfatsil-the sodium, 0,02% furacilin, 0,4% «Ekor-Forte», 3% «Alaminol», 0,25% «Lizophormine - 3000», «Ahdez 3000» (solution, ready to use) or a solution «Pjurgjavel» (solution prepared by dissolving 1 "sparkling" tablets «Pjurgjavel» in 1 liter of water) at pH 2.0 purulent masses remain thick, viscous, sticky and viscous, while after the interaction with these same solutions, but at pH 12.0 viscosity festering mass decreases several times, and viscous thick pus becomes liquid and gain good fluidity.

In parallel, we have studied the ability of water and solutions of medicines thin thick pus and delete it out in the conditions of mechanical mixing of environments through standardized shaking model cavity filled with commensurate amounts of pus and investigated solutions. It is shown that alkalization to pH 12,0 water for injection, solutions 0.9% and 10% sodium chloride, 10% and 20% sodium sulfatsil, 0,02% furatsilin, «Ahdez 3000», 0,4% «Ekor-Forte», 3% «Alaminol», 0,25% «Lizophormine» and 2% «Pjurgjavel» at a temperature of +24°C and +42°C with speeds up the process of thinning and removal of thick pus in 4 and 5 times, respectively, compared with their original wash activity. However, active shake for 30 minutes model cavities containing comparable amounts of thick pus and the above solutions do not lead to absolutely complete dissolution of all pus clots.

Then, we have examined the state of dense pus through 15 minutes of its location at a temperature of +24°C in solutions of 2.4% and 24% of aminophylline (issued in the dosage form «solutions for injections» with a pH about 9.0 and 12.0 respectively in accordance with the requirement of the Pharmacopoeia article) solutions 4% and 10% sodium bicarbonate (have a value of pH 8.0), or in solutions 5% and 20% glucose (issued in the dosage form «solution for injections» with a pH of about 3,5 in accordance with the requirement of the Pharmacopoeia article). It is established, that after 15 minutes of interaction thick, purulent mass with acidic solutions 5% and 20% glucose rheology, appearance and microstructure of purulent masses remains almost unchanged, and in 15 minutes of interaction purulent masses with alkaline solutions 2.4% and 24% aminofillina or 4% and 10% sodium bicarbonate thick pus almost fully loses its viscosity, becomes liquid and is very fluid. The biggest intensity of the dilution of the pus was observed when placed in a solution of 10% sodium bicarbonate. Following this, we have conducted research in the state of dense pus when mixing it with a solution of 4% sodium bicarbonate through standardized shaking model cavity filled with equal volumes of interacting environments. It is shown that when +24°C solution 4% sodium bicarbonate leads to complete the liquefaction of pus from the formation of a homogeneous fluid liquid after 15 minutes, at + 37° C - through of 12.5 minutes, and at +42°C - through of 12 minutes of interaction. The study of the microstructure of pus showed that after thinning he is a homogenous transparent intercellular environment in which very rarely are «scattered» non-differential cellular elements that have preserved their structures and their congestions appear «loose» (non-dense).

Then we studied the dynamics of the movement purulent masses after infusion of water in them and different solutions. The results obtained showed that pus sinks in water and solutions of traditional antiseptic and disinfectant means, but it floats on the surface of a saturated solution of sodium bicarbonate. Exploring the quality of thick pus obtained from the pleural cavity of patients pleural empyema and abdominal patients with peritonitis, we have determined that it represents a heavy, viscous and acidic isotonic biomass. The acidity of pus is within pH 6.0, osmotic activity of pus within 300 мОсмоль/l of water, and the share of pus is within 1,035 ± 0.005 g/cm$^3$. In turn, solutions of sodium bicarbonate in concentrations above 5% a value pH of 8.0, osmotic activity is above 500 мОсмоль/l of water, and the density is higher 1,050 g/cm$^3$. These data allowed to explain that it is due to the gravitational force and the difference between the weights of purulent masses emerge in solutions of 5 - 10% sodium bicarbonate up, situated in the upper layer of the solution, and sink in water, as well as in all the «easy» solutions of medicines.

After that we decided to increase the concentration of gases in a solution of 4% sodium bicarbonate, turning it into carbonated mineral water, hoping thus to strengthen its aggressive action up to the ability to form foam and geyser. To do this, we added a soda solution in the first case, hydrogen peroxide, and in the second case - carbonic gas under high pressure of 0.2 ATM like sparkling drinking water.

It is shown that injections of warm solution of 3% hydrogen peroxide and 4% sodium bicarbonate in isolated fragments thick pus in 5 minutes transformation of two interacting environments in one turbid liquid with a fluid properties, microstructure which is a divorced translucent colloidal intercellular environment with «scoured out» cellular elements united in groups of up to 20 cells.

In a laboratory model of the purulent fistula filled with serous tissue, taken from the pleural cavity of the patient destructive pneumonia, it is shown that a single injection of the bottom part of the channel solution of 4% sodium bicarbonate carbonated carbon dioxide for to excess pressure of 0.2 ATM and heated up to +42°C provides instant rapid expulsion from fistula of almost all purulent contents foaming like a geyser [18]. Found that warm solution of sodium bicarbonate and hydrogen peroxide has the ability to dissolve pus by alkaline burn intercellular colloid environment catalyzed by its heating and interstitial application only boiling (blasting). Alkaline burn intercellular colloid environment softens clot for by hydrolysis of proteins and saponification of fats, accelerated by heat and massive «explosions» of the surface layer of pus on the boundary separating the media bubbles of oxygen released from hydrogen peroxide under the action of the enzyme catalase.

In experiments with ultrasound carbonated solutions in plastic and rubber packages it is shown that the appearance of bubbles gas in solution of 3% hydrogen peroxide and 4% sodium bicarbonate, and the water from the tap, or in solution of 0.9% sodium chloride or 4% sodium bicarbonate when their carbonated carbon dioxide under excessive pressure of 0.2 ATM, «makes» the movement of solution more visible. It is established that the presence of gas bubbles in solutions provides visualization not only of the solutions of a modified their ultrasonic density, but almost every bubble in them, as well as the direction of its movement in the solution. In addition it is shown that visualised using ultrasound direction of movement of gas bubbles in the solution allows you to monitor, and change the location of the cavity in space allows you to change the direction of movement of gas bubbles, as well as the solution and liquid colloidal environment inside the cavity [10,11].

## 4. Conclusion

It is shown that the rheology of solutions studied plasma substitutes drugs and antiseptic before and after mixing with purulent masses mostly depends on the volume of drug solution, the temperature, alkalinity, osmotic activity, total concentration of the ingredients, the specific weight and carbonation. It was found that the rheology of liquid and viscous and dense biological tissues improve drug having the following physicochemical characteristics: hypertermia, hyperalkalization and hypergazation. Found that the leader rheology improving drugs and biological tissues is sodium bicarbonate, hydrogen peroxide and carbon dioxide introduced into the pressurized drug (similar food carbonated beverages). The data allowed to develop a new hygienic medicine designed to liquefy thick purulent masses with pleural empyema, peritonitis, rhinitis, sinusitis, conjunctivitis, osteomyelitis, and thinning of cerumen and dissolution of the lacrimal stones. New hygiene medicament comprises an aqueous solution of 0.5 - 3% hydrogen peroxide and 0.5 - 10% sodium bicarbonate, carbonated with carbon dioxide to a pressure of 0.2 ATM and heated up to +39 - +42° C [3,8,12,18,20,21,22].

## References

[1] Urakova N A et al. 2009. *Medical Assistance.* **1** pp 46-48
[2] Urakov A L et al. 2011. *Ural Med. J.* **6** pp 142-145
[3] Urakov A L et al. 2011. *Herald of Ural Med. Acad. Science.* **1** (33) pp 84–87
[4] Urakov A L and Urakova N A 2013. *Modern Problems of Science and Education.* **1** URL: http://www.science-education.ru/107-7607.
[5] Urakov A L, Urakova N A and Kasatkin A A 2013. *Thrombosis Research.* **131** Suppl. 1 pp S79
[6] Urakov A L 2013. *Successes of Modern Natural Sciences.* **12** pp 8 - 12
[7] Urakov A L 2013. *Successes of Modern Natural Sciences.* **11** pp 32-36
[8] Urakov A L 2013. *Fundamental Research.* **9(3)** pp 486-490; URL: www.rae.ru/fs/?section=content&op=show_article&article_id=10001589
[9] Urakov A L and Urakova N A 2013. *Modern Problems of Science and Education.* **1** URL: http://www.science-education.ru/107-7607
[10] Urakov A.L. 2000. *Economic Herald of Pharmacy.* **8** pp 101-104
[11] Urakov A L et al. 2011. *Recipe.* **6(80)** pp 26-31
[12] Strelkov N S et al. 2007. *Inventions. Utility Models.* **30** Patent RF № 2308894
[13] Chereshnev V A et al. 2008. *Inventions. Utility Models.* **23** Patent RF № 2331441
[14] Chereshnev V A et al. 2008. *Inventions. Utility Models.* **30** Patent RF № 2336833
[15] Mihailova N A et al. 2008. *Inventions. Utility Models.* **18** Patent RF № 2327471
[16] Urakov A L et al. 2009. *Inventions. Utility Models.* **19** Patent RF № 2360685
[17] Urakov A L et al. 2010. *Inventions. Utility Models.* **13** Patent RF № 2389013
[18] Urakov A L et al. 2009. *Inventions. Utility Models.* **26** Patent RF № 2367375
[19] Reshetnikov A P et al. 2009. *Inventions. Utility Models.* **30** Patent RF № 2371532
[20] Urakov A L et al. 2009. *Inventions. Utility Models.* **12** Patent RF № 2387465
[21] Urakov A L et al. 2012. *Inventions. Utility Models.* **18** Patent RF № 2468776
[22] Bondarenko L B et al. 2012. *Inventions. Utility Models.* **16** Patent RF № 2452478

# AUTHOR INDEX

# AUTHOR INDEX

**Institute of Physics**
Dirac House, Temple Back
Bristol BS1 6BE UK

CURRAN ASSOCIATES INC
proceedings
.com